AF443434

Series Editor
John M. Walker
School of Life and Medical Sciences
University of Hertfordshire
Hatfield, Hertfordshire, AL10 9AB, UK

For further volumes:
http://www.springer.com/series/7651

Somatic Stem Cells

Methods and Protocols

Second Edition

Edited by

Shree Ram Singh

Basic Research Laboratory, Stem Cell Regulation and Animal Aging Section, National Cancer Institute, Frederick, MD, USA

Pranela Rameshwar

Department of Medicine-Hematology/Oncology, Rutgers New Jersey Medical School, Newark, NJ, USA

 Humana Press

Editors
Shree Ram Singh
Basic Research Laboratory
Stem Cell Regulation and Animal Aging Section
National Cancer Institute
Frederick, MD, USA

Pranela Rameshwar
Department of Medicine-Hematology/Oncology
Rutgers New Jersey Medical School
Newark, NJ, USA

ISSN 1064-3745   ISSN 1940-6029   (electronic)
Methods in Molecular Biology
ISBN 978-1-4939-8696-5   ISBN 978-1-4939-8697-2   (eBook)
https://doi.org/10.1007/978-1-4939-8697-2

Library of Congress Control Number: 2018949604

Cover illustration: Immunofluorescence image of induced neural stem cells under proliferation and after differentiation. From Chapter 2, Figure 2B—Luis Azmitia and Philipp Capetian.

This Humana Press imprint is published by the registered company Springer Science+Business Media, LLC part of Springer Nature.
The registered company address is: 233 Spring Street, New York, NY 10013, U.S.A.

Preface

Stem cells reside in almost all organisms. There are several types of stem cells used in medical research with both promise and limitations. Stem cells have recently attracted significant attention because of their therapeutic potential in regenerative medicine and for developing anticancer therapies to eliminate cancer stem cells. Understanding the mechanisms regulating stem cell proliferation and differentiation is a very hot topic in developmental biology and stem cell medicine. However, the mechanism of stem cell self-renewal and differentiation remains elusive. Recent identification of induced pluripotent stem cells further strengthens the field of drug development and modeling human diseases.

This new volume of *Somatic Stem Cells* is intended to present selected genetic, molecular, and cellular techniques used in somatic stem cell research and its clinical application. Composed in the highly successful *Methods in Molecular Biology* series format, each chapter contains a brief introduction, step-by-step methods, a list of necessary materials, and a notes section which shares tips on troubleshooting to avoid known pitfalls. The chapters mainly focus on the isolation, characterization, purity, plasticity, and clinical uses of somatic stem cells from a variety of human and animal tissues. *Somatic Stem Cells* provides fundamental techniques to cell and molecular biologists, developmental biologists, tissue engineers, geneticists, clinicians, and students and postdocs working in the various disciplines of stem cell research and its potential application in regenerative medicine.

We would like to thank Prof. John M. Walker and the staff at Humana+Springer for their invitation, editorial guidance, and assistance throughout preparation of the book for publication. I also would like to express my sincere appreciation and gratitude to the contributors for sharing their precious laboratory expertise with the stem cell community. Finally, we would like to thank our family members for their continued support.

Frederick, MD, USA   *Shree Ram Singh*
Newark, NJ, USA   *Pranela Rameshwar*

Contents

Contributors

POONAM AGGARWAL • *Basic Research Laboratory, Stem Cell Regulation and Animal Aging Section, National Cancer Institute, Frederick, MD, USA*

CRISTOPHER AGUILAR • *Basic Research Laboratory, Stem Cell Regulation and Animal Aging Section, National Cancer Institute, Frederick, MD, USA*

YASUYUKI AMOH • *Department of Dermatology, Kitasato University, Sagamihara, Japan*

LUIS AZMITIA • *Department of Neurosurgery, University of Kiel, Kiel, Germany*

ALEKSANDAR BAJIC • *Department of Human and Molecular Genetics, Baylor College of Medicine, Houston, TX, USA; Jan and Dan Duncan Neurological Research Institute at Texas Children's Hospital, Houston, TX, USA*

DODGE L. BALUYA • *MD Anderson Cancer Center, University of Texas, Houston, TX, USA*

IVAN BERTONCELLO • *Department of Pharmacology and Therapeutics, Lung Health Research Centre, University of Melbourne, Parkville, VIC, Australia*

RADOVAN BOROJEVIC • *Centro de Medicina Regenerativa, Faculdade de Medicina de Petrópolis–FASE, Rio de Janeiro, Brazil*

PHILIPP CAPETIAN • *Department of Neurology, University of Würzburg, Würzburg, Germany; Institute of Neurogenetics, University of Lübeck, Lübeck, Germany*

GIANNI CARRARO • *Department of Medicine, Lung and Regenerative Medicine Institutes, Cedars-Sinai Medical Center, Los Angeles, CA, USA*

BEYONCÉ CARRINGTON • *Basic Research Laboratory, Stem Cell Regulation and Animal Aging Section, National Cancer Institute, Frederick, MD, USA*

CHING-CHENG CHEN • *Stem Cell & Leukemia Research, Gher Family Center for Leukemia Research, Beckman Research Institute of City of Hope, Duarte, CA, USA*

KEVIN CHEN • *Department of Biosciences, Rice University, Houston, TX, USA*

TIFFANY CHEN • *Basic Research Laboratory, Stem Cell Regulation and Animal Aging Section, National Cancer Institute, Frederick, MD, USA*

XI CHEN • *Department of Plastic and Reconstructive Surgery, Shanghai Tenth People's Hospital, TongJi University School of Medicine, Shanghai, People's Republic of China*

WILLIAM T. CHOI • *Jan and Dan Duncan Neurological Research Institute at Texas Children's Hospital, Houston, TX, USA; Program in Developmental Biology, Baylor College of Medicine, Houston, TX, USA; Medical Scientist Training Program, Baylor College of Medicine, Houston, TX, USA*

PETER CONATY • *Division of Hematology/Oncology, Department of Medicine, New Jersey Medical School, Rutgers School of Biomedical Health Science, Newark, NJ, USA*

ALEXANDRA CONDÉ-GREEN • *Division of Plastic Surgery, Department of Surgery, New Jersey Medical School, Rutgers School of Biomedical Health Science, Newark, NJ, USA*

CATHARINE DIETRICH • *Basic Research Laboratory, Stem Cell Regulation and Animal Aging Section, National Cancer Institute, Frederick, MD, USA*

CARLOS ANTÔNIO DO NASCIMENTO SANTOS • *Instituto de Biofísica Carlos Chagas Filho, Universidade Federal do Rio de Janeiro, Cidade Universitária, Rio de Janeiro, Brazil*

MAYRA GARCIA • *Stem Cell & Leukemia Research, Gher Family Center for Leukemia Research, Beckman Research Institute of City of Hope, Duarte, CA, USA*

QIANPING GUO • *Biomaterials and Cell Mechanics Laboratory (BCML), Orthopedic Institute, Soochow University, Suzhou, China*

ROBERT M. HOFFMAN • *AntiCancer, Inc., San Diego, CA, USA; Department of Surgery, University of California San Diego, San Diego, CA, USA*

XINGBIN HU • *Stem Cell & Leukemia Research,Gher Family Center for Leukemia Research, Beckman Research Institute of City of Hope, Duarte, CA, USA; Department of Transfusion Medicine, Xijing Hospital, Fourth Military Medical University, Xi'an, PR China*

MADHURI KANGO-SINGH • *Department of Biology, University of Dayton, Dayton, Ohio, USA*

BON-KYOUNG KOO • *IMBA Institute of Molecular Biotechnology, Stem Cells and Organoids, Vienna, Austria*

VASANTH S. KOTAMARTI • *New Jersey Medical School, Rutgers School of Biomedical Health Science, Newark, NJ, USA*

BIJENDER KUMAR • *Department of Radiation Oncology, City of Hope Medical Center, Duarte, CA, USA*

EDWARD S. LEE • *Division of Plastic Surgery, Department of Surgery, New Jersey Medical School, Rutgers School of Biomedical Health Science, Newark, NJ, USA*

BIN LI • *Biomaterials and Cell Mechanics Laboratory (BCML), Orthopedic Institute, Soochow University, Suzhou, China*

YUHONG LI • *Department of Cell Biology, Third Military Medical University, Chongqing, China*

GUANGPENG LIU • *Department of Plastic and Reconstructive Surgery, Shanghai Tenth People's Hospital, TongJi University School of Medicine, Shanghai, People's Republic of China*

SRIDESHIKAN SARGUR MADABUSHI • *Department of Radiation Oncology, City of Hope Medical Center, Duarte, CA, USA*

CHRISTINA M. MAEDATAKIYA • *Instituto de Biofísica Carlos Chagas Filho, Universidade Federal do Rio de Janeiro, Cidade Universitária, Rio de Janeiro, Brazil*

SCOTT T. MAGNESS • *Department of Medicine, Center for Gastrointestinal Biology and Disease, University of North Carolina at Chapel Hill, Chapel Hill, NC, USA; NC State/UNC Joint Department of Biomedical Engineering, University of North Carolina at Chapel Hill, Chapel Hill, NC, USA; Department of Cell Biology and Physiology, University of North Carolina at Chapel Hill, Chapel Hill, NC, USA*

MIRJANA MALETIC-SAVATIC • *Jan and Dan Duncan Neurological Research Institute at Texas Children's Hospital, Houston, TX, USA; Department of Pediatrics and Neuroscience, Program in Developmental Biology, Baylor College of Medicine, Houston, TX, USA*

JONATHAN L. MCQUALTER • *School of Health and Biomedical Sciences, RMIT University, Bundoora, VIC, Australia*

PARAS KUMAR MISHRA • *Department of Cellular and Integrative Physiology, University of Nebraska Medical Center, Omaha, NE, USA; Department of Anesthesiology, University of Nebraska Medical Center, Omaha, NE, USA*

CAITLYN A. MOORE • *Division of Hematology/Oncology, Department of Medicine, University of Medicine and Dentistry of New Jersey-Rutgers—New Jersey Medical School, Newark, NJ, USA*

YAHAIRA NAALDIJK • *Division of Hematology/Oncology, Department of Medicine, New Jersey Medical School, Rutgers School of Biomedical Health Science, Newark, NJ, USA; Institute of Chemistry, University of Sao Paulo, Sao Paulo, Brazil*

LUIZ EURICO NASCIUTTI • *Instituto de Ciências Biomédicas, Universidade Federal do Rio de Janeiro, Cidade Universitária, Rio de Janeiro, Brazil*

NATHAN PINTO • *Basic Research Laboratory, Stem Cell Regulation and Animal Aging Section, National Cancer Institute, Frederick, MD, USA*

PRANELA RAMESHWAR • *Department of Medicine-Hematology/Oncology, Rutgers New Jersey Medical School, Newark, NJ, USA*

LEIGH A. SAMSA • *Department of Medicine, Center for Gastrointestinal Biology and Disease, University of North Carolina at Chapel Hill, Chapel Hill, NC, USA*

NILOY N. SHAH • *Department of Medicine, Division of Hematology/Oncology, University of Medicine and Dentistry of New Jersey-New Jersey Medical School, Newark, NJ, USA*

LAUREN S. SHERMAN • *Division of Hematology/Oncology, Department of Medicine, New Jersey Medical School, Rutgers Biomedical and Health Sciences, Newark, NJ, USA*

SHREE RAM SINGH • *Basic Research Laboratory, Stem Cell Regulation and Animal Aging Section, National Cancer Institute, Frederick, MD, USA*

RACHIT SINHA • *National Cancer Institute, Frederick, MD, USA*

CAROLINE P. SMITH • *Division of Hematology/Oncology, Department of Medicine, University of Medicine and Dentistry of New Jersey-Rutgers—New Jersey Medical School, Newark, NJ, USA*

ALEXANDRA STOLZING • *Centre for Biological Engineering, Loughborough University, Loughborough, UK*

CHRISTINE TANG • *Department of Biosciences, Rice University, Houston, TX, USA*

HENNING ULRICH • *Institute of Chemistry, University of Sao Paulo, Sao Paulo, Brazil*

JAMES H.-C. WANG • *MechanoBiology Laboratory, Departments of Orthopaedic Surgery and Bioengineering, University of Pittsburgh School of Medicine, Pittsburgh, PA, USA*

XIAOYANG WANG • *Department of Cell Biology, Third Military Medical University, Chongqing, China; Department of Developmental and Cell Biology, University of Irvine, Irvine, CA, USA*

LIHONG WENG • *Stem Cell & Leukemia Research, Gher Family Center for Leukemia Research, Beckman Research Institute of City of Hope, Duarte, CA, USA*

IAN A. WILLIAMSON • *Department of Medicine, Center for Gastrointestinal Biology and Disease, University of North Carolina at Chapel Hill, Chapel Hill, NC, USA; NC State/ UNC Joint Department of Biomedical Engineering, University of North Carolina at Chapel Hill, Chapel Hill, NC, USA*

YIZHAN XING • *Department of Cell Biology, Third Military Medical University, Chongqing, China*

SANTOSH KUMAR YADAV • *Department of Cellular and Integrative Physiology, University of Nebraska Medical Center, Omaha, NE, USA*

JIANYING ZHANG • *MechanoBiology Laboratory, Departments of Orthopaedic Surgery and Bioengineering, University of Pittsburgh School of Medicine, Pittsburgh, PA, USA*

XINPING ZHANG • *School of Medicine and Dentistry, University of Rochester Medical Center, Rochester, NY, USA*

PINGHUI ZHOU • *Biomaterials and Cell Mechanics Laboratory (BCML), Orthopedic Institute, Soochow University, Suzhou, China*

Part I

Introduction

Chapter 1

An Update on the Therapeutic Potential of Stem Cells

Pranela Rameshwar, Caitlyn A. Moore, Niloy N. Shah, and Caroline P. Smith

Abstract

The seeming setbacks noted for stem cells underscore the need for experimental studies for safe and efficacious application to patients. Both clinical and experimental researchers have gained valuable knowledge on the characteristics of stem cells, and their behavior in different microenvironment. This introductory chapter focuses on adult mesenchymal stem cells (MSCs) based on the predominance in the clinic. MSCs can be influenced by inflammatory mediators to exert immune suppressive properties, commonly referred to as "licensing." Interestingly, while there are questions if other stem cells can be delivered across allogeneic barrier, there is no question on the ability of MSCs to provide this benefit. This property has been a great advantage since MSCs could be available for immediate application as "off-the-shelf" stem cells for several disorders, tissue repair and gene/drug delivery. Despite the benefit of MSCs, it is imperative that research continues with the various types of stem cells. The method needed to isolate these cells is outlined in this book. In parallel, safety studies are needed; particularly links to oncogenic event. In summary, this introductory chapter discusses several potential areas that need to be addressed for safe and efficient delivery of stem cells, and argue for the incorporation of microenvironmental factors in the studies. The method described in this chapter could be extrapolated to the field of chimeric antigen receptor T-cells (CAR-T). This will require application to stem cell hierarchy of memory T-cells.

Key words Stem cell, Mesenchymal stem cells, CAR-T

1 Introduction

The area of stem cells, although studied for decades, has mainly focused on the hematopoietic stem cells. This has wrongly led to the premise that the bone marrow is a unique organ since it contained due to the presence of stem cells. The field has since evolved to demonstrate the existence of stem cells in all organs. This has led to the proposition of the cancer stem cell theory, which is a field that began over a decade (discussed in [1, 2]). Another evolving field of stem cell is the memory T-cells, which are now placed in hierarchy [3].

As the area of stem cell biology expands the field of medicine is forced to change. As examples, as the field of cancer stem cell

Shree Ram Singh and Pranela Rameshwar (eds.), *Somatic Stem Cells: Methods and Protocols*, Methods in Molecular Biology, vol. 1842, https://doi.org/10.1007/978-1-4939-8697-2_1, © Springer Science+Business Media, LLC, part of Springer Nature 2018

expands, the drug developing area are involved in studies to generate targeting drugs. As one would expect, this field has affected other areas of medicine such as radiation oncology. The lingering question is how radiation would be used in cancer biology when the cancer stem cells appear to resist radiation [4]. Also, other research studies have demonstrated benefit to patients with combinations of drugs and radiation, with respect to targeting of cancer stem cells [5]. This may lead to a serious consideration of how the training field of oncology would progress going forward. It may be necessary to merge the medical and radiation oncology programs, perhaps by beginning to combine the training programs. The latter will prepare the oncologists with evolving methods to delivery treatments from both fields of medicine.

Another area of changes worth noting is the field of chimeric antigen receptor (CAR-T). As scientists begin to develop hierarchy of T-cells within the memory T-cell space this might benefit the CAR-T space. If the ongoing trials indicate a more prolonged existence of CAR-T the field may determine to engineer to memory T-cells, or the cell types below with respect to maturation to limit lifelong presence of the CAR-T cells.

Since the hematopoietic stem cells are the source of immune and blood cells, there was no controversy on the existence of stem cells in bone marrow. As would be expected for any type of research, there were challenges in studies to identify hematopoietic stem cells and lineage differentiation. Despite the time spent on studies to understand lineage maturation of hematopoietic stem cells, new reagents continue to identify additional pathways in the development of hematopoietic stem cells. If the longest studied stem cells still require investigation, this underscores the complex mechanisms in the maturation of stem cells to mature cells.

There is little to no doubt among investigators that a small subset of bone marrow cells can reconstitute lethally irradiated animals. The very low frequency of this stem cells in an individual can be explained by the self-renewal mechanism reported by Drs Till and McCullock [6]. Other investigators reported on chromosomal preservation as a method to reduce the frequency of mutation in the stem cells. These findings, mostly determined for hematopoietic stem cells, can be extrapolated to stem cells in other organs and tissues.

The identification of the major histocompatibility complex (MHC)-Class I and II molecules have been most valuable in cause transplantation of hematopoietic stem cells across allogeneic barrier. This has been met with remarkable success for hematological disorders, solid tumors and autoimmune disease. The success of these transplants is met with the attitude of improving protocols with the incorporation of the new science. As an example, during the early time of transplantation to repopulate the immune system, relatively young individuals were considered as qualified for the procedure. As science and the clinical practice gain more experience with transplants, the procedure is indicated for older subjects. Of note is that the field of stem cells is being developed for indications in older individuals. This point is important because science must determine if the data with young stem cells will benefit the information needed for the aging population.

During the past few years, stem cells have been identified in all organs. Although it is unclear about the physiological functions of these stem cells, studies on their differentiation to specialized cells indicate that the endogenous stem cells in the various organs might be required for daily replacement of damaged cells. This has led to vast investigations across the globe on cell replacement and protection by stem cells. The growing literature across the globe indicates that stem cells and their products could be among the new wave of therapy for cell replacement and repair, as well as other indications.

The field has undergone years of research including clinical trials. Thus far, there is no evidence of cell replace or long-term sustenance of stem cells. The new and growing field of extracellular vesicles might explain the benefit of stem cells. Since these vesicles are identified as vehicles for intercellular communication, their contents need to be dissected to better "harness" stem cells for future treatment.

The experimental studies are not limited to repair tissues of stem cells from the same organs, but to use stem cells of another other. In addition, scientists have focused on the immature nature of stem cells to explore their plasticity, which is defined as the ability of stem cells to generate cells of another germ layer. As an example, stem cells of mesodermal origin can form cells of ectodermal origin. Based on the vast number of publications on this subject, it is obvious that adult stem cells are functionally plastic. The current challenge is to utilize the most efficient method to translate the experimental studies. Unlike the experimental studies, stem cells, when placed in patients could be under stress due to the different microenvironment and can behave differently. This chapter discusses some of the challenges and identifies some solutions that could be explored to succeed in the translation of stem cells.

2 Challenges

The categories of stem cells are broadly divided into embryonic stem cells, fetal stem cells and adult stem cells. Since the cord blood and placenta contains fetal stem cells, these types are considered as subsets of fetal stem cells. Induced pluripotent stem cells (iPS) can be generated from adult stem cells and differentiated somatic cells. Thus, the iPS cells are placed in its own category. Despite embryonic stem cells being the most primitive cells, if the ethical issues are dissected, the science of these stem cells needs to take a pause on its future in medicine. Simply, embryonic stem cells need to be MHC matched. There are few reports that showed an immune suppressive role of embryonic stem cells when tested as third-party cells to two-way mixed lymphocyte reaction [7, 8]. At this time, the type of immune response of embryonic stem cells is still unresolved since others have reported on their ability to induce both humoral and cellular responses [9]. In the best case, if embryonic stem cells can be transplanted across allogeneic barrier through their ability to be immune suppressive, this does not address the instability of these stem cells to form tumors [10]. There are insights into the future of embryonic stem cells since some trials have been initiated with embryonic stem cell-derived differentiated cells for macular degeneration and diabetes [11].

The developing hematopoietic stem cells are well studied in the embryo and fetus, which brings insights into how the adult stem cells could respond during insults. In parallel with the renewed interest in stem cells, scientists continue to examine stem cells from fetus; in particular from tissues that are discarded: umbilical cord blood and the placenta [12]. Stem cells from other regions of the fetus have been studied. These include, but are not limited to multipotential progenitors, nonhematopoietic stem cells from human fetal livers [13]. Amniotic stem cells can be obtained from human and mice [14]. In several laboratories have isolated amniotic stem cells, expanded them and, induce them to generate cells of all three germ layers. Amniotic fluid is an excellent source of stem cells due to the large number of women who required amniocentesis. This source is not easily available because new methods have been developed to analyze the fetal DNA with a blood draw. A small amount of amniotic fluid following clinical analyses would be sufficient to acquire adequate amount of stem cells for therapy. Indeed, stem cells from amniotic fluid continue to be a major subject of research for several clinical disorders [15, 16].

Adult stem cells are found in all organs. Each organ has its own unique microenvironment and requires that the stem cells are studied in the context of the milieu within the organ, or at least to develop experimental models that recapitulate the organ. These types of investigations, as well as evolving novel methods will achieve the following: (1) acquire these stem cells in cases where they would be discarded, such as during surgery. If they can be expanded and shows plasticity ex vivo, they would be excellent source of stem cells for tissue repair. (2) By studying the stem cells within their own microenvironment, it will be possible to stimulate them endogenously during tissue insult to replace damaged tissue.

As an example, in the case of tissue insults such as wounds, it might be possible to induce the endogenous skin stem cells to repair the damaged cells [17]. (3) If the environment of a tissue is changed by an inflammatory process, investigative studies will determine if, instead, of targeting the stem cells, the microenvironment could be regulated by drugs to allow repair by the endogenous stem cells. The premise is that changing the microenvironment will cause the endogenous stem cells to mature into desired cell types. (4) It is necessary to keep in mind that several genes linked to pluripotency can exert oncogenic and tumor suppressor functions. The experimental evidence indicate that tumors can be generated from stem cells, referred to as tumor within stem cells [18]. Indeed, recent studies showed higher incidence of tumor formation in tissue with highly proliferating stem cells [19]. Based on the above discussion in this section, it could be concluded that an understanding of how stem cells interact with the microenvironment during self-renewal and differentiation would provide insights on malignancies, and aid in developing efficacious treatment strategies.

3 Status of Phenotypic Evaluation of Stem Cells

This section will not discuss the specific markers on stem cells since this area continues to be evolving. Also, it is important to evaluate the various types of stem cells with multiple designations to determine if they could be functionally similar. As examples, the phenotype and functions of stem cells from placenta, adipose tissue and amniotic fluid strongly indicate that they could be functionally similar to mesenchymal stem cells (MSCs). The question is why these stem cells are discussed as if they distinct stem cells. The problem with the field is the disagreement among scientists on nomenclature. This could be due to spread in stem cell discussions at different forums. Thus, an international team is needed to develop consensus terms. However, such endeavor might need to wait until further scientific research dictates the subsets of cells in cultures.

The widespread use of terms for stem cells with similar phenotype might be attributed to the lack of inclusion among scientists with a long history of related research. This obstacle should be overcome by broad societies, such as those specific for stem cells. It is expected that time will solve these obstacles as the field is gaining clarity on how stem cells from different organs interact with the microenvironment. Another consideration for the use of distinct designations could be due to claims on intellectual property for organ-specific stem cells. The challenges discussed in this section are more than academic as they might slow down the progress in the field as new scientists enter the field.

The field of stem cell biology should always combine phenotype and function, with new or published markers. An investigator should perform functional assays by using the published literature only as guide. If this approach is made, it is likely that a small group of laboratories will add significantly to the literature to the field of stem cell biology. Contrary to this approach could result in numerous publications in the wrong directions of the field. This is demonstrated in the field of cancer stem cells, expressing biomarkers such as CD133, CD24, and CD44, were touted as the gold-standard for isolating cancer stem cells [20–22]. Scientists can still use the published markers of stem cells until proven otherwise but acknowledge the limitations. The data could provide clarity in retrospective analyses as the literature provides clarity on the subset of stem cells.

In addition to phenotype, other common functions in multipotency or pluripotency should be applied in characterizing stem cells. These include self-renewal properties, lineage differentiation and, the expression of genes linked to stem cells such as *Oct4*, *Nanog*, and *Sox2* [23]. The genetic characterization should also include genes linked to cell differentiation. Ideally, assays could be done to show that the stem cells can reconstitute an organ. As an example, stem cells from the mammary gland should reconstitute an organ to generate cells of all lineages [24].

4   Mesenchymal Stem Cells (MSCs)—Overview

As discussed above, the use of embryonic stem cells to reach the clinic has been slow. Adult and fetal stem cells have taken a more rapid path to the clinic. Since placenta, amniotic and umbilical cord stem cells seem to be functionally similar to MSCs, this section will discuss MSCs as potential for patient care. It is important to conduct other studies if the information here is extrapolated to other similar stem cells.

In culture MSCs are generally considered as heterogeneous and can be found several adult and fetal tissues [16–18]. Similar heterogeneity may be noted in vivo since the bone marrow contains cells with overlapping functions as MSCs. These include CXCL12-abundant reticular cells (CAR) cells and pericytes, both of which surround the vascular system [25, 26]. Even after more than a decade of active laboratory research involving MSCs, the question of heterogeneity remains a subject of investigation.

The adult bone marrow is the major organ of MSCs. In bone marrow, MSCs surround blood vessels and are in contact with the trabeculae as well as close to the endosteum [27, 28]. The frequency of MSCs varies, depending on the source. For example, the frequency is low in umbilical cord blood, but high in adult bone marrow and adipose tissues [22, 29]. In contrast to umbilical cord blood, the perivascular region and Wharton Jelly of the cord have a higher frequency of MSCs [30, 31]. As for other stem cells, MSCs show multipotency by their ability to differentiate along multiple lineages. As examples, when given the appropriate cues, MSCs can generate fibroblasts, adipocytes, chondrocytes, osteogenic cells, and cartilage [32].

The growing body of literature suggests that the efficiency of MSCs to differentiate into a particular cell type might depend on the anatomical source of the MSC [33]. The literature suggests that subsets of cultured MSCs express genes which suggests their ability to generate specialized cells [34]. For example a population of CD146-expressing cells have been shown to generate osteogenic cells and fibroblasts, suggesting that CD146(+) MSCs could be osteoprogenitors [35]. In other functions, the CD146-expressing cells can be transplanted to form a supporting hematopoietic microenvironment [35]. This information is significant for specific application of MSCs. If specific markers can be identified to select a subset of stem cells, this will benefit medicine. Going forward, it may be possible to select subsets of MSCs to treat specific diseases. MSCs have been reported to express neural markers, such as neural ganglioside, GD2 [36–38]. It will be interesting to determine if this marker is expressed on a subset of MSCs to generate neural cells, such as neurons [39–41].

Regarding techniques to characterize MSCs, these stem cells are morphological symmetrical with fibroblastoid appearance [42]. Phenotypically, MSCs can be characterized by the expression of several markers such as CD44, CD29, CD105, CD73, and CD166 and lack markers of hematopoietic lineage, CD45 [43, 44]. The markers seem to be expanding with CD200 added as a marker with immunomodulatory property [34]. Human MSCs have been suggested to be perivascular, also referred to as pericytes [45, 46]. Pericytes have been isolated from different human organs, express CD146, NG2, and PDGF receptor type 1, and form myogenic cells [45]. As discussed above, there is growing evidence that the pericytes might be a subset of MSCs.

A discussion regarding the origin of embryonic origin of MSCs is relevant to an understanding of their ability to generate different cell types. The origin of MSCs have been reported to be mesodermal and neuroepithelial [32, 47]. They show smooth muscle type structures that make them stem cells of mesodermal origin. The neuroepithelial origin of MSCs would explain the ease by which they form functional neurons and other neuronal cells [39–41, 48, 49]. As mesodermal stem cells, the generation of functional neurons indicates that the MSCs have crossed germ layer by forming cells of ectodermal origin, validating the plasticity of MSCs. On the other hand, if they are neuroepithelial stem cells, they should be able to generate ectodermal cells and this might deter their ability to be designated as plastic cells in their generation of functional neurons [39, 41, 48, 50].

5   MSCs-Safety and Early Trials

A major issue of safety is to evaluate the immune response of transplanted MSCs. This is particularly relevant to MSCs because these cells are transplanted across allogeneic barrier, as evident by the large number of clinical trials listed on clinicaltrial.gov [51]. MSCs when placed in an inflammatory microenvironment can be induced/licensed to act as immune suppressor, resulting in blunting functions such those by dendritic, natural killer, and T- and B-cells [42, 44, 52–55].

The safety of MSCs has been determined in several clinical trials, including those in which MSCs are used as third-party cells for graft versus host disease [43, 56]. As far as we are aware, there is no report on adverse effect, including the formation of tumors. Thus far, there is no report contradicting the use of allogeneic MSCs or otherwise known as "off-the-shelf" MSCs. As third-party cells, MSCs show promise as cellular therapy for graft versus host diseases by bone marrow transplantation as well as for organ transplant. MSCs could also be applied to patients with other inflammatory disorders such as asthma, inflammatory bowel disorders, and skeletal disorders [54, 57–59]. This list of disorders indicated for MSCs is exhaustive and cannot be discussed in this brief introductory chapter.

The results of early clinical trials with MSCs are mixed, with limited explanation on the contradictory outcomes. There are some discussion on the discordance shown by these clinical trials and the experimental studies [52]. The clinical trial fails to sustain long-term immune suppression. It is unclear if long-term immune suppression is desirable and if so, what can be done to achieve this sustained effect. Insights into this question require a closer look at the ability of MSCs to establish functional cross talk with mediators within tissue microenvironment.

Experimental studies indicate that transplanted allogeneic MSCs do not survive for long period in the recipient. Despite this, MSCs still have beneficial effects on the targeted pathology. This suggests that the mechanism of tissue repair/regeneration could be caused by the MSC secretome. Presently, one cannot make assumptions regarding the need to retain MSCs for long period in the transplanted recipient. The scientific community has however learnt much from the trials and this has led to robust research to examine the secretomes and microvesicles.

MSCs express specific receptors for chemokines and cytokines that are expected at sites of tissue injury [60–62]. These studies underscore the ability of MSCs to establish functional cross talk with cytokines when they are placed within a milieu of inflammatory mediators. Although MSCs have been linked to immune suppressor functions, they can also exert immune-enhancing functions such as antigen presentation and autoimmune responses [63–66]. Therefore, when MSCs are placed within a milieu of inflammatory responses, it might be difficult to predict the outcome. The dual immune responses of MSCs are consistent with the lack luster results of clinical trials [52].

The ability of MSCs to establish functional cross talk with immune mediators within a microenvironment of cytokines is not limited to undifferentiated stem cells. MSC-derived neurons also express cytokine receptors such as IL-1, IL-2, IL-6, and TNFα [61]. More importantly, the question that lingers is whether cells, differentiated from MSCs, or any other stem cells can be rejected by the host. To explain this further, MSCs are considered desirable for transplantation as off-the-shelf stem cells, indicating their delivery across allogeneic barrier. The initial delivery can be safe due to the immune suppressive properties of MSCs [67]. However, if future application shows that MSCs can replace endogenous tissues, the new cells will express class I MHC of the donor. The question is whether the new cells will be rejected or if tolerance will be developed. Regardless, this should be important investigations, in parallel to current work with MSCs. Also, the new cells, through the expression of cytokine receptor, would be able to establish cross talk with the microenvironment of tissue injury. The mixed trials with adult stem cells for cardiac disease have provided valuable information on the complex network of activated cells and soluble factors that stem cells have to accommodate to repair damaged tissue [68]. Together, this section describes the complex biology that is not mutually exclusive in the efficiency of MSCs, and other stem cells to protect and repair damaged tissues.

6   Immune Biology of MSCs

This section provides an in-depth discussion on the immune responses of MSCs since such information will be fundamental to the future clinical trials of these stem cells. MSCs show functional plasticity with regard to their immune properties by exerting both immune suppressor and enhancer functions [69]. In addition, MSCs might be an instructive cells to macrophages as a mechanism of tissue repair [70]. MSCs produce cytokines that can mediate autocrine and/or paracrine stimulation [27]. Others have suggested that MHC-II expression should be included among the minimal requirements MSC designation [71, 72]. Several reports indicated the isolation of pluripotent cells with properties consistent for MSCs, but undetectable MHC-II. These cells exhibit plastic adherence and can be differentiated along multiple lineages. It is possible that not all subsets of MSCs express MHC-II and that its expression does not require prior stimulation with inflammatory mediators [42, 66]. However, the expression of MHC-II on MSCs is a major consideration for stem cell therapy. MHC-II expression provides the cells with the ability to act as antigen presenting cells (APCs) [42, 63, 73–75]. Although MHC-II might not be expressed in unstimulated MSCs, its expression can be induced by interferon gamma (IFNγ) [66]. The significance of MHC-II expression, whether constitutive or induced, is significant to transplantation studies because if MSCs begin to serve as APCs, this might create confounds to the treatment. Therefore, the analyses of MHC-II and the consequence to cell therapy should be carefully examined, going forward to achieve efficient cell therapy. Another relevant property of MSCs is their ability to endocytose particles [63]. Again, this is highly significant for tissue repair where vast amounts of necrotic cells are likely to be present. Therefore, by placing MSCs in these regions of tissue injury, the cells could begin to engulf necrotic tissues to initiate an immune response.

Although MSCs express MHC-II and act as APCs, they differ from other professional APCs such as macrophage as regards *MHC-II* gene expression [63, 76]. IFNγ, which is a major proinflammatory inducer of MHC-II, shows a bimodal effect on MHC-II expression on MSCs [63]. MSCs produce baseline IFNγ that maintain MHC-II expression [63]. However, when MSCs are exposed to high level of IFNγ, its expression is decreased. If this observation is extrapolated to in vivo transplantation for acute inflammation, the MHC-II will be decreased and the stem cells will exert immune suppressive function. This will be desirable until the IFNγ level is decreased in MHC-II will be reexpressed.

The bimodal expression of MHC-II on MSCs has been attributed to the differential effects of IFNγ on the master regulator of MHC-II, CIITA [77]. At high IFNγ levels, the CIITA is retained in the cytosol, thereby preventing MHC-II transcription [77]. This is a highly relevant finding since the retention of CIITA in the cytosol could be explored for effective treatment to prevent the reexpression of MHC-II on MSCs.

Not only is the effect of CIITA relevant for stem cells, but the mechanism appears to be similar for neurons derived from MSCs that are exposed to IFNγ [78]. This further underscores the significance of CIITA as future targets for the efficiency of transplanting MSCs as off-the-shelf stem cells. The expression of MHC-II in cells that were derived from allogeneic MSCs is likely to occur long after their delivery to replace damaged tissue. The immune system could develop tolerance for the Class I MHC, which is normally considers as a weak antigen for allogeneic response. However, if the host is subjected to an infection, there will be high levels of IFNγ that could cause MHC-II to be expressed. This could result in late rejection of the replaced cells.

7   Cytokines in Stem Cell Responses

A thorough understanding on the roles of cytokines in the behavior of stem cells require a separate, but lengthy review due to the exhaustive literature on cytokines and chemokines as well as other immune mediators such as hormones and extracellular matrices. At sites of tissue injury, the microenvironment is expected to encompass a complex network of multiple soluble and insoluble mediators as well as several immune cell subsets. Another level of complexity is the timeline changes of the factors as well as the altered tissue microenvironment at sites close to the area of tissue insult. For example, there could be timeline changes in cytokine levels, following tissue injury. In addition, at a specific time, the cytokines could show a gradient concentration from the site of injury. These changes make it difficult to predict how MSCs, or any other stem cells, should be implanted.

This review briefly discusses three cytokines with broad, but opposing functions: interleukin-1α (IL-1α), the proinflammatory tumor necrosis factor α (TNFα), and transforming growth factor-beta (TGF-β). IL-1 is selected because it can regulate other cytokines with positive and negative effects. TGF-β is discussed due to its role as a proinflammatory and anti-inflammatory mediator. Another reason to discuss TGF-β is due to its association with oncogenesis. The placement of any stem cell within a milieu of inflammatory mediators could predispose the cell to transformation. At the time of designing any trial with stem cells, one needs to consider that the genes associated with pluripotency are also linked to oncogenesis. TNFα is a proinflammatory cytokine and its role as mediators in cross talk between the microenvironment and stem cells is prototypical of several cytokines. We also briefly discuss the role of small noncoding RNA (miRNA). The latter topic is not addressed with any depth since it requires a review article on its own.

IL-1α belongs to the family of cytokines that are central to inflammation and host defense [79]. IL-1α and IL-1β appear to exhibit similar effects through the type I IL-1 receptor. The type II receptor subtype lacks an intracellular signaling domain. IL-1 could be significant in understanding how stem cells respond to tissue factors. For example, IL-1 induces the expression of other inflammatory mediators in the stem cells as well as neighboring cells. Since it is most likely that IL-1 will be present at an area of tissue injury, it is expected that this cytokine could initiate a network of other cytokines that will cause cross-communication between stem cells and other immune cells [79]. In this regard, IL-1 could mediate stem cell responses through direct and indirect mechanisms [80, 81]. IL-1 would be able to affect cells derived from MSCs since its receptor is expressed on MSC-derived neurons [79]. In the case of neurons, IL-1 can induce the production of neurotransmitters, which could cause a cascade reaction to stimulate immune cells at the site of tissue injury, and perhaps indirectly expand to distant organs through the movement of activated immune cells [81–84].

TGF-β1 belongs to a superfamily of proteins including the activins, inhibins, and bone morphogenic proteins [85]. TGF-β receptors are ubiquitously expressed on normal and malignant cells [86, 87]. TGF-β1 interacts with subtypes I and II where the type 1 form is activated by type II [88, 89]. Type I signaling activates four members of Smad transcription factors [90–92]. TGF-β is involved in development during embryogenesis and neurogenesis [85, 93]. TGF-β1 also modulates immune responses and, inhibits cell proliferation, differentiation, and apoptosis [94, 95]. TGF-β1 exerts both tumor suppressor and oncogenic properties [96].

TNFα, along with other members of its family, interact with the TNF receptor superfamily [97]. TNFα is produced by activated macrophages and monocytes to induce cell death, and is involved in inflammatory disorders such as arthritis [98]. Since MSCs are suggested for inflammatory responses, the role of TNFα could be significant in tissue repair. It is unclear if MSCs can produce TNFα, but these cells are responsive through specific receptors [99, 100]. This suggests that TNFα, which is a proinflammatory cytokine, when present at areas of tissue injury, could immediately initiate cross-communication between MSCs and the microenvironment.

TNFα enhances the adhesion of MSCs [101, 102]. This role is interesting since it is possible that TNFα might be involved in mediating the attachment between MSCs and tissues. In cases of TNFα decrease in the tissue environment, MSCs are able to exert immune suppressive functions [70]. The enhanced immune function could be an advantage or a disadvantage. If TNFα is required for the MSCs to adhere to tissue for their retention within the damaged tissue, the decrease could be a disadvantage. This section underscores the complex points that could occur when cytokine levels are changed within the tissue.

An understanding of the role of cytokines in mediating cross talk between stem cells and the microenvironment of tissue injury requires discussions of miRNAs. They are single strand RNA of approximately 19–25 nucleotides [103]. MiRNAs are derived from precursor hairpin-shaped transcripts and most act as a guide in post-transcriptional gene silencing by forming base pair structures with mRNA [103, 104]. Although miRNAs repress translational of mRNA they do not prevent mRNA from docking to polyribosomes and do not block the initiation of translation [94, 104–106]. MiRNAs are involved in the development from stem cells along distinct lineages [107–109]. The role of miRNAs in stem cell biology has been established as their roles in neurogenesis, including MSC-derived neurons has been reported [110–112]. Taken together, cytokines are involved in the behavior of stem cells and miRNAs are involved in development. Therefore, studies are required to determine how cytokines affect stem cells through the regulation of miRNA.

8   Interferon Gamma (IFNγ) in MSC Functions

IFNγ has been shown to be central in the immune response of MSCs, and perhaps other stem cells, including those from allogeneic sources. Type 1 IFN-γ is important in viral protection as well as the general proinflammatory responses [113]. IFN-γ is produced by T-cells, NK cells and MSCs, and through the type I receptor (IFNγRI), activates JAK1 and JAK2, resulting in the phosphorylation and dimerization of STAT1α [63, 113–115]. Additionally, IFNγ can also active cellular responses through intracrine mechanism [116–118].

Regarding the regulation of MHC-II, IFNγ shows a bimodal effect in MSCs. At high levels, MHC-II is decreased, partly through the retention of the master regulator CIITA in the cytoplasm without any change in the IFNγ receptor [63, 77]. This method of IFNγ-mediated expression of MHC-II in MSCs is different from macrophages where the effect of IFNγ is dose-dependent [63, 119, 120]. The reduction in MHC-II by high IFN-γ level correlates with the immune suppressive function of MSCs [121, 122]. In line with this effect of MSCs is the report showing suppressive effect of third party MSCs during graft versus host disease, with high levels of IFN-γ [123]. There is always a need to balance benefit with potential risk. As an example, although MSCs can benefit graft-versus-host disease, one should keep in mind that MSCs might compromise graft versus tumor effect [124]. Such untoward effect could occur because of the immune suppressive effects of MSCs. As graft versus host disease subsides the level of IFNγ will decrease, resulting in reexpression of MHC-II on the surviving MSCs. The discussion in this paragraph needs to be carefully studied with animal models since these investigations are needed for efficient application of MSCs.

Additional questions are needed to understand the mechanism by which high levels of IFNγ cause immune suppression of MSCs. It is likely that IFN-γ could induce the production of the immune suppressor TGF-β1 [121, 122]. Also, IFNγ can induce the release of other immune inhibitors in MSCs such as indoleamine 2,3-dioxygenase [125]. The role of IFNγ on MSC function is complex, which was underscored by studies showing cross-presentation of antigens in MSCs [126]. Thus, high levels of IFNγ will not always exert immune suppressive properties. Interestingly, TGFβ1 did not alter this cross-presentation despite its changes in the intracellular molecules linked to antigen loading to the MHC molecule [126].

9   Cytokines/miRNAs in Stem Cells

The above discussion focuses on the role of cytokines in immune suppression of MSCs and as mediators for intercellular interactions. This section discusses whether cytokines can change the expression of genes linked to pluripotency. The discussion focuses on *REST* (repressor element-1 silencing transcription factor), also known as NRSF (neuron restrictive silencing factor). REST is a DNA-binding protein that exerts both tumor suppressor and oncogenic properties [127]. REST assembles a repressor complex to modify histone acetylation, chromosomal methylation, and DNA phosphorylation in promoter regions of a wide array of genes [128–134]. The cofactors utilized by REST depend on the type of cell [128, 135–137], suggesting that there could be differences in the mechanisms by which *REST* is regulated in MSCs, and perhaps other stem cells. If there are differences by which *REST* is expressed in stem cells this would indicate that stem cells, through the expression of REST, would cause varied outcomes in response to inflammatory mediators. This is particularly relevant to neurogenesis since REST represses the transcription of neuronal genes in nonneural tissues [130–132, 134].

REST has been linked to multipotency in cancer and healthy stem cells [138, 139]. It is possible that *REST* expression could be changed by cytokines when the MSCs or other stem cells are placed at sites of tissue injury. Since *REST* is also expressed in tumors, including the cancer stem cells, its regulation in stem cell therapy needs to be considered with respect to safety. The same argument can be made for the other stem cell-associated gene, *Oct4*, which is linked to both oncogenesis and pluripotency [140].

IL-1 can regulate the expression of other cytokines. One of these cytokines, TGF-β1, can negatively affect inflammatory responses. IL-1α has been shown to cause a rapid decrease in *REST* expression in MSCs and their neuronal-induced cells [141]. While it is an advantage to have a pluripotent gene decrease for differentiation, rapid decrease in *REST* expression could predispose the cell to transformation, based on the reports that REST could have a tumor suppressor role [142–144].

In addition to cytokines, *REST* is also linked to other molecules during the development of stem cells to specialized cells. MiRNAs have emerged as a central family of mediators in the developmental of stem cells to specialized cells. It is expected that their roles will be tightly linked to other genes such as *REST*, *Oct-4* and inflammatory mediators. As an example, during the development of MSCs to neurons there is a decrease in *REST* expression, which correlates with an increase in miR124a to promote neurogenesis [145]. The regulated expression of *REST*, in consort with miR124a expression, are examples of the mechanisms by which different categories of molecules can coordinate to suppress the expression of nonneuronal genes in neurons while enhancing the expression of neural genes. It is highly likely that the link between miR124 and REST depends on *Oct4*. Computer analyses have determined the presence of multiple REST sites in the 5′ regulatory region of *Oct4* and vice versa. This suggests that *Oct4* and *REST* could regulate the expression of each other. The inclusion of miRNAs is not expected to be independent of the cytokines. For example, IL-1 stimulation can cause a decrease in *REST* [110]. Reduced level of REST could lead to decreased Oct4 and increase in miR124. While these parameters are established as a simplified network, this interaction is complex with other mediators and interactions with nearby cells. Ultimately, the cross talk between the stem cells and mediators within the microenvironment could affect the functional outcome of stem cells such as MSCs. We propose that robust research studies are needed to examine how cytokines, microvesicles, and noncoding RNA such as miRNA can affect multipotency of stem cells.

To explain how cross talk between stem cells and microenvironmental factors affect the fate of stem cells, we discuss two neurotransmitter genes in the context of *REST*. Stem cells such as MSCs, are expected to repress neural genes but this would be reversed as the stem cells mature to neurons. *REST* expression is among the critical genes involved in the expression of neural and nonneural genes. The 5′ noncoding region of the neurotransmitter *TAC1* gene has one functional REST binding site while the *tyrosine hydroxylase* gene has three sites [141, 146]. As expected, REST acts as a repressor for *TAC1* transcription in nonneuronal cells [130, 131, 141]. During the development of MSCs to neurons, *REST* expression is gradually decreased, leading to *TAC1* expression [141]. IL-1 stimulation of MSCs or the early differentiated MSCs toward neurogenesis causes a rapid decrease in REST with concomitant increase in *TAC1* expression [141].

This increase in the neurotransmitter gene is consistent with a repressor function of REST. If this finding is placed in perspective with MSCs at sites of tissue injuries, the MSCs would be exposed to multiple cytokines. This would facilitate cross talk between the MSCs and cytokines. This indicates that it would be difficult to predict the responses of stem cells within a changing microenvironment. The methods described in this book should be applied to begin to address the complex signaling that could occur in stem cells within an inflammatory environment.

A significant question for stem cells in a microenvironment of inflammation is to study if the *Oct4* gene is regulated by cytokines, as well as by other mediators. If cytokines can regulate *Oct4* expression, this would lead to insights on the behavior of stem cells through changes in *Oct4* expression not only upon implantation, but also during the development of stem cells to mature specialized cells. *REST* and *Oct4* are among the genes that could explain the complex functions of stem cells and/or the cells that they generate. Several questions remain unanswered regarding the mechanisms by which cytokines affect the expression of these genes, and/or if there are indirect effects on one to affect the expression of the other.

Oct4 and REST interact with other stem cell-related genes [138]. Thus, it is expected that the molecular network developed by the stem cell genes could be maintained or compromised within an inflammatory milieu with various cytokines. It is highly possible that activators of cytokines could be involved in regulating *Oct-4* and/or *REST* in stem cells. Consequently, there could be responses by the stem cells to remain as pluripotent cells or to generate specialized cells. The outcome might depend on the milieu that is expected of injury. Although IL-1 and TGFβ are discussed as inflammatory mediators in the model of cross talk with stem cells, IFNγ and TNFα have been studied in the immune biology of MSCs.

10 Representative Application

This section discusses the use of MSCs and/or their generated dopamine (DA) neurons for diseases such as Parkinson's (PD) or traumatic brain injury (TBI), both of which are associated with defects in the dopaminergic system. PD and TBI are selected as examples to represent examples of cross talk between an environment and stem cells because it is expected that the microenvironment could represent distinct milieu of inflammatory mediators. Since PD is a chronic disease, the pathology is likely to be different from TBI during the acute phase and even the beginning of a more chronic phase when there will be an abundance of inflammatory cells. In this case, one would expect the cross talk between the implanted cells and microenvironment to be different.

In the substantia nigra, DA neurons are required for motor control, and hence their association with Parkinson's disease [147, 148]. At present, it is unclear if brain disorders will be treated with MSCs and/or their generated DA-producing cells. Since this type of cellular treatment will be an alternative to fetal cells, other issues of interactions with inflammatory mediators and the possibility of rejection will need in-depth analyses before the data could be translated to patients [48, 147, 149].

DA is a phenylethylamine neurotransmitter. Its synthesis requires two enzymatic steps [149]. Tyrosine hydroxylase converts tyrosine to L-DOPA, followed by decarboxylation to DA [147, 149]. DA is stored in synaptic vesicles and upon release, it interacts with any of five related G-protein coupled receptors [150]. Similar to MSC-derived peptidergic neurons, MS-derived DA neurons also express receptors for inflammatory mediators [151]. It is these receptors that support communication with the regions of injuries. In fact, studies with nonhuman primates that were subjected to chemically induced PD reported a promising outcome, and suggest cross talk between the stem cells and factors in the microenvironment [152].

Another area of interest is the relevance of the microenvironment and stem cells in subjects with spinal cord injury (SCI). Several groups, including ours have generated efficient methods to develop MSCs into functional neurons, discussed above. However, lacking are robust studies to determine if inflammatory mediators could benefit the behavior of stem cells in SCI to induce axonal regrowth. This question might not be answered at the molecular level in a complex system such as animal models. An in vitro system would allow us to answer this question in an autologous system in which neurons are placed in contact with skeletal muscle.

In an in vitro system, it would be possible to axotomize a neuron by microdissection to recapitulate injured neurons in SCI patients. Cross talk of injured neurons with inflammatory mediators can be studied by adding cytokines to the system. The system would allow for studies before and after injury on the responses of the injured nerve within a milieu of tissue factors. By establishing an efficient system, the research could identify factors that are relevant to damage and repair. This information would allow for targeted translation of the science to patients where the treatment would take advantage of expected outcome from the development of cross talk between the stem cells and tissue factors at the site of tissue injury. The model would determine whether physicians can repair SCI through cross talk between neurons and other stem cells. Through such models, research could lead to an understanding of nerve regrowth, and synapse formation with skeletal muscle.

11   Conclusion

This review summarizes the complex interactions among cytokines and stem cells, using MSCs as an example. Despite the focus on this single type of stem cells, the information can be extrapolated to other stem cell types. This introductory chapter attempts to bring attention to the deficits in the field of stem cell biology. The methods described in this book could aid in developing experimental studies to contribute to the vast body of existent literature. This will bring stem cells safely to patients, and with efficacious outcome. It is paramount to consider cytokines, other proinflammatory and anti-inflammatory mediators as well as resident cells that could establish a cross talk with the implanted stem cells or their differentiated cells. The discussion with few cytokines such as IL-1 and TGF-β was meant to show how the tissue milieu can cause molecular and developmental changes in stem cells. MSCs as well as other stem cells express multiple receptors for cytokines. This adds to the complexity of stem cell therapy since it would be difficult to predict which receptors are expressed at a given time. This would indicate that the cross talk between the cells and mediators in a microenvironment could change rapidly, depending on the rate of differentiation. Furthermore, if the stem cells are dispersed within the site of tissue injury, there will be lack of synchrony with regard to the developmental stage of the stem cells and, the type of receptors on each cell. Therefore, cross talk between the cells and mediators within the microenvironment of tissue damage would vary within particular regions of tissue damage.

The promise for successful therapy by MSCs is great even if their use would not require additional immune suppressive therapies to prevent rejection [153]. The plasticity of MSCs is evident from studies that show their ability to be preconditioned by microenvironmental factors [154]. MSCs have been reported to integrate in brain regions of animals [155, 156]. Despite the immune suppressive properties of MSCs [42], it appears that this might not be harmful to patients because these individuals are likely to clear a viral infection [114].

Although this chapter does not discuss the use of MSCs in drug/miRNA delivery, the literature suggests that these stem cells could be used in cellular and targeted delivery of drugs. The use of MSCs in drug delivery can also occur in the brain [51]. The efficient and safe use of MSCs in drug delivery requires the information on stem cell-microenvironmental cross talk. It is important to understand if there are other effects of stem cells within tissues as the drug is delivered.

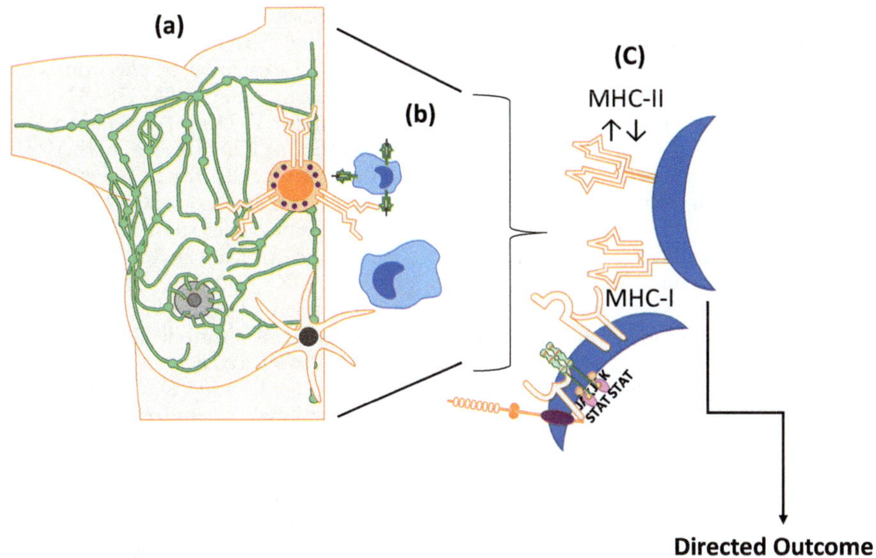

Fig. 1 Diagram depicts the cross talk between the lymphatic system and stem cells. The lymphatic system is shown (**a**) releasing immune cells. The stem cells are shown expressing MHC-I with MHC-II, up or down (**c**, top). Also shown, is the expression of stem cells capable of responding to inflammatory mediators through the expression of specific receptors such as those for cytokines (**b**). Additionally, the cells also express of the innate immune system such as TLR (**c**, bottom). Arrows point to specific outcome, depending on the interaction between the stem cells and the microenvironment

In all clinical outcomes, a common thread is the interaction between the immune system and the therapeutic stem cells. Figure 1 shows an activated lymphatic system capable of interacting with stem cells, which express multiple receptors for cytokines and those of the innate immune system such as the toll receptors (TLR). It is expected that the interaction would dictate the desired outcome. Thus, robust research with established and new stem cell techniques must be applied to understand these interactions.

Acknowledgments

This work is supported by a grant awarded by F.M Kirby Foundation.

References

1. Sell S (2010) On the stem cell origin of cancer. Am J Pathol 176:2584–2594
2. Sell S (2004) Stem cell origin of cancer and differentiation therapy. Crit Rev Oncol Hematol 51:1–28
3. Roberto A, Castagna L, Zanon V et al (2015) Role of naive-derived T memory stem cells in T-cell reconstitution following allogeneic transplantation. Blood 125:2855–2864

4. Frank NY, Schatton T, Frank MH (2010) The therapeutic promise of the cancer stem cell concept. J Clin Invest 120(1):41–50

5. Baumann M, Krause M, Hill R (2008) Exploring the role of cancer stem cells in radioresistance. Nat Rev Cancer 8:545–554

6. McCulloch EA, Till JE (2005) Perspectives on the properties of stem cells. Nat Med 11:1026–1028

7. Drukker M, Katchman H, Katz G et al (2006) Human embryonic stem cells and their differentiated derivatives are less susceptible to immune rejection than adult cells. Stem Cells 24:221–229

8. Li L, Baroja ML, Majumdar A et al (2004) Human embryonic stem cells possess immune-privileged properties. Stem Cells 22:448–456

9. Swijnenburg RJ, Schrepfer S, Govaert JA et al (2008) Immunosuppressive therapy mitigates immunological rejection of human embryonic stem cell xenografts. Proc Natl Acad Sci 105:12991–12996

10. Blum B, Benvenisty N (2008) The tumorigenicity of human embryonic stem cells. In: George FV (ed) Advances in cancer research, vol 100. Academic Press, Cambridge, pp 133–158

11. Ilic D, Ogilvie C (2017) Concise review: human embryonic stem cells-what have we done? What are we doing? Where are we going? Stem Cells 35:17–25

12. Parolini O, Alviano F, Bagnara GP et al (2008) Concise review: isolation and characterization of cells from human term placenta: outcome of the first international workshop on placenta derived stem cells. Stem Cells 26:300–311

13. Dan YY, Riehle KJ, Lazaro C et al (2006) Isolation of multipotent progenitor cells from human fetal liver capable of differentiating into liver and mesenchymal lineages. Proc Natl Acad Sci 103:9912–9917

14. De Coppi P, Bartsch G, Siddiqui MM et al (2007) Isolation of amniotic stem cell lines with potential for therapy. Nat Biotechnol 25:100–106

15. Perin L, Sedrakyan S, Giuliani S et al (2010) Protective effect of human amniotic fluid stem cells in an immunodeficient mouse model of acute tubular necrosis. PLoS One 5(2):e9357

16. Siegel N, Rosner M, Unbekandt M et al (2010) Contribution of human amniotic fluid stem cells to renal tissue formation depends on mTOR. Hum Mol Genet 19:3320–3331

17. Fuchs E (2008) Skin stem cells: rising to the surface. J Cell Biol 180:273–284

18. Takebe N, Ivy SP (2010) Controversies in cancer stem cells: targeting embryonic signaling pathways. Clin Cancer Res 16:3106–3112

19. Tomasetti C, Vogelstein B (2015) Cancer etiology. Variation in cancer risk among tissues can be explained by the number of stem cell divisions. Science 347:78–81

20. Clement V, Dutoit V, Marino D et al (2009) Limits of CD133 as a marker of glioma self-renewing cells. Int J Cancer 125:244–248

21. DA Cruz PA, Lopes C (2017) Implications of different cancer stem cell phenotypes in breast cancer. Anticancer Res 37:2173–2183

22. Lee OK, Kuo TK, Chen WM et al (2004) Isolation of multipotent mesenchymal stem cells from umbilical cord blood. Blood 103:1669–1675

23. Greco SJ, Liu K, Rameshwar P (2007) Functional similarities among genes regulated by Oct4 in human Mesenchymal and embryonic stem cells. Stem Cells 25:3143–3154

24. Woodward WA, Chen MS, Behbod F et al (2005) On mammary stem cells. J Cell Sci 118:3585–3594

25. Sugiyama T, Nagasawa T (2012) Bone marrow niches for hematopoietic stem cells and immune cells. Inflamm Allergy Drug Targets 11:201–206

26. de Souza LE, Malta TM, Kashima Haddad S et al (2016) Mesenchymal stem cells and pericytes: to what extent are they related? Stem Cells Dev 25:1843–1852

27. Castillo M, Liu K, Bonilla LM et al (2007) The immune properties of mesenchymal stem cells. Intl J Biomed Sci 3:100–104

28. Sakaguchi Y, Sekiya I, Yagishita K et al (2004) Suspended cells from trabecular bone by collagenase digestion become virtually identical to mesenchymal stem cells obtained from marrow aspirates. Blood 104:2728–2735

29. Meliga E, Strem BM, Duckers HJ et al (2007) Adipose-derived cells. Cell Transplant 16:963–970

30. Baksh D, Yao R, Tuan RS (2007) Comparison of proliferative and multilineage differentiation potential of human mesenchymal stem cells derived from umbilical cord and bone marrow. Stem Cells 25:1384–1392

31. Troyer DL, Weiss ML (2008) Wharton's jelly-derived cells are a primitive stromal cell population. Stem Cells 26:591–599

32. Caplan AI (1994) The mesengenic process. Clin Plast Surg 21:429–435

33. Fernandez-Moure JS, Corradetti B, Chan P et al (2015) Enhanced osteogenic potential of mesenchymal stem cells from cortical bone: a comparative analysis. Stem Cell Res Ther 6:203

34. Delorme B, Ringe J, Gallay N et al (2008) Specific plasma membrane protein phenotype of culture-amplified and native human

bone marrow mesenchymal stem cells. Blood 111(5):2631–2635

35. Sacchetti B, Funari A, Michienzi S et al (2007) Self-renewing osteoprogenitors in bone marrow sinusoids can organize a hematopoietic microenvironment. Cell 131:324–336

36. Deng J, Petersen BE, Steindler DA et al (2006) Mesenchymal stem cells spontaneously express neural proteins in culture and are neurogenic after transplantation. Stem Cells 24:1054–1064

37. Martinez C, Hofmann TJ, Marino R et al (2007) Human bone marrow mesenchymal stromal cells express the neural ganglioside GD2: a novel surface marker for the identification of MSCs. Blood 109:4245–4248

38. Padovan CS, Jahn K, Birnbaum T et al (2003) Expression of neuronal markers in differentiated marrow stromal cells and CD133+ stem-like cells. Cell Transplant 12:839–848

39. Greco SJ, Zhou C, Ye JH et al (2007) An interdisciplinary approach and characterization of neuronal cells transdifferentiated from human mesenchymal stem cells. Stem Cells Dev 16:811–826

40. Trzaska KA, King CC, Li KY et al (2009) Brain-derived neurotrophic factor facilitates maturation of mesenchymal stem cell-derived dopamine progenitors to functional neurons. J Neurochem 110:1058–1069

41. Trzaska KA, Reddy BY, Munoz JL et al (2008) Loss of RE-1 silencing factor in mesenchymal stem cell-derived dopamine progenitors induces functional maturity. Mol Cell Neurosci 39:285–290

42. Potian JA, Aviv H, Ponzio NM et al (2003) Veto-like activity of mesenchymal stem cells: functional discrimination between cellular responses to alloantigens and recall antigens. J Immunol 171:3426–3434

43. Le Blanc K, Frassoni F, Ball L et al (2008) Mesenchymal stem cells for treatment of steroid-resistant, severe, acute graft-versus-host disease: a phase II study. Lancet 371:1579–1586

44. Pittenger MF, Mackay AM, Beck SC et al (1999) Multilineage potential of adult human mesenchymal stem cells. Science 284:143–147

45. Crisan M, Yap S, Casteilla L et al (2008) A perivascular origin for mesenchymal stem cells in multiple human organs. Cell Stem Cell 3:301–313

46. Caplan AI (2008) All MSCs are pericytes? Cell Stem Cell 3:229–230

47. Takashima Y, Era T, Nakao K et al (2007) Neuroepithelial cells supply an initial transient wave of MSC differentiation. Cell 129:1377–1388

48. Trzaska KA, Kuzhikandathil EV, Rameshwar P (2007) Specification of a dopaminergic phenotype from adult human mesenchymal stem cells. Stem Cells 25:2797–2808

49. Brohlin M, Mahay D, Novikov LN et al (2009) Characterisation of human mesenchymal stem cells following differentiation into Schwann cell-like cells. Neurosci Res 64:41–49

50. Cho KJ, Trzaska KA, Greco SJ et al (2005) Neurons derived from human mesenchymal stem cells show synaptic transmission and can be induced to produce the neurotransmitter substance P by interleukin-1{alpha}. Stem Cells 23:383–391

51. Sherman LS, Shaker M, Mariotti V et al (2017) Mesenchymal stromal/stem cells in drug therapy: new perspective. Cytotherapy 19:19–27

52. Ben JJ, Steven JM (2008) Immunosuppression by mesenchymal stromal cells: from culture to clinic. Exp Hematol 36:733–741

53. Bocelli-Tyndall C, Barbero A, Candrian C et al (2006) Human articular chondrocytes suppress in vitro proliferation of anti-CD3 activated peripheral blood mononuclear cells. J Cell Physiol 209:732–734

54. Nemeth K, Keane-Myers A, Brown JM et al (2010) Bone marrow stromal cells use TGF-beta to suppress allergic responses in a mouse model of ragweed-induced asthma. Proc Natl Acad Sci 107:5652–5657

55. Zhao S, Wehner R, Bornhañuser M et al (2010) Immunomodulatory properties of mesenchymal stromal cells and their therapeutic consequences for immune-mediated disorders. Stem Cells Dev 19:607–614

56. Wagner J, Kean T, Young R et al (2009) Optimizing mesenchymal stem cell-based therapeutics. Curr Opin Biotechnol 20:531–536

57. Ko IK, Kim BG, Awadallah A et al (2010) Targeting improves MSC treatment of inflammatory bowel disease. Mol Ther 18:1365–1372

58. Lim PK, Patel SA, Gregory LA et al (2010) Neurogenesis: role for microRNAs and mesenchymal stem cells in pathological states. Curr Med Chem 17:2159–2167

59. Richardson SM, Hoyland JA, Mobasheri R et al (2010) Mesenchymal stem cells in regenerative medicine: opportunities and challenges for articular cartilage and intervertebral disc tissue engineering. J Cell Physiol 222:23–32

60. da Silva ML, Caplan AI, Nardi NB (2008) In search of the in vivo identity of mesenchymal stem cells. Stem Cells 26:2287–2299

61. Greco SJ, Rameshwar P (2007) Enhancing effect of IL-1 on neurogenesis from adult

human mesenchymal stem cells: implication for inflammatory mediators in regenerative medicine. J Immunol 179:3342–3350

62. Schinkothe T, Bloch W, Schmidt A (2008) In vitro secreting profile of human mesenchymal stem cells. Stem Cells Dev 17:199–206

63. Chan JL, Tang KC, Patel AP et al (2006) Antigen-presenting property of mesenchymal stem cells occurs during a narrow window at low levels of interferon-gamma. Blood 107:4817–4824

64. Campeau PM, Rafei M, Francois M et al (2008) Mesenchymal stromal cells engineered to express erythropoietin induce anti-erythropoietin antibodies and anemia in allorecipients. Mol Ther 17:369–372

65. Heng BC, Cowan CM, Davalian D et al (2009) Electrostatic binding of nanoparticles to mesenchymal stem cells via high molecular weight polyelectrolyte chains. J Tissue Eng Regen Med 3:243–254

66. Romieu-Mourez R, Francois M, Boivin MN et al (2007) Regulation of MHC class II expression and antigen processing in murine and human Mesenchymal stromal cells by IFN-{gamma}, TGF-beta, and cell density. J Immunol 179:1549–1558

67. Pistoia V, Raffaghello L (2010) Potential of mesenchymal stem cells for the therapy of autoimmune diseases. Expert Rev Clin Immunol 6:211–218

68. Rameshwar P, Qiu H, Vatner SF (2010) Stem cells in cardiac repair in an inflammatory microenvironment. Minerva Cardioangiol 58:127–146

69. Patel SA, Sherman L, Munoz J et al (2008) Immunological properties of mesenchymal stem cells and clinical implications. Arch Immunol Ther Exp 56:1–8

70. Kim J, Hematti P (2009) Mesenchymal stem cell-educated macrophages: a novel type of alternatively activated macrophages. Exp Hematol 37:1445–1453

71. Dominici M, Le BK, Mueller I et al (2006) Minimal criteria for defining multipotent mesenchymal stromal cells. The International Society for Cellular Therapy position statement. Cytotherapy 8:315–317

72. Dominici M, Paolucci P, Conte P et al (2009) Heterogeneity of multipotent mesenchymal stromal cells: from stromal cells to stem cells and vice versa. Transplantation 87:S36–S42

73. Fujiwara N, Kobayashi K (2005) Macrophages in inflammation. Curr Drug Targets Inflamm Allergy 4:281–286

74. Pfeifer JD, Wick MJ, Roberts RL et al (1993) Phagocytic processing of bacterial antigens for class I MHC presentation to T cells. Nature 361:359–362

75. Stagg J (2007) Immune regulation by mesenchymal stem cells: two sides to the coin. Tissue Antigens 69:1–9

76. Herrero C, Sebastian C, Marquqs L et al (2002) Immunosenescence of macrophages: reduced MHC class II gene expression. Exp Gerontol 37:389–394

77. Tang KC, Trzaska KA, Smirnov S et al (2008) Down-regulation of MHC-II in mesenchymal stem cells at high IFN- can be partly explained by cytoplasmic retention of CIITA. J Immunol 180:1826–1833

78. Castillo MD, Trzaska KA, Greco SJ et al (2008) Immunostimulatory effects of mesenchymal stem cell-derived neurons: implications for stem cell therapy in allogeneic transplantations. Clin Transl Sci 1:27–34

79. Allan SM, Tyrrell PJ, Rothwell NJ (2005) Interleukin-1 and neuronal injury. Nat Rev Immunol 5:629–640

80. Laver J, Moore MAS (1989) Clinical use of recombinant human hematopoietic growth factors. J Natl Cancer Inst 81:1370–1382

81. Moore MA (2002) Cytokine and chemokine networks influencing stem cell proliferation, differentiation, and marrow homing. J Cell Biochem Suppl 38:29–38

82. Bagby GC (1989) Interleukin-1 and hematopoiesis. Blood Rev 3:152–161

83. Dinarello CA (1996) Biologic basis for interleukin-1 in disease. Blood 87:2095–2147

84. Dinarello CA (2005) Blocking IL-1 in systemic inflammation. J Exp Med 201:1355–1359

85. Roberts AB, Sporn MB (1993) Physiological actions and clinical applications of transforming growth factor-beta (TGF-beta). Growth Factors 8:1–9

86. Massague J, Andres J, Attisano L et al (1992) TGF-beta receptors. Mol Reprod Dev 32:99–104

87. Massague J, Weis-Garcia F (1996) Serine/threonine kinase receptors: mediators of transforming growth factor beta family signals. Cancer Surv 27:41–64

88. Shi Y, Massague J (2003) Mechanisms of TGF-[beta] signaling from cell membrane to the nucleus. Cell 113:685–700

89. Wrana JL (2000) Regulation of Smad activity. Cell 100:189–192

90. Massague J, Wotton D (2000) Transcriptional control by the TGF-beta/Smad signaling system. EMBO J 19:1745–1754

91. Mehra A, Wrana JL (2002) TGF-beta and the Smad signal transduction pathway. Biochem Cell Biol 80:605–622

92. Sato M, Muragaki Y, Saika S et al (2003) Targeted disruption of TGF-beta1/Smad3 signaling protects against renal tubulointerstitial fibrosis induced by unilateral ureteral obstruction. J Clin Invest 112(10):1486–1494

93. Golestaneh N, Mishra B (2005) TGF-beta, neuronal stem cells and glioblastoma. Oncogene 24:5722–5730

94. Downing JR (2004) TGF-{beta} signaling, tumor suppression, and acute lymphoblastic leukemia. N Engl J Med 351:528–530

95. Sokol JP, Schiemann WP (2004) Cystatin C antagonizes transforming growth factor {beta} signaling in normal and cancer cells. Mol Cancer Res 2:183–195

96. Kim SJ, Letterio J (2003) Transforming growth factor-beta signaling in normal and malignant hematopoiesis. Leukemia 17:1731–1737

97. Tracey D, Klareskog L, Sasso EH et al (2008) Tumor necrosis factor antagonist mechanisms of action: a comprehensive review. Pharmacol Ther 117(2):244–279

98. Locksley RM, Killeen N, Lenardo MJ (2001) The TNF and TNF receptor superfamilies: integrating mammalian biology. Cell 104:487–501

99. Romieu-Mourez R, Francois M, Boivin MN et al (2009) Cytokine modulation of TLR expression and activation in mesenchymal stromal cells leads to a proinflammatory phenotype. J Immunol 182:7963–7973

100. van den Berk LCJ, Jansen BJII, Siebers-Vermeulen KGC et al (2009) Mesenchymal stem cells respond to TNF but do not produce TNF. J Leukoc Biol 87:283–289

101. Fu X, Han B, Cai S et al (2009) Migration of bone marrow-derived mesenchymal stem cells induced by tumor necrosis factor-alpha and its possible role in wound healing. Wound Repair Regen 17:185–191

102. Kim YS, Park HJ, Hong MH et al (2009) TNF-alpha enhances engraftment of mesenchymal stem cells into infarcted myocardium. Front Biosci 14:2845–2856

103. Kim VN (2005) MicroRNA biogenesis: coordinated cropping and dicing. Nat Rev Mol Cell Biol 6(5):376–385

104. Bartel DP (2004) MicroRNAs: genomics, biogenesis, mechanism, and function. Cell 116:281–297

105. Kim J, Krichevsky A, Grad Y et al (2004) Identification of many microRNAs that copurify with polyribosomes in mammalian neurons. Proc Natl Acad Sci 101:360–365

106. Zamore PD, Haley B (2005) Ribo-gnome: the big world of small RNAs. Science 309(5740):1519–1524

107. Chen CZ, Li L, Lodish HF et al (2004) MicroRNAs modulate hematopoietic lineage differentiation. Science 303:83–86

108. Harfe BD (2005) MicroRNAs in vertebrate development. Curr Opin Genet Dev 15:410–415

109. Ramkissoon SH, Mainwaring LA, Ogasawara Y et al (2006) Hematopoietic-specific microRNA expression in human cells. Leuk Res 30:643–647

110. Greco SJ, Rameshwar P (2007) miRNAs regulate synthesis of the neurotransmitter substance P in human mesenchymal stem cell-derived neuronal cells. Proc Natl Acad Sci U S A 104:15484–15489

111. Miska E, varez-Saavedra E, Townsend M et al (2004) Microarray analysis of microRNA expression in the developing mammalian brain. Genome Biol 5:R68

112. Schaefer A, O'Carroll D, Tan CL et al (2007) Cerebellar neurodegeneration in the absence of microRNAs. J Exp Med 204:1553–1558

113. Kobayashi M, Fitz L, Ryan M et al (1989) Identification and purification of natural killer cell stimulatory factor (NKSF), a cytokine with multiple biologic effects on human lymphocytes. J Exp Med 170:827–845

114. Kang HS, Habib M, Chan J et al (2005) A paradoxical role for IFN-[gamma] in the immune properties of mesenchymal stem cells during viral challenge. Exp Hematol 33:796–803

115. Stark GR, Kerr IM, Williams BRG et al (1998) How cells respond to interferons. Annu Rev Biochem 67:227–264

116. Ahmed CMI, Burkhart MA, Mujtaba MG et al (2003) The role of IFN{gamma} nuclear localization sequence in intracellular function. J Cell Sci 116:3089–3098

117. Subramaniam PS, Green MM, Larkin J III et al (2001) Nuclear translocation of IFN-gamma is an intrinsic requirement for its biologic activity and can be driven by a heterologous nuclear localization sequence. J Interf Cytokine Res 21:951–959

118. Subramaniam PS, Johnson HM (2004) The IFNAR1 subunit of the type I IFN receptor complex contains a functional nuclear localization sequence. FEBS Lett 578:207–210

119. Bergmann CC, Parra B, Hinton DR et al (2003) Perforin-mediated effector function within the central nervous system requires IFN-{gamma}-mediated MHC up-regulation. J Immunol 170:3204–3213

120. Boss JM (1997) Regulation of transcription of MHC class II genes. Curr Opin Immunol 9:107–113

121. English K, Barry FP, Mahon BP (2008) Murine mesenchymal stem cells suppress dendritic cell migration, maturation and antigen presentation. Immunol Lett 115(1):50–58

122. English K, Ryan JM, Tobin L et al (2009) Cell contact, prostaglandin E(2) and transforming growth factor beta 1 play non-redun-

dant roles in human mesenchymal stem cell induction of CD4+CD25(high) forkhead box P3+ regulatory T cells. Clin Exp Immunol 156:149–160

123. Polchert D, Sobinsky J, Douglas G et al (2008) IFN-gamma activation of mesenchymal stem cells for treatment and prevention of graft versus host disease. Eur J Immunol 38:1745–1755

124. Capitini CM, Herby S, Milliron M et al (2009) Bone marrow deficient in IFN-{gamma} signaling selectively reverses GVHD-associated immunosuppression and enhances a tumor-specific GVT effect. Blood 113:5002–5009

125. DelaRosa O, Lombardo E, Beraza A et al (2009) Requirement of IFN-gamma-mediated Indoleamine 2,3-Dioxygenase expression in the modulation of lymphocyte proliferation by human adipose-derived stem cells. Tissue Eng A 15:2795–2806

126. Francois M, Romieu-Mourez R, Stock-Martineau S et al (2009) Mesenchymal stromal cells cross-present soluble exogenous antigens as part of their antigen-presenting cell properties. Blood 114(13):2632–2638

127. Coulson JM (2005) Transcriptional regulation: cancer, neurons and the REST. Curr Biol 15:R665–R668

128. Belyaev ND, Wood IC, Bruce AW et al (2004) Distinct RE-1 silencing transcription factor-containing complexes interact with different target genes. J Biol Chem 279:556–561

129. Bruce AW, Donaldson IJ, Wood IC et al (2004) Genome-wide analysis of repressor element 1 silencing transcription factor/ neuron-restrictive silencing factor (REST/NRSF) target genes. Proc Natl Acad Sci 101:10458–10463

130. Fiskerstrand CE, Newey P, McGregor GP et al (2000) A role for Octamer binding protein motifs in the regulation of the proximal preprotachykinin-A promoter. Neuropeptides 34:348–354

131. Quinn JP, Bubb VJ, Marshall-Jones ZV et al (2002) Neuron restrictive silencer factor as a modulator of neuropeptide gene expression. Regul Pept 108:135–141

132. Su X, Kameoka S, Lentz S et al (2004) Activation of REST/NRSF target genes in neural stem cells is sufficient to cause neuronal differentiation. Mol Cell Biol 24:8018–8025

133. Wood IC, Belyaev ND, Bruce AW et al (2003) Interaction of the repressor element 1-silencing transcription factor (REST) with target genes. J Mol Biol 334:863–874

134. Yeo M, Lee SK, Lee B et al (2005) Small CTD phosphatases function in silencing neuronal gene expression. Science 307:596–600

135. Greenway DJ, Street M, Jeffries A et al (2007) RE1 silencing transcription factor maintains a repressive chromatin environment in embryonic hippocampal neural stem cells. Stem Cells 25:354–363

136. Sun YM, Greenway DJ, Johnson R et al (2005) Distinct profiles of REST interactions with its target genes at different stages of neuronal development. Mol Biol Cell 16:5630–5638

137. Zhang P, Pazin MJ, Schwartz CM et al (2008) Nontelomeric TRF2-REST interaction modulates neuronal gene silencing and fate of tumor and stem cells. Curr Biol 18:1489–1494

138. Nahas GR, Murthy RG, Patel SA et al (2016) The RNA-binding protein Musashi 1 stabilizes the oncotachykinin 1 mRNA in breast cancer cells to promote cell growth. FASEB J 30:149–159

139. Thakore-Shah K, Koleilat T, Jan M et al (2015) REST/NRSF knockdown alters survival, lineage differentiation and signaling in human embryonic stem cells. PLoS One 10(12):e0145280

140. Trosko JE (2006) From adult stem cells to cancer stem cells: Oct-4 gene, cell-cell communication, and hormones during tumor promotion. Ann N Y Acad Sci 1089:36–58

141. Greco SJ, Smirnov S, Rameshwar P (2007) Synergy between RE-1 silencer of transcription (REST) and NFêB in the repression of the neurotransmitter gene *Tac1* in human mesenchymal stem cells: Implication for microenvironmental influence on stem cell therapies. J Biol Chem 282:30039–30050

142. Weissman AM (2008) How much REST is enough? Cancer Cell 13:381–383

143. Majumder S (2006) REST in good times and bad: roles in tumor suppressor and oncogenic activities. Cell Cycle 5:1929–1935

144. Reddy BY, Greco SJ, Patel PS et al (2009) RE-1-Çosilencing transcription factor shows tumor-suppressor functions and negatively regulates the oncogenic TAC1 in breast cancer cells. Proc Natl Acad Sci 106:4408–4413

145. Conaco C, Otto S, Han JJ et al (2006) Reciprocal actions of REST and a microRNA promote neuronal identity. Proc Natl Acad Sci 103:2422–2427

146. Kim SM, Yang JW, Park MJ et al (2006) Regulation of human tyrosine hydroxylase gene by neuron-restrictive silencer factor. Biochem Biophys Res Commun 346:426–435

147. Perrier AL, Studer L (2003) Making and repairing the mammalian brain – in vitro production of dopaminergic neurons. Semin Cell Dev Biol 14:181–189

148. Lange KW, Mecklinger L, Walitza S et al (2006) Brain dopamine and kinematics

of graphomotor functions. Hum Mov Sci 25:492–509

149. Trzaska KA, Rameshwar P (2007) Current advances in the treatment of Parkinson's disease with stem cells. Curr Neurovas Res 4:99–109

150. Shigetomi S, Fukuchi S (1994) Recent aspect of the role of peripheral dopamine and its receptors in the pathogenesis of hypertension. Fukushima J Med Sci 40:69–83

151. Tansey MG, McCoy MK, Frank-Cannon TC (2007) Neuroinflammatory mechanisms in Parkinson's disease: potential environmental triggers, pathways, and targets for early therapeutic intervention. Exp Neurol 208:1–25

152. Sanberg PR (2007) Neural stem cells for Parkinson's disease: to protect and repair. PNAS 104:11869–11870

153. Blanc KL, Pittenger MF (2005) Mesenchymal stem cells: progress toward promise. Cytotherapy 7:36–45

154. Gregory CA, Ylostalo J, Prockop DJ (2005) Adult bone marrow stem/progenitor cells (MSCs) are preconditioned by microenvironmental "niches" in culture: a two-stage hypothesis for regulation of MSC fate. Sci STKE 2005:e37

155. Honma T, Honmou O, Iihoshi S et al (2006) Intravenous infusion of immortalized human mesenchymal stem cells protects against injury in a cerebral ischemia model in adult rat. Exp Neurol 199:56–66

156. Shyu WC, Lee YJ, Liu DD et al (2006) Homing genes, cell therapy and stroke. Front Biosci 11:899–907

Part II

Stem Cell Study in Model Organisms and Human

Chapter 2

Single-Step Plasmid Based Reprogramming of Human Dermal Fibroblasts to Induced Neural Stem Cells

Luis Azmitia and Philipp Capetian

Abstract

Reprogramming of somatic cells to induced pluripotent stem cells (iPSC) and subsequent differentiation opened up the opportunity of deriving cell types in vitro which (like neurons) had a very restricted accessibility in the past. However, cell culture protocols for iPSC reprogramming, neural induction and differentiation tend to be labor and time intensive, costly and commonly depend on viral vector delivery. Single-step reprogramming to induced neural stem cells (iNSC) avoids many of the necessary intermediate steps of the aforementioned method but yields a cell type that proliferates over longer time spans and readily differentiates to mature neurons when required. Here we describe a plasmid based reprogramming protocol employing defined, commercially available components for induction and proliferation of iNSC, followed by a defined, small molecule based differentiation step toward mature neurons. The described method might be of particular interest for groups with limited resources and/or restricted access to higher biosafety level facilities required for viral transduction, but also for groups requiring a high throughput for dealing with large numbers of cell lines.

Key words Induced neural stem cells, Plasmid transfection, Direct neural reprogramming, Small molecule based differentiation, Human dermal fibroblasts, Human neurons, Small molecules

1 Introduction

Induced pluripotency introduced the possibility of reprogramming human somatic cells to a state of stemness that theoretically allows subsequent differentiation to any cell type of the human body with the help of the appropriate protocol [1]. While groups with a scientific focus on the basic properties of pluripotent stem cells or the early steps of induction and differentiation to mature cell types inevitably require the cultivation of iPSC, groups solely interested in the mature cell type would rather experience the time and labor consuming protocols of maintaining pluripotent cells in vitro rather as a "necessary evil." These considerations lead to the development of protocols allowing the reprogramming of somatic cells to mature cell types (e.g., neurons) without the intermediate culti-

Shree Ram Singh and Pranela Rameshwar (eds.), *Somatic Stem Cells: Methods and Protocols*, Methods in Molecular Biology, vol. 1842, https://doi.org/10.1007/978-1-4939-8697-2_2, © Springer Science+Business Media, LLC, part of Springer Nature 2018

vation as iPSC [2–4]. However, the low yield and inability of terminally differentiated cells to proliferate limit the practicability of this paradigm. Generating proliferative neural precursors circumvents these drawbacks. In general, two paradigms for this approach have been established: The first one employs reprogramming factors also suitable for reprogramming toward iPSC, but addition of the appropriate medium and factors promotes the formation of neural instead of pluripotent stem cells [5]. The second uses overexpression of transcription factors associated with neural stem cells [6]. The first approach carries the theoretical risk of generating fully pluripotent stem cells and thus deriving nonneural cell types. Indeed, one study could demonstrate the transient presence of pluripotent intermediates during the reprogramming for murine cells [7]. For human cells, our group could demonstrate a temporary expression of endogenous transcription factors associated with pluripotency (Oct3/4) [8]. However, since the long-term culture conditions in a serum-free environment preclude the growth of iPSC, we feel that this objection has no practical consequence for groups focused on the final differentiated neurons.

The protocol presented on the following pages is a detailed description of a reprogramming method our group recently established [8]. It is based on plasmid transfection of three reprogramming vectors by electroporation, followed by a one-step neural reprogramming process to induced neural stem cells (iNSC), mediated by addition of a neural induction medium. Colonies are identified by their morphology and selected by manual picking, followed by propagation as an adherent monolayer. Differentiation is achieved by a defined neural culture medium with the addition of a small molecule inhibitor of the notch pathway (DAPT). All components are commercially available and a facility offering biosafety level S1 is sufficient. Furthermore, the schedule of cell culture maintenance can easily be performed in a 5-day workweek and thus requires no weekend duties (as usually required for iPSC culture). The protocol might therefore be particularly suitable for groups interested in patient derived neurons without the necessary funds and work force required for iPSC cultivation. The regional identity of the differentiated neurons is mostly cortical, but GABAergic and dopaminergic neurons are also present in culture. However, scientists in need of an enriched particular regional phenotype have either to perform selection steps or refer to appropriate directed differentiation protocols from iPSC.

2   Materials

2.1   Consumables

1. 24-well polystyrene clear multiple well plates.

2. 6-well polystyrene clear multiple well plates.

3. Amaxa NHDF nucleofector kit (Lonza, Cat. No. VPD-1001, *see* **Note 1**) and Amaxa Nucleofector II device.

4. Accutase.

5. DAPT (also known as GSI-IX or LY-374973).

6. Donkey anti-mouse IgG, highly crossabsorbed, Alexa Fluor 488 (ThermoFisher Scientific A-21202).

7. Donkey anti-rabbit IgG, highly crossabsorbed, Alexa Fluor 568 (ThermoFisher Scientific A10042).

8. Gilson Pipette Tips 200 μL.

9. hEGF (human Epidermal Growth Factor).

10. hFGF2 (human Fibroblast Growth Factor 2).

11. hFGF4 (human Fibroblast Growth Factor 4).

12. Matrigel growth factor reduced matrix.

13. Mouse anti-MAP2 monoclonal antibody (Millipore, Cat. No. MAB378, 1:800).

14. Mouse anti-nestin monoclonal antibody (BD Biosciences, Cat. No. 55639, 1:2000).

15. Nalgene™ Rapidflow™ 0.2 μm 500 mL filter unit.

16. NHDF Nucelofector™ Kit (100 RCT) (Lonza, Cat. No. VVPD-1001, *see* **Note 1**).

17. Nunclon 6-well plates.

18. PFA (paraformaldehyde).

19. Plasmid pCXLE-hSK.

20. Plasmid pCXLE-hOCT 3/4-shp53-F.

21. Plasmid pCXLE-hUL.

22. Plasmid pCXLE-EGFP.

23. Rabbit anti-bIII-tubulin polyclonal antibody.

24. Rabbit anti-GFAP polyclonal antibody.

25. Round scalpel no. 10.

26. STEMDiff™ Neural Induction Medium.

27. STEMDiff™ Neural Progenitor Medium.

28. Sterile filter unit 500 mL.

29. T75 Red Cap Flasks.

30. TrypLE.

31. Y27632 Stemolecule.

2.2 Laboratory Equipment

1. Cell centrifuge.

2. Cell culture incubator 37 °C 5% CO_2.

3. Cell culture bench.

4. Confocal laser scanning microscope.

5. Horizontal orbital shaker.

6. Light microscope.

7. Magnetic heating plate.

8. Nucelofector™ 2b device.

9. Tabletop microcentrifuge.

2.3 Cell Culture Medium and Staining Solutions for Immunofluorescence

1. Basic fibroblast medium: DMEM is enriched with 15% FCS, 1% GlutaMAX, 1% HEPES buffer, 1% antibiotic–antimycotic final concentrations and filtered through a filter unit. Medium can be stored at 4 °C for up to 1 month.

2. N2B27 medium: Mix DMEM/F12 with Neurobasal medium to 1:1. Add 1% N2 supplement, 2% B27 supplement, 1% NEAA, 1% GlutaMAX, and 1% antibiotic–antimycotic. Filter through a filter unit and store at 4 °C for up to 1 month (*see* **Note 2**).

3. DAPI solution: DAPI is dissolved at a final concentration of 2 μg/mL in PBS and stored at 4 °C protected from light.

4. 5% Donkey Serum Blocking buffer: Donkey serum is added to PBS at a final concentration of 5%, Triton X-100 at 0.1%, and NaN_3 at 0.01%. Store at 4 °C.

5. 1% Donkey Serum Incubation buffer: Donkey serum is added to PBS at a final concentration of 1%, Triton X-100 at 0.1%, and NaN_3 at 0.01%. Store at 4 °C.

6. 4% PFA: Dissolve PFA into PBS at 65 °C at a final concentration of 4%. The mixture is transferred to a fume hood and maintained on a heating plate at 60 °C under continuous stirring until dissolved. The pH is adjusted to 7.2 by addition of NaOH and the final volume adjusted to 1 L.

7. 0.1% Triton X-100 Washing buffer: Dissolve Triton X-100 into PBS at room temperature at a final concentration of 0.1%.

3 Methods

3.1 Transfection of Human Dermal Fibroblasts (HDF)

1. Before commencing, prepare Matrigel coated 6-well plates: Thaw one 300 μL aliquot of Matrigel per 6-well plate on ice. Meanwhile, let the 6-well plates and the necessary plastic ware cool down in an −20° lab freezer. Aliquot 6 mL of ice-cold DMEM medium per aliquot in a precooled 50 mL falcon tube. By using precooled pipette tips, transfer the Matrigel to the DMEM and mix by pulse vortexing or shaking. Put the mixture back on ice. Add 1 mL of the mixture to each well of a cold 6-well plate with a cold pipette and immediately shake the plate to cover the bottom evenly. In order to polymerize

the Matrigel before further use, leave the plates for 30 minutes at room temperature under the hood. Unused plates can be stored in the fridge at 4 °C for up to 3 weeks. However, plates need to rewarm for 30 min at room temperature when taken out of the freezer.

2. Collect fibroblasts at an early passage (below passage 10, *see* **Note 3**) by enzymatic digestion with TrypLE for 10–15 min. Spin down at 75 × *g* for 5 min and gently resuspend the pellet in 1 mL basic fibroblast medium.

3. Count the cell density of the suspension, spin down two aliquots of 500 K cells for every cell line in a tabletop centrifuge (75 × g for 5 min).

4. Discard the supernatant as thoroughly as possible without disturbing the pellet.

5. Add 100 µL of the Nucleofector solution and the plasmids as described below:

Control transfection:

   1 µg pCXLE-EGFP plasmid.

Reprogramming transfection:

   1 µg pCXLE-hSK plasmid.

   1 µg pCXLE-hOCT3/4-shp53-F plasmid.

   1 µg pCXLE-hUL plasmid.

6. Resuspend the pellets by gently pipetting up with the provided pipette and down without introducing air bubbles.

7. Transfer the mixture by the designated plastic pipettes to the cuvettes. Electroporate with the Nucleofector device with program U-023 (*see* **Note 1**).

8. Immediately after electroporation, use 2 mL of warm basic fibroblast medium to "wash" the cell solution out of the cuvettes and transfer everything to a Matrigel covered well after removing the DMEM supernatant. Gently rock the plate to distribute the cells evenly and move the plate to the cell incubator.

9. Leave the cells with basic fibroblast medium for 2 days.

10. Follow up your transfection efficiency by checking the pCXLE-EGFP expression under the fluorescence microscope. In case of sufficient transfection efficiency, wells expressing the control plasmid can be discarded to save medium.

3.2 Neural Induction and Expansion of the Reprogrammed Cells

1. On day 2, start the stepwise transfer to N2B27-medium, by increasing the concentration by 25% every other day. At the same time point start supplementing the growth factors: hFGF2 at a final concentration of 20 ng/mL, hFGF4 10 ng/mL,

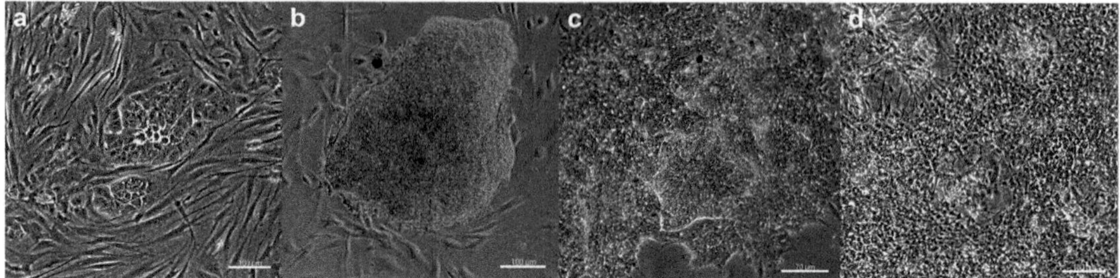

Fig. 1 Morphology of the reprogramming and differentiation process: Roughly 10 days after the transfection, fibroblasts will lose their characteristic morphology and form colonies with tightly packed, small round cells (**a**). Those colonies will eventually continue growing up to a considerable size that allows for picking them mechanically (**b**). After replating those colonies, cells will continue growing in high densities and (especially during early passages) form "neural rosettes" (**c**). Differentiation by omitting growth factors and adding DAPT will lead to a change of morphology of cells spreading out a dense network of neurites (**d**)

and hEGF 10 ng/mL (*see* **Note 4**). We recommend 2 mL of medium per 6-well plate (*see* **Note 2**).

2. After day 8, change the medium three times per week (e.g., Monday, Wednesday, and Friday).

3. Between day 10 and day 30 morphological changes will show among the transfected fibroblasts (Fig. 1a): Cells start losing their processes, acquire rounded shapes, and form colonies. Observe individual colonies for at least 1 week to ensure a sufficient growth with a densely packed, "bulging" morphology (Fig. 1b, *see* **Notes 5** and **6**).

3.3 Picking and Propagation of Reprogrammed Induced Neural Stem Cells (iNSC)

1. Before picking the cells, change them to fresh N2B27 medium and add Y27632 at a final concentration of 10 µM. Be sure to thoroughly disinfect your working area and microscope, because of the particular high risk of contamination during this step.

2. Cut the selected iNSC colonies with a round sterile scalpel no. 10 through a chopping movement. Cut every colony into 10–20 small pieces depending on their size. Dislodge the chunks by a lifting movement into the medium and collect the entire medium containing the cells with a pipette. Replate the cells on a new well of a 6-well plate coated with Matrigel. Continue the process until all selected colonies have been transferred. Add a final amount of fresh N2B27-medium to the source plates and keep them for further pickings of new colonies at later time points (*see* **Note 7**).

3. After 1–2 weeks in culture, outgrowing cells will form a uniform monolayer of polygonal cells, growing in a characteristic "palisading" manner and (at earlier passages) form "neural rosettes" [9] (Fig. 1c). Once a sufficient density (~100–120%) is achieved, split cells with 1 mL accutase for 5–8 min inside

the incubator, take up the cell suspension in an additional amount of 4 mL DMEM and spin down for 1 min at 1000 RPM. Discard the supernatant and resuspend the cells thoroughly in 1 mL of N2B27-medium. Cells can either be replated at a fixed ratio (1:5 — 1:10 works in most cases) or 250 000 cells per new Matrigel coated well of a 6-well plate containing 2 mL N2B27-medium +20 ng/mL hFGF2, 10 ng/mL hFGF4, 10 ng/mL hEGF, and 10 µM Y27632. From then on, cultures are kept under proliferative conditions and split when reaching high confluency. iNSC grow particularly well under high density without detaching for achieving a high cell yield per well and should not be plated to sparse (*see* **Note 8**).

3.4 Differentiation of iNSC

1. Coat a desired number of wells of a 24-well plate with Matrigel as described before.

2. When splitting proliferating iNSC, count the cell suspension and plate 50 K of cells per each well of the 24-well plate an add 500 µL of N2B27-medium + 20 ng/mL hFGF2, 10 ng/mL hFGF4, 10 ng/mL hEGF, and 10 µM Y27632 to each well.

3. Cultivate the plates until all plates reach a sub confluent cell density. At this step, it is important not to let the cells overgrow as this can lead to detachment of the monolayer during the differentiation and not let them be too sparse, as this will impair the process and cells will eventually die off (*see* **Note 9**). Once they reach the appropriate density, add N2B27-medium without bFGF, FGF4, and EGF but instead add DAPT at a final concentration of 10 µM. Change medium three times a week (e.g., Mon, Wed, and Fri). This will start the differentiation process. Cells should start changing morphology within a week: The cell density will decrease; cell bodies will acquire a more rounded morphology and neurites will grow out (Fig. 1d). Usually after 1 month, cells will reach a sufficient maturity for staining with adult neuronal and glial marker. However, if you are planning experiments requiring a more mature phenotype (e.g., electrophysiology) you should give them at least 2–3 months of differentiation time or even longer.

3.5 Immunofluorescence

The 24-well format has proven in our hands as the most suitable for immunofluorescence. We would therefore advise to plate cells undergoing the staining procedure on this format beforehand. If not otherwise stated use 500 µL of solution per well. Of course, other plates can be used as well, scale volumes up or down accordingly. All washing and incubation steps are performed on a horizontal orbital shaker. If not stated otherwise, perform steps at RT.

1. Fix proliferating iNSC reaching 80–90% confluency and iNSC after 30 days of differentiation with 4% PFA for 15 min.

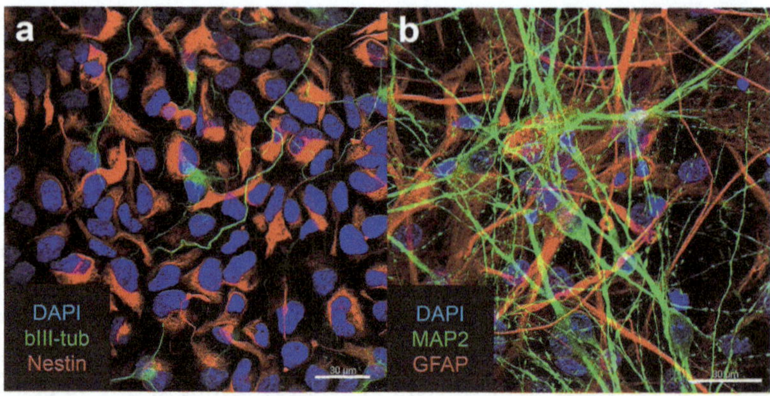

Fig. 2 Immunofluorescence of iNSC under proliferation and after differentiation: Under proliferative conditions, the majority of iNSC will be positive for the neural stem cell marker nestin, with interspersed neuronal progenitors positive for bIII-tubulin (**a**). Multipotent iNSC will differentiate into MAP2-positive neurons and GFAP-positive astrocytes (**b**)

2. Wash three times for 5 min with 1% PBS.

3. Add blocking buffer for 45 min to block unspecific binding sites.

4. Dilute the primary antibody according to the dilutions mentioned in the consumables list in dilution buffer (200 µL per sample). Combine nestin with bIII-tubulin and MAP2 with GFAP. Incubate overnight at 4 °C.

5. The next day, wash three times for 15 min with washing buffer.

6. Dilute the secondary antibody according to the dilutions mentioned in the consumables list in dilution buffer (200 µL per sample). Combine donkey anti-mouse 488 and donkey anti-rabbit 568. Incubate for 2 h.

7. Wash three times for 15 min with washing buffer.

8. Incubate with DAPI solution for 10 min.

9. Wash three times with PBS for 5 min.

10. Analyze cells on a fluorescent microscope or confocal laser-scanning microscope (CLSM): Most proliferating cells should express nestin with interspersed positivity of bIII-tubulin (Fig. 2a). After differentiation, there should be MAP2-positive cells with long and complex neurites on a basal layer of GFAP-positive astrocytes (Fig. 2b).

4 Notes

1. In our hands, successful reprogramming was only achieved by electroporation of fibroblasts in contrast to chemical reagents. When using other electroporation devices, the transfection protocol and even the amount of plasmid needs to be adapted since overloading cells can cause cell death and failure of reprogramming.

2. As an alternative to growth factor supplemented N2B27-medium, we were successful with using commercially available neural induction medium for reprogramming, followed by neural progenitor medium for proliferation (both from STEMCELL Technologies). However, differentiation was still carried out in N2B27-medium with DAPT. These media were originally developed for neural induction and cultivation of iPSC derived neural progenitors. Using them reduces the workload of mixing culture medium and producing aliquots of growth factors. Furthermore, they can increase consistency of the reprogramming process.

3. The reprogramming success depends on the condition of the fibroblasts. Fibroblasts should be below passage 10. If you generate your own primary fibroblast lines, reprogramming cells before freezing/thawing greatly enhances the efficiency.

4. Store your growth factors at workable concentrations and in small amounts, so you can add them to the medium just before adding the medium to the wells and to avoid prolonged storage at 4 °C (best is to use them up within 1 month).

5. Colonies can only arise from a dense fibroblast layer supporting the reprogramming process. Since fibroblasts stop proliferating in serum-free medium, give the cells enough time to grow before switching to N2B27-medium.

6. If you do not observe some changes in cell morphology or at least some small colonies by day 14 after transfection, discard and start again with fibroblasts of a lower passage. However, keep in mind that some lines are easier to transfect than others for no obvious reason. Some lines will never work whatsoever.

7. Keep in mind that cells reprogrammed by this protocol transit through an intermediate pluripotent step. The neural factors and medium greatly enhance neural progeny; however, the yield is not 100%. Therefore, check on outgrowing cells after replating and discard all clones that either show a flat morphology with large nuclei and very prominent nucleoli (fully or partially pluripotent stem cells, fig. 3a) or any other aberrant morphology (e.g., epithelial cells showing a "cobblestone"-like growth pattern, fig. 3b).

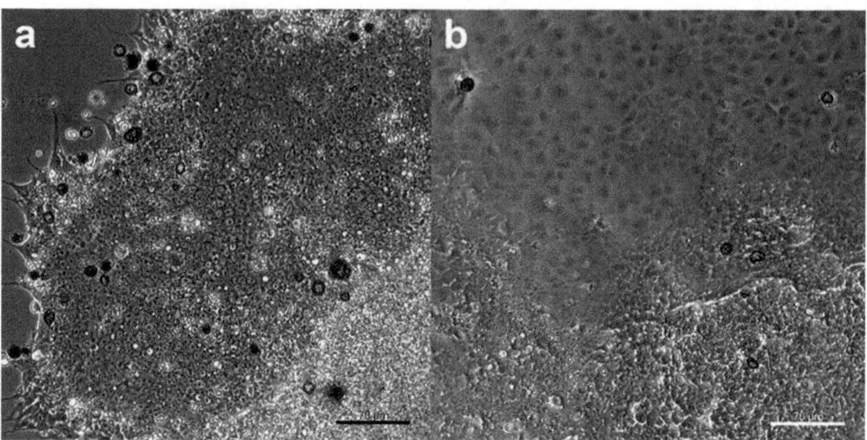

Fig. 3 Undesirable results: After picking and replating colonies, outgrowing cells might show aberrant morphologies. Cells with a large round nucleus and prominent nuclei growing in flat colonies represent partially or fully pluripotent stem cells (**a**). In general, any cell type showing a different growth pattern than neural stem cells are suspicious. For example flat cells growing in "cobblestone" patterns represent epithelial cells (**b**)

8. iNSC can be kept for 15–20 passages in culture without losing their ability to proliferate and differentiate. It is a robust and rather fail-safe protocol. However, as this is a FGF-based protocol, those cells are not immortal and will lose their potential at some point. The idea behind this protocol is deriving neurons in-vitro from patients with a limited amount of workload in a short time. If you need very early and distinct stages of neural precursors that show very little changes during passaging, you should rather refer to other protocols available [10].

9. If your iNSC start dying in between media changes, you will have to consider changing the media every day (instead of every other day) or increasing the amount of media per well.

References

1. Takahashi K, Tanabe K, Ohnuki M et al (2007) Induction of pluripotent stem cells from adult human fibroblasts by defined factors. Cell 131:861–872. https://doi.org/10.1016/j.cell.2007.11.019

2. Pfisterer U, Kirkeby A, Torper O et al (2011) Direct conversion of human fibroblasts to dopaminergic neurons. Proc Natl Acad Sci U S A 108:10343–10348. https://doi.org/10.1073/pnas.1105135108

3. Caiazzo M, Giannelli S, Valente P et al (2015) Direct conversion of fibroblasts into functional astrocytes by defined transcription factors. Stem Cell Reports 4(1):25–36. https://doi.org/10.1016/j.stemcr.2014.12.002

4. Halliwell RF (2016) Electrophysiological properties of neurons derived from human stem cells and iNeurons in vitro. Neurochem Int 106:37–47. https://doi.org/10.1016/j.neuint.2016.10.003

5. Kim J, Efe JA, Zhu S et al (2011) Direct reprogramming of mouse fibroblasts to neural progenitors. Proc Natl Acad Sci 108:7838–7843. https://doi.org/10.1073/pnas.1103113108

6. Lujan E, Chanda S, Ahlenius H et al (2012) Direct conversion of mouse fibroblasts to self-renewing, tripotent neural precursor cells. Proc Natl Acad Sci U S A 109:2527–2532. https://doi.org/10.1073/pnas.1121003109

7. Maza I, Caspi I, Zviran A et al (2015) Transient acquisition of pluripotency during somatic cell transdifferentiation with iPSC reprogramming factors. Nat Biotechnol 33(7):769–774. https://doi.org/10.1038/nbt.3270

8. Capetian P, Azmitia L, Pauly MG et al (2016) Plasmid-based generation of induced neural stem cells from adult human fibroblasts. Front Cell Neurosci 10:245. https://doi.org/10.3389/fncel.2016.00245

9. Elkabetz Y, Panagiotakos G, Shamy GA et al (2008) Human ES cell-derived neural rosettes reveal a functionally distinct early neural stem cell stage. Genes Dev 22:152–165. https://doi.org/10.1101/gad.1616208

10. Reinhardt P, Glatza M, Hemmer K et al (2013) Derivation and expansion using only small molecules of human neural progenitors for neurodegenerative disease modeling. PLoS One 8(3):e59252. https://doi.org/10.1371/journal.pone.0059252

Isolation and Analysis of Mesenchymal Progenitors of the Adult Hematopoietic Niche

Mayra Garcia, Lihong Weng, Xingbin Hu, and Ching-Cheng Chen

Abstract

Mesenchymal stromal cells are an important component of the adult hematopoietic stem cell niche. They are a diverse population of cells that include a hierarchy of primitive, intermediate, and mature osteoprogenitors that support HSCs and supply the bone with matrix producing osteoblast. To understand the different roles played by individual types of progenitors, it is necessary to separate individual populations and analyze them in a controlled environment. Here we describe two transplantation models, an ectopic bone forming assay and an intravenous injection assay, in which niche components can be isolated and manipulated to dissect their individual properties.

Key words Mesenchymal stromal progenitors, Hematopoietic stem cell niche, In vivo transplantation

1   Introduction

The hematopoietic stem cell (HSC) niche is composed of a complex set of stromal support cells that maintain HSCs and promote normal hematopoiesis in the bone marrow (BM) [1–4]. Mesenchymal stromal cells, along with other support cells such as endothelial cells, CXCL12-abunant reticular (CAR) cells, and osteoblasts, are thought to compose the HSC niche within the BM. These support cells provide key signals that help maintain self-renewal and regulate differentiation of HSCs [1–4]. The mesenchymal stromal cell population is composed of a diverse set of cells that include mesenchymal stromal progenitors and osteoprogenitors. The main obstacle in studying the role of mesenchymal stromal cells in HSC maintenance is the heterogeneous nature of the population which makes it difficult to target specific stromal cells.

Several investigators have employed markers such as Osterix, LepR, Nestin, PDGFR-α, Sca1, and CD166 to subdivide the mesenchymal stromal cells to better understand how they themselves

Shree Ram Singh and Pranela Rameshwar (eds.), *Somatic Stem Cells: Methods and Protocols*, Methods in Molecular Biology, vol. 1842, https://doi.org/10.1007/978-1-4939-8697-2_3, © Springer Science+Business Media, LLC, part of Springer Nature 2018

are maintained and the roles they play in HSC maintenance [5–9]. Recently, we identified a hierarchy of mesenchymal stromal progenitors using Sca1 (a marker expressed by HSCs and other stem cells), CD146 (a marker for human osteoprogenitors), and CD166 (an adhesion molecule expressed by both HSCs and mesenchymal stromal cells) to subdivide the mesenchymal stromal population (CD45⁻Ter119⁻CD31⁻) [6, 10, 11]. We identified a CD146⁻CD166⁻Sca1⁺ (Sca1⁺; primitive progenitor) population which gave rise to CD166⁻CD146⁺ (CD146⁺; intermediate progenitor) and CD146⁻CD166⁺ (CD166⁺ mature osteoprogenitor) cells [12]. All three progenitors could support HSCs and we observed that the Sca1⁺ progenitors are a significant source of KitL, a key ligand for HSC self-renewal and maintenance [12, 13]. Compared to previous markers we found that all three populations were LepR positive suggesting that our markers can be used to subdivide this population [7, 12]. In contrast, only CD146⁺ and CD166⁺ cells contained Nestin suggesting that Nestin is a marker for more differentiated stromal cells [5, 12]. Here we present the in vivo bone forming assay and intravenous (IV) injection assay we used to conduct our studies. These assays are powerful for HSC niche studies as they provide a means to target specific niche components that would otherwise not be accessible for in vivo investigations.

2 Materials

2.1 Isolation of Progenitors

1. Four adult mice (> 8 weeks).
2. Dissection scissors and forceps.
3. E14-E15 staged mouse embryos.
4. Phosphate-buffered saline.
5. Glass mortar and pestle.
6. 20 G1 needle.
7. 3 mL syringe.
8. 15 mL tube.
9. 50 mL tube.
10. Collagenase Mix: 1 mg/mL Collagenase D, 100 ng/mL DNaseI, 1 mg/mL BSA, 20 mM Hepes in DMEM.
11. Red Cell Lysis Buffer: 150 mM NH_4Cl, 10 mM $KHCO_3$, 0.1 mM EDTA.
12. CD45 MicroBeads, mouse (Miltenyi Biotec).
13. LD columns (Miltenyi Biotec).
14. Falcon Cell Strainers-Mesh Size: 40 μM.
15. 16 G1 needle.

16. 5 mL syringe.

17. 5 mL FACS tube.

18. CD16/32 antibody (1:100 BioLegend 101301).

19. CD45-Alexa700 (1:100 BioLegend 103128).

20. CD31-PE-Cy7 (1:100 BioLegend 102418).

21. Ter119-PE-Cy5 (1:500 BioLegend 116210).

22. Sca1-BV605 (1:20 BioLegend 108133).

23. CD146-Fitc (1:25 BioLegend 134705) or CD146-Dylight650 (1:25 BioLegend 134702-Conjugated to Dylight650).

24. CD166-PE (1:25 R&D Systems FAB1172P).

25. LIVE/DEAD® Fixable Aqua Dead Cell Stain Kit, for 405 nm excitation (ThermoFisher).

26. 0.1% BSA.

27. Swing bucket centrifuge.

28. Incubator.

29. Nutator or rocker.

30. FACS sorter.

31. Dissection microscope.

2.2 shRNAi Manipulation of Progenitors

1. Lentivirus: we prepare using VSV.G and CMVDR8.74 packaging plasmids and jetPrime transfection reagent.

2. Supplemented MEM α medium: MEM α medium, 15% FBS, 1 mM sodium pyruvate, 1× MEM nonessential amino acids, 50 μM β-mercaptoethanol, 10 mM hepes, and 1× Pen-Strep.

3. 96-well cell culture plates; round bottom.

4. Trypsin.

5. 0.1% BSA.

6. Incubator set at 37 °C, 5% CO_2, 2% O_2.

7. Swing bucket centrifuge.

8. FACS sorter.

2.3 Transplantation of Progenitors

1. BD Matrigel (BD Bioscience).

2. Sterile phosphate-buffered saline.

3. Immunodeficient mouse.

4. Sterilized surgical tools: scissors, forceps, suture kit, skin staples.

5. Isoflurane, USP or another anesthetic.

6. Ophthalmic ointment/sterile artificial tears ointment.

7. Povidone–iodine.

8. 70% isopropanol.

9. Hamilton® Gastight® syringe, 1700 series, 25 µL 22 s ga bevel tip (Sigma-Aldrich).

10. Sterile cotton swabs.

11. Postoperative analgesic such as Buprenex.

12. Insulin syringe.

13. UBC-creERT2/iDTR mice.

14. 20 mg/mL tamoxifen in corn oil.

15. 1 ng/µL diphtheria toxin in PBS.

16. Swing bucket centrifuge.

17. Irradiator for mice.

2.4 Post-transplantation Analysis

1. Phosphate-buffered saline.

2. Dissection scissors and forceps.

3. 60 mm tissue culture dish.

4. 15 mL tube.

5. Collagenase Mix: 1 mg/mL collagenase D, 100 ng/mL DNaseI, 1 mg/mL BSA, 20 mM Hepes in DMEM.

6. Nutator or rocker.

7. 40 µM mesh cell strainer.

8. Cryomolds.

9. Tissue-Tek® OCT compound.

10. Cryostat.

11. Fluorescence dissection microscope.

12. Swing bucket centrifuge.

13. FACS analyzer.

14. Fluorescence compound or confocal microscope.

3   Methods

3.1 Isolation of Progenitors

3.1.1 Isolation of Adult Mesenchymal Stromal Progenitors

1. Dissect hind limbs from four adult mice (> 8 weeks). Completely remove the muscle from the bones keeping the bone including the knee cap intact (Fig. 1a) (*see* **Note 1**).

2. Place the bones in a glass mortar add 3–5 mL cold PBS. Gently, crush the bones with a pestle. The marrow will be released into the PBS.

3. Remove the PBS/marrow with a 20 G1 needle and 3 mL syringe be careful to not take up the bone (*see* **Note 2**).

4. Repeat **steps 2** and **3** until the marrow is removed from the bones (Fig. 1b).

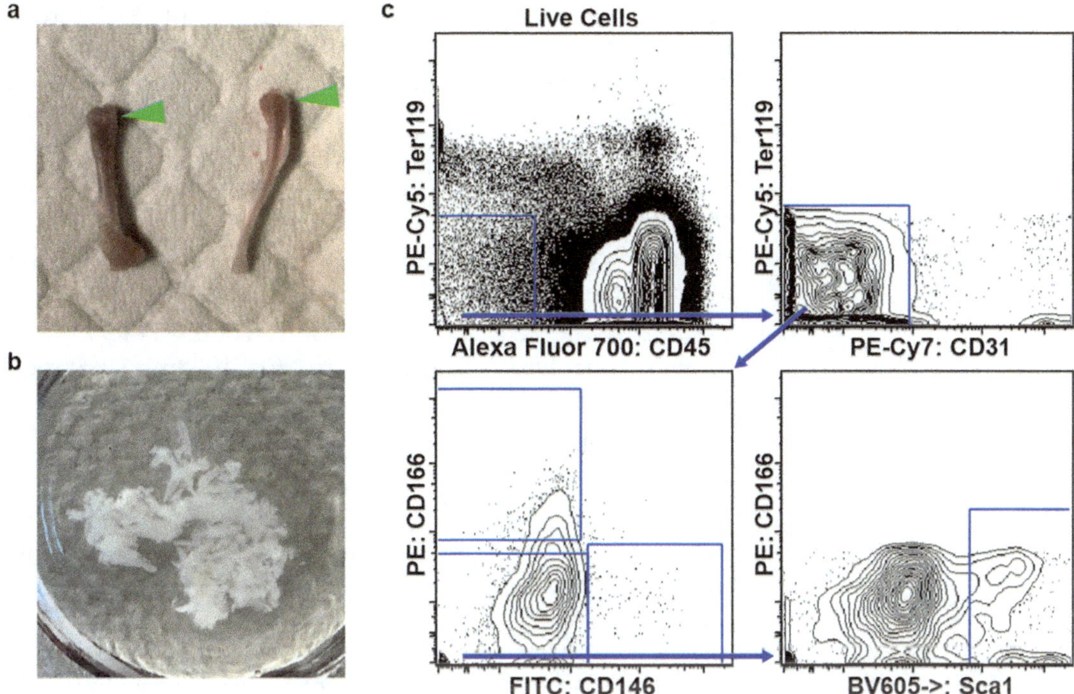

Fig. 1 Isolation of adult mesenchymal stromal cells from the bone. (**a**) Hind limb bones harvested from an adult mouse. The intact knee cap is shown by the green arrowheads. (**b**) Crushed bone from one hind limb is shown with the marrow removed. (**c**) FACS profile of cells isolated from the hind limb bones of four mice, only 2.5 million events were recorded. In this case the CD45 cells were not depleted prior to sorting. The starting panel contains only live cells. Previous FSC and SSC gates as well as live cell gating are not shown. The hematopoietic cells are gated out by collecting the Ter119 negative and CD45 negative population. Then the endothelial cells are gated out by taking the CD31 negative population. The stromal cells are then subdivided using CD166, CD146 and Sca1

5. Transfer the bone fragments to a 15-mL tube and add 5 mL of warm collagenase mix. Place the tube on a rocker at 37 °C agitate for 45 min (*see* **Note 3**).

6. Cut the tip of a 1 mL pipet tip and homogenize the bone collagenase mixture by pipetting up and down ten times.

7. Allow the bone to settle and transfer the liquid (bone disassociated cells will be in the liquid fraction) to a 50-mL tube passing through a 40 μm mesh strainer.

8. Add 5 mL of PBS to the settled bone and repeat **steps 6** and **7** twice.

9. Pellet the cells by centrifugation at 300 × *g* at 4 °C.

10. Resupsend the cell pellet in 1 mL of Red Cell Lysis Buffer. Incubate at room temperature for 5 min. Then add 10 mL of PBS and pellet the cells by centrifugation at 200 × *g* at 4 °C.

11. Deplete hematopoietic cells using CD45 microbeads and LD columns per the manufactures protocol (*see* **Note 4**).

12. Pellet the cells at 300 × g at 4 °C in a 5 mL FACS tube. Resuspend the cells in PBS with anti-CD16/32 antibody to block the Fc receptors. Incubate for 5 min on ice. Then stain by adding labeled monoclonal antibodies against CD45-Alexa700, CD31-PE-Cy7, Ter119-PE-Cy5, Sca1-BV605, CD146-Fitc or CD146-DyLight650 if using a GFP mouse and CD166-PE. Incubate for 20 min on ice (Fig. 1c).

13. Wash the cells with 500 µL of PBS then pellets the cells at 300 × g at 4 °C.

14. Resuspend the cells in 100 µL of 1/1000 LIVE/DEAD® Fixable Aqua Dead Cell Stain Kit, for 405 nm excitation. Incubate for 10 min on ice.

15. Wash the cells with 1 mL of PBS then pellet the cells at 300 × g at 4 °C.

16. Resuspend the cells in 150 µL of PBS with 0.1% BSA. Sort the cells into PBS with 0.1% BSA on an FACS sorter. If available, use a larger nozzle we used the 85 µm nozzle on the Aria III sorter.

3.1.2 Isolation of Fetal Bone Cells

1. Dissect the fore and hind limbs from E14-E15 stage fetuses by gently pinching and pulling with forceps and suspend in 15 mL cold PBS buffer in a 50-mL tube. The embryos from one or two pregnant females can be pooled. A dissection scope may be used for easier visualization of the embryo.

2. Pass the suspension through a 16 G1 needle with a 5 mL syringe several times until the bones are completely separated from the skin and muscles.

3. Pellet the bones by short centrifugation at 180 × g, 4 °C, discard the supernatant.

4. Resuspend the bones in cold PBS buffer by passing through 16 G1 syringe several times.

5. Pellet the bones by short centrifugation at 180 × g, 4 °C, discard the supernatant.

6. Repeat **steps 4** and **5** until the bones are clean and the buffer is cleared.

7. Warm the Collagenase Mix and add 5 mL to the collected bones, close the tube tightly, and place on a rocker, 37 °C, agitating constantly for 20 min.

8. After digestion, apply the suspension to a 40 µm mesh cell strainer placed in a 50-mL corning tube. The undigested materials left on strainer can be gently homogenized with the end of a pipette tip.

9. Wash the cell strainer with PBS buffer.

10. Discard the cell strainer and add PBS buffer to a final volume 50 mL.

11. Centrifuge cell suspension at $300 \times g$ for 5 min, 4 °C. Aspirate supernatant completely.

12. Resuspend the cells in PBS with 0.1% BSA.

3.2 shRNAi Manipulation of Progenitors

1. Prepare the lentivirus using standard procedures [14], we used VSV.G and CMVDR8.74 packaging plasmids along with jetPrime transfection reagent.

2. Culture 1000–5000 sorted stromal cells in 100 μL of supplemented MEM α medium in a 96 well round bottom plate at 37 °C, 5% CO_2, 2% O_2. Add the lentivirus immediately or culture overnight, allowing the cells to adhere to plate, then add the virus.

3. Incubate the cells with the virus overnight and then remove the media and add fresh media. Culture the cells in fresh media for 48 h.

4. Harvest the cells by removing the media and adding warm trypsin. If the cells do not detach easily gently pipet up and down. Pellet the cells at $300 \times g$ and resuspend in PBS with 0.1% BSA. FACS-sort for GFP+ or another fluorescent label carried by your transduced vector. At this point you can transplant the cells as in Subheading 3.3 or you can culture to expand the cells depending on your yield. Do not culture the sorted cells for longer than 1 week as they may differentiate and lose progenitor properties.

3.3 Transplantation of Progenitors

3.3.1 Ectopic Bone Forming Assay

This protocol is modified from a previous protocol used to identify a fetal osteochondral progenitor [12, 15, 16]. For easy visualization of the transplanted cells or to distinguish adult cells from transplanted fetal cells it is best to use adult cells sorted from a GFP mouse or cells transduced with a GFP or other fluorescently labeled shRNA or overexpression vector.

1. Prepare the cells for transplantation by pelleting at $300 \times g$ for 5 min, 4 °C; either 5000 adult stromal cells alone or 5000 adult stromal cells mixed with 30,000 fetal bone cells.

2. Thaw Matrigel on ice. Resuspend the cell pellet in 10 μL of thawed Matrigel and keep it on ice. The Matrigel will harden if the temperature is increased.

3. Select an immunodeficient mouse as the recipient.

4. Anesthetize the mouse with isoflurane.

5. Apply ophthalmic ointment on the eyes of the mouse to protect them from the light and over drying.

6. After the anesthetic has taken effect, shave the left flank and back of the mouse.

7. Sterilize the skin with povidone–iodine and then wipe off with 70% isopropanol. Repeat the step at least twice.

8. At this point, sterile/aseptic technique should be followed. The surgery site/sterile area should be isolated and maintained.

9. Make a small incision in the skin on the flank, about 1.3 cm from the rib cage, perpendicular to the spine. Expose the peritoneum (Fig. 2a).

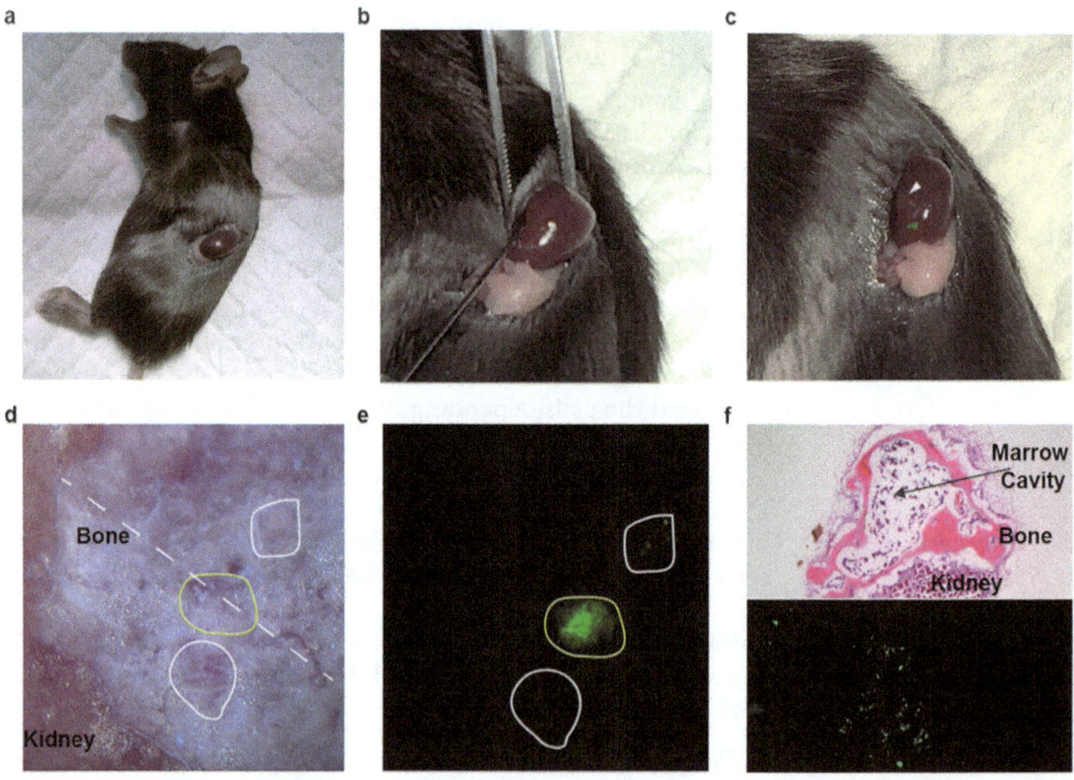

Fig. 2 Ectopic bone forming assay. (**a**) Exposed kidney shown in relation to the mouse body to highlight the location. (**b**) The kidney is stabilized with the tweezers and the needle is inserted underneath the capsule. (**c**) The needle insertion site is highlighted by the green arrowhead. The location of the transplanted cells is highlighted by the white arrowhead. The lighter color on the kidney surface is the transplanted cells. (**d**) One month after transplantation with fetal bone cells and Sca1+GFP+ adult cells, the kidney was harvested. Bright-field image of the bone graft sitting on top of the kidney. Two marrow cavities are outlined in white and one in yellow. The dashed line shows the sectioning plane. (**e**) The marrow cavity highlighted in yellow has a large population of the adult Sca1+ transplanted cells, which were harvested from a GFP mouse and are fluorescent when imaged under a GFP filter. The marrow cavities highlighted in white are out of focus and only have small populations of adult Sca1+ cells. (**f**) H&E stained section (upper panel) of the bone graft shows the bone stained in pink and marrow cavity sitting on top of kidney cells. Much of the kidney was removed prior to imbedding for sectioning. The fluorescent image (lower panel) shows the Sca1+GFP+ adult cells in the marrow cavity

10. Make a small incision in the peritoneum to expose the kidney, which is located directly adjacent to the spleen on the posterior side (Fig. 2a).

11. Gently squeeze either side of the incision to raise the kidney out of the peritoneal cavity.

12. Clean the blood with sterile dry cotton swab and moisten the kidney with sterile PBS buffer or sterile saline solution.

13. Draw 10 μL of the prepared cells using a precooled 25 μL Hamilton syringe with beveled needle.

14. Insert the Hamilton needle underneath the kidney capsule with the beveled end facing up. Having the beveled end up will allow you to see the bulge made by the transplanted cell–Matrigel mix. You can use tweezers to stabilize the kidney while inserting the needle, but be careful not to squeeze the kidney (Fig. 2b).

15. Move around carefully to make a pocket between the capsule and renal parenchyma.

16. Slowly inject the cells into the far end of the pocket (to avoid the leakage of cells out of the insertion site) (Fig. 2c).

17. Carefully remove the Hamilton syringe.

18. Clean the blood with dry cotton swab. Moisten the kidney with PBS buffer. If done properly bleeding if any should be minimal.

19. Gently place the kidney back to the peritoneal cavity.

20. Suture the peritoneum and staple the skin, respectively. Administer analgesic to mouse and follow standard postoperative procedures.

21. Allow the graft to grow for at least a month.

3.3.2 Intravenous Injection Assay

For this assay, we sorted Sca1$^+$ cells from UBC-creERT2/iDTR mice. When we injected the mice with tamoxifen 3 weeks after transplantation, the diphtheria toxin receptor (DTR) was expressed in the transplanted cells. We were thus able to ablate the transplanted cells by injecting the mice with diphtheria toxin. This allowed us to specifically target and analyze the loss of Sca1$^+$ cells. This assay is useful for targeting specific cells that lack a specific Cre for conditional targeting. If analyzing a new cell type homing should be assayed first by transplanting fluorescently labeled cells.

1. Pellet 5000 stromal cells at 300 × g; 4 °C sorted from donor mouse (i.e., Sca1$^+$ cells from UBC-creERT2/iDTR mice). Resuspend the cells in 100 μL of sterile PBS.

2. Sublethally irradiate the recipient mouse with 600 rad doses.

3. Anesthetize the recipient mouse with isoflurane. Load 100 μL of cells into an insulin syringe and inject the cells into the retro-orbital vein of the recipient mouse.

4. After 3 weeks inject the recipient mice with 125 mg/kg of tamoxifen (tamoxifen dissolved in corn oil at 37 °C for 6 h). This activates the expression of the DTR in the transplanted Sca1$^+$ cells.

5. After 1 week inject the recipient mice with 4 ng/g of diphtheria toxin. This will ablate the transplanted Sca1$^+$ cells.

6. Analyze the mice at desired time point. We analyzed 1 month after ablation.

3.4 Post-transplantation Analysis

3.4.1 Processing of Bone Graft

1. One month after transplantation. Euthanize the recipient mouse and harvest the host kidney. The graft may be visible without magnification. It will appear as a white growth on the surface of the kidney.

2. Immerse the kidney in cold PBS in a culture dish. Make sure that the kidney is completely covered in PBS. Then visualize the kidney under a fluorescent dissecting microscope equipped with a camera. This will allow you to perform initial documentation of the graft (Fig. 2d, e). You may now analyze the graft in one of the two ways; FACS or sectioning/microscopy.

3. For FACS analysis, place the kidney on a clean 60 mm tissue culture dish cover without any PBS use a scalpel or razor blade to cut the graft away from the kidney. Try to remove as much of the kidney as possible while maintaining the graft. This is more easily done while looking through a dissection microscope.

4. Place the graft in glass mortar with 1 mL of PBS and crush the bone with glass pestle. Transfer the bone and PBS to a 15-mL tube. Wash the mortar with PBS and transfer the wash to the same 15 mL tube. Pellet the bone and cells at $300 \times g$; 4 °C for 5 min. Resuspend the pellet in 1 mL of collagenase mix. Rotate at 37 °C for 15 min.

5. After digestion, apply the suspension to a 40 μm mesh cell strainer placed in a 50-mL corning tube. The undigested materials left on strainer can be gently homogenized with the end of a pipette tip.

6. Wash the 15-mL tube and cell strainer with 3 mL of PBS.

7. The cells can now be stained and analyzed by FACS. Process a piece of un-transplanted kidney in a similar way to use as a negative control.

8. For sectioning, place the kidney in a clean culture dish without any PBS use a scalpel or razor blade to cut the kidney and create a flat surface. Imbed the kidney with the graft in OCT in an appropriate cryomold for your cryostat. Four micrometers to eight micrometers sections can be made and mounted on slides for microscopy using standard staining procedures such as hematoxylin and eosin (H&E), direct visualization of GFP or immunohistochemistry (Fig. 2f).

3.4.2 Analysis of IV Transplanted Mouse

1. One month after ablation of the transplanted cells the bone can be harvested and analyzed following the same protocol used Subheading 3.1.1.

2. Instead of FAC-sorting, the samples can be analyzed on an FACS analyzer.

3. To analyze the BM hematopoietic cell populations, the marrow will be processed as in **Note 2**. Follow the same procedure in Subheading 3.1.1 minus the collagenase-related steps. The samples will be stained using hematopoietic specific antibodies [12].

4 Notes

1. Each mouse will have around 1000–2000 Sca1$^+$ cells, 1000–2000 CD166$^+$ cells, and 500–1000 CD146$^+$ cells. Thus, you will need to batch mice to obtain enough cells. We routinely sort from four mice.

2. Alternatively, a pipet can be used to remove the supernatant/wash. If you wish to analyze the marrow as well, you may keep the supernatant/wash and pass the marrow through the syringe several times until there are no more marrow clumps.

3. If you are processing one mouse at a time you can use 1 mL of collagenase mix instead.

4. This step is optional. It will decrease sorting time and increase the yield.

Acknowledgments

This study was supported in part by grants to C.-C.C from the ThinkCure! Foundation, the Margaret E. Early Medical Research Trust, the STOP Cancer Foundation and the American Cancer Society (Grant 128766-RSG-15-162). X.H. was supported by a California Institute for Regenerative Medicine (CIRM) training grant.

References

1. Boulais PE, Frenette PS (2015) Making sense of hematopoietic stem cell niches. Blood 125(17):2621–2629

2. Morrison SJ, Scadden DT (2014) The bone marrow niche for haematopoietic stem cells. Nature 505(7483):327–334

3. Moore KA, Lemischka IR (2006) Stem cells and their niches. Science 311(5769):1880–1885

4. Ehninger A, Trumpp A (2011) The bone marrow stem cell niche grows up: mesenchymal stem cells and macrophages move in. J Exp Med 208(3):421–428

5. Méndez-Ferrer S, Michurina TV, Ferraro F, Mazloom AR, MacArthur BD, Lira SA, Scadden DT, Maaposayan A, Enikolopov GN, Frenette PS (2010) Mesenchymal and haematopoietic stem cells form a unique bone marrow niche. Nature 466(7308):829–834

6. Ohneda O, Ohneda K, Arai F, Lee J, Miyamoto T, Fukushima Y, Dowbenko D, Lasky LA,

Suda T (2001) ALCAM (CD166): its role in hematopoietic and endothelial development. Blood 98(7):2134–2142

7. Zhou BO, Yue R, Murphy MM, Peyer JG, Morrison SJ (2014) Leptin-receptor-expressing mesenchymal stromal cells represent the main source of bone formed by adult bone marrow. Cell Stem Cell 15(2):154–168

8. Greenbaum A, Hsu YM, Day RB, Schuettpelz LG, Christopher MJ, Borgerding JN, Nagasawa T, Link DC (2013) CXCL12 in early mesenchymal progenitors is required for haematopoietic stem-cell maintenance. Nature 495(7440):227–230

9. Morikawa S, Mabuchi Y, Kubota Y, Nagai Y, Niibe K, Hiratsu E, Suzuki S, Miyauchi-Hara C, Nagoshi N, Sunabori T, Shimmura S, Miyawaki A, Nakagawa T, Suda T, Okano H, Matsuzaki Y (2009) Prospective identification, isolation, and systemic transplantation of multipotent mesenchymal stem cells in murine bone marrow. J Exp Med 206(11):2483–2496

10. Bradfute SB, Graubert TA, Goodell MA (2005) Roles of Sca-1 in hematopoietic stem/progenitor cell function. Exp Hematol 33(7):836–843

11. Sacchetti B, Funari A, Michienzi S, Di Cesare S, Piersanti S, Saggio I, Tagliafico E, Ferrari S, Robey PG, Riminucci M, Bianco P (2007) Self-renewing osteoprogenitors in bone marrow sinusoids can organize a hematopoietic microenvironment. Cell 131(2):324–336

12. Hu X, Garcia M, Weng L, Jung X, Murakami JL, Kumar B, Warden CD, Todorov I, Chen CC (2016) Identification of a common mesenchymal stromal progenitor for the adult haematopoietic niche. Nat Commun 7:13095

13. Ding L, Saunders TL, Enikolopov G, Morrison SJ (2012) Endothelial and perivascular cells maintain haematopoietic stem cells. Nature 481(7382):457–462

14. Metz M, Piliponsky AM, Chen C-C, Lammel V, Abrink M, Pejler G, Tsai M, Galli SJ (2006) Mast cells can enhance resistance to snake and honeybee venoms. Science (New York, NY) 313(5786):526–530

15. Chan CKF, Chen C-C, Luppen CA, Kim J-B, DeBoer AT, Wei K, Helms JA, Kuo CJ, Kraft DL, Weissman IL (2009) Endochondral ossification is required for haematopoietic stem-cell niche formation. Nature 457(7228):490–494

16. Weng L, Hu X, Kumar B, Garcia M, Todorov I, Jung X, Marcucci G, Forman SJ, Chen CC (2016) Identification of a CD133-CD55- population functions as a fetal common skeletal progenitor. Sci Rep 6:38632

Chapter 4

Identification and Isolation of Mice and Human Hematopoietic Stem Cells

Bijender Kumar and Srideshikan Sargur Madabushi

Abstract

Hematopoietic stem cells (HSCs) are multipotent cells capable of differentiating into all types of blood cells. The important feature of the HSCs is their ability to repopulate the complete blood cells after BM ablation. For clinical application, cord blood derived HSCs and G-CSF mobilized peripheral blood HSCs are good alternative to bone marrow HSCs. For immunological and hematological studies the obvious choice of model organism is Mouse. Therefore, understanding HSCs in murine model is important. In this chapter, we describe the common/currently used methods to isolate and identify human and mouse HSCs.

Key words Hematopoietic stem cells, Bone marrow, Mice, Human

1 Introduction

Hematopoietic stem cells (HSCs) have huge therapeutic potential and have been harnessed in the clinic for many years in context of hematological disorders and transplantation. Multipotent long term hematopoietic stem cells (LT-HSCs) reside in bone marrow microenvironment where the different niche components help in preserving the HSCs at steady state. The HSCs self-renew by dividing asymmetrically to maintain a pool of HSCs or can give rise to short term HSC (ST-HSCs)/multipotent progenitors (MPP) and lineage restricted oligopotent progenitors which undergo proliferation and differentiation to give rise to functionally different mature hematopoietic cells [1, 2]. The frequency of HSCs in bone marrow is about 0.005–0.01% of total nucleated hematopoietic cells. Though very low in percentage they must persist for lifespan of the model organism to replenish the hematopoietic system throughout the life. Due to their very low frequency in BM cells it is of utmost importance to identify the phenotype of HSC and

Bijender Kumar, Srideshikan Sargur Madabushi contributed equally to this work.

Shree Ram Singh and Pranela Rameshwar (eds.), *Somatic Stem Cells: Methods and Protocols*, Methods in Molecular Biology, vol. 1842, https://doi.org/10.1007/978-1-4939-8697-2_4, © Springer Science+Business Media, LLC, part of Springer Nature 2018

studies in mice have been going on for quite some time. Over the last 20–25 years the different study groups have pioneered the HSC biology field and contributed mainly in identification of HSCs and other progenitors through applying functional approaches and transcriptome analysis.

The HSCs are enriched in lineage marker-negative CD34$^{-/low}$ cKit$^+$ Sca1$^+$ (CD34$^{-/low}$ LSK), CD34$^{-/low}$ Flt3$^-$ LSK cells and CD48$^-$ CD150$^+$ LSK cells [3–6]. Several long-term transplantation experiments have been carried out by many researchers and CD34$^-$ LSK cells has showed the highest reconstitution potential. The CD34$^-$ LSK cells are usually Flt3$^-$ CD48$^-$ CD150$^{+/low}$ cells. Therefore, CD34$^-$ LSK cells may be pure HSCs but adding SLAM markers CD150 and CD48 could enhance the purity of HSCs. The SLAM CD150$^+$ CD48$^-$ LSK cells have highest reconstitution potential and multilineage differentiation capacity [7].

Recently Morrison group further divided the CD150$^+$ cells based on CD229 and CD244 expression pattern. They showed that LSK CD150$^+$ CD48$^-$ CD229$^-$ CD244$^-$ have more long-term reconstitution potential than LSK CD150$^+$ CD48$^-$ CD229$^+$ CD244$^-$ and CD229 positive population within CD150$^+$ compartment represents a slightly differentiated population [8] Table 1. The molecular regulation of specific HSCs properties such as long term self- renewal has been elucidated in murine HSCs. However, the biology of human HSCs remains poorly studied because of their rarity and lack of methods and cell surface markers to distinguish HSCs from multipotent progenitors. Over the last 20 years CD34$^+$ positivity was considered the sole criterion for HSC selection in clinic and would be used for mobilization and transplantation studies [9]. The bulk of HSCs are CD34$^+$ substantiated by human transplantation in clinics [10, 11], immune-deficient mice xenografts repopulation assays and in-vitro colony formation assays. However, most CD34$^+$ cells are lineage restricted progenitors and HSCs fraction in it remains very low [12, 13]. The discovery of CD90 (Thy1) on HSCs advanced enrichment of HSCs properties with long term reconstitution potential within CD34$^+$CD38$^-$ compartment [13–15]. Loss of CD90 expression in lineage depleted cord blood CD34$^+$ CD38$^-$ CD45RA$^-$ was sufficient to distinguish HSCs from multipotent progenitors. The studied CD49f surface expression on HSCs and analyzed HSC could be delineated using CD49F expression status [16] (Table 2).

The CD90$^+$ cells were divided into CD49f$^+$(CD90$^+$CD49f$^+$) and CD49f$^-$(CD90$^+$CD49f$^-$) fractions and evaluated for their long-term multilineage chimerism capability in intrafemoral injected immunodeficient NSG mice recipients. The CD49f$^+$ HSC had about sevenfold increased engraftment ability compared to CD49f negative fraction cells suggesting true functional HSCs existed in

Table 1
Hierarchy and classification of mice HSC and other progenitors used by different study groups

Mice cell types	Cell surface marker expressed
LSK	Lin− c-kit+ Sca1+
LT-HSC, Weissman model	LSK CD135− Thy1^{low}
ST-HSC, Weissman model	LSK CD135+Thy1^{low}
LT-HSC, Nakauchi/Jacobsen model	LSK CD34− CD135−
LT-HSC Goodell Model	LSK SP CD150+
ST-HSC, Jacobsen model	LSK CD34+ CD135−
LT-HSC, Morrison model 2005	LSK CD150+ CD48−CD135−
HSC-1, Morrison model 2013	LSK CD150+ CD48$^{-/low}$ CD229$^{-/low}$ CD244−
HSC-2, Morrison model 2013	LSK CD150+CD48$^{-/low}$ CD229+ CD244−
MPPs, Morrison model 2013	LSK CD150− CD48 $^{-/low}$ CD229 $^{-/low}$ CD244− LSK CD150−CD48 $^{-/low}$ CD229 + CD244− LSK CD150−CD48 $^{-/low}$ CD229 + CD244+
MPPs, Jacobsen model	LSK CD135+ CD34+
HSPCs	Lin− c-kit+ Sca1−
CLP	Lin− C-kit^{low} Sca1+ CD127+ CD135+
CMP	Lin− c-kit+ Sca1− CD127− CD16/32− CD34+
GMP	Lin− c-kit+ Sca1− CD127− CD16/32+ CD34+
MEP	Lin− c-kit+ Sca1− CD127− CD16/32− CD34−

LT-HSC long-term hematopoietic stem cells, *HSC-1* hematopoietic stem cell 1, *HSC-2* hematopoietic stem cell-2, *ST-HSC* short term hematopoietic stem cells, *LSK* lin-Sca1+Kit+, *MPP* multipotent progenitors, *HSPCs* hematopoietic stem/progenitor cells, *CLP* common lymphoid progenitor, *GMP* granulocyte macrophage progenitor, *MEP* megakaryocyte erythrocyte progenitor

Table 2
Hierarchy and classification of human HSCs and other progenitors used by different study groups

Human cell types	Cell surface marker expressed
HSC, Weissman model 2007	Lin− CD34+ CD38− CD90+ CD45RA−
HSC, Dick J model 2011	Lin− CD34+ CD38− CD90+ CD45RA− CD49f+
MPP, Weissman model 2007	Lin− CD34+ CD38− CD90− CD45RA−
MLP, Weissman model 2007	Lin− CD34+ CD38− CD135− CD123− CD45RA+
CMP, Weissman model 2007	Lin− CD34+ CD38+ CD10− CD135+ CD123^{low} CD45RA−
GMP, Weissman model 2007	Lin− CD34+ CD38+ CD10− CD135+ CD123^{low} CD45RA+
MEP, Weissman model 2007	Lin− CD34+ CD38+ CD10− CD135− CD123− CD45RA−

CD49f+ fraction. Limiting dilution assays also confirmed that CD49f+ CD90+ had higher frequency of HSCs compared to CD49f− CD90+ cells [16].

1.1 Dye Exclusion Methods for Hematopoietic Stem Cells Identification

Apart from cell surface antigen expression, primitive hematopoietic stem cells from mice, human and other species have been identified based on the expression of multidrug resistance related drug transporter and their ability to efflux some florescent dyes. Their ability to efflux dyes are attributed to high expression of p-glycoproteins and ABC transporter gene family members which are highly expressed in hematopoietic stem cells that remain in quiescent stage when compared to proliferating multipotent stem cells. Side population (SP) cells have been identified in human cord blood and adult bone marrow of monkeys [17]. Based on HSC/HSPC's efflux properties, they form a characteristic cluster of events to the lower left side in dual wavelength FACS profile of Hoechst stained cells [18].

Goodell MA group Identified HSCs in side populations (SP) based on Red and Blue double staining. The double negative side population (SP) is enriched in CD150+ enriched progenitor cells [19]. Further, the lower fraction within SP is enriched for even more CD150+ HSCs. In human, mitochondrial specific dye Rhodamine 123(Rho123) can be used to characterize HSPCs (lin−CD34+CD38−) populations with long term reconstitution potential [20, 21].

2   Materials

1. 40 and 70-μ strainers.

2. 4 mL EDTA vials.

3. Ficoll Hypaque or LSM (lymphocyte separation medium) can also be used.

4. 21G needle and 27G 1 mL heparin injection syringe.

5. 5 mL and 10 mL syringes.

6. Pestle and mortar.

7. 15 mL and 50 mL falcon tubes.

8. Plastic 10 mL transfer pipette.

9. DPBS (pH 7.4) without Ca++ and Mg++ salts.

10. Staining buffer: DPBS 1× without calcium and magnesium 0.1% BSA (deionized BSA). Heat-inactivated FBS can also be used instead of BSA.

11. ACK Lysing Buffer or prepare by dissolving 8.25 g ammonium chloride, 1 g potassium bicarbonate, and 37.2 mg disodium EDTA salt in milliQ autoclaved H_2O to make up final volume to 1000 mL and adjust pH to 7.2–7.4.

12. Complete IMDM media containing 10% heat-inactivated FBS,1% glutamine,1% HEPES,1% penicillin–streptomycin,1% nonessential amino acid mix, and 0.1% β-mercaptoethanol (β-ME).

13. Human CD34 MicroBead enrichment kit.

14. Mouse CD117 MicroBeads enrichment kit.

15. LS or MS column.

16. Mini MACS or MidiMACS separator and MACS MultiStand.

17. MACS buffer—0.5%BSA and 2 mM EDTA final conc. In DPBS.

18. Centrifuge.

19. 1.5 mL microcentrifugation tubes.

20. 1.2 mL and 4 mL FACS tubes.

21. Purified rat anti-mouse CD16/CD32 (mouse BD fc block).

22. Human BD fc block™.

23. Anti-mouse Ig, κ/negative control compensation particles set.

24. Anti-rat and anti-hamster Ig κ/negative control compensation particles set.

25. A FACS machine with four lasers (violet, blue, green, and red) capable of analyzing 10+ colors such as BD Fortessa and BD ARIAIII.

3　Methods

3.1　Isolation of Mice Bone Marrow Cells

1. Sacrifice the mouse according to the accepted protocol as per the institutes animal care and Use committee. Harvest Femurs and tibia from mouse and place them in sorting buffer.

2. Bone marrow cells can be harvested either by flushing using 27 Gauge needles and 5 mL syringes with staining buffer or by crushing using a sterile mortar and pestle.

3. When using a pestle and mortar, crush and grind the bones in 1 mL PBS for 2–3 min. Wash the pestle with 5 mL Staining buffer (cold). Grinding usually breaks up the clumps in the bone marrow cells to single cells, which are then filtered through a 40-μ nylon filter.

4. For flushing, cut the ends of the bones and using 27 gauge needles flush the bone marrow cells using 5–10 mL staining buffer.

5. Centrifuge the bone marrow cells at $290 \times g$ at 4 °C for 5 min.

6. Decant the supernatant and suspend in 4 mL fresh staining buffer (room temperature)/IMDM 10% FBS medium. Layer 3 mL of LSM (Lymphocyte Separation Medium) on the top of the medium containing cells. Alternately, we can pipette the LSM into the bottom of the 15 mL falcon tubes

containing cells, so that the 3 mL LSM is below the 4-mL media containing cells.

7. Centrifuge at $453 \times g$ for 30 min at 20 °C in swinging bucket rotor without any brakes. Temperature is critical here, if temperature is lower, the separation of cells (buffy coat) between the two layers will be affected.

8. Remove the buffy coat or mononuclear cells (MNCs) between the LSM and medium. RBCs and polymorphonuclear granulocytes are settled at the bottom of the tube.

9. Wash and filter the MNCs with excess staining buffer, $290 \times g$ for 7 min 4 °C and resuspend in 100 µL of MACS buffer (containing 0.5% BSA and 2 mM EDTA in PBS) and proceed to C-kit enrichment step.

10. Add 5 µL of CD117 microbead for every 10^7 cells used and incubate for 20 min at 4 °C in dark (*see* **Note 1**).

11. Wash the cells with 1 mL MACS buffer and centrifuge at $290 \times g$ for 5 min, remove supernatant and resuspend in 500 µL MACS buffer (*see* **Note 2**).

12. Load the cells suspension on the column and collect the CD117 negative unlabeled cells flow through the column in a collection tube. Load 1 mL of MACS buffer to completely remove the unlabeled cells.

13. Remove the column from the MACS system magnet and place it on top of 15 mL falcon tube. Load 1 mL PBS flush out the labeled CD117$^+$ cells from column by pushing the plunger in to the column and repeat the procedure to completely remove any remaining cells.

14. Centrifuge the cells at $290 \times g$ for 5 min at 4 °C and resuspend the CD117$^+$ enriched cell pellet in 50 µL of staining buffer and add 0.1 µL purified FcR blocking antibody and incubate for 5 min.

15. Suspend cells in staining buffer containing antibodies to stain HSC, HSPC and Lineage antibody cocktail. The lineage cocktail contains B220, CD19, CD3, CD4, CD8, Mac1, Gr1, and Ter119. It will save time if we use lineage antibodies (primary antibodies) conjugated to a fluorochrome like Pacific Blue. Otherwise, incubate cells in biotinylated lineage cocktail for 25–30 min on ice. Centrifuge, discard supernatant, and wash with staining buffer twice, $290 \times g$, 5 min 4 °C.

16. Suspend cells in antibody cocktail containing viability dye, CD117(c-KIT), Sca1, CD 34, CD16/32 (FCR), CD150, CD135, CD48, and anti lineage antibody. Incubate on ice for 45–60 min (mix cells by tapping every 15 min to prevent cells from settling at the bottom). Fig. 1. For good staining of CD34 and other stem cell markers, at least 60 min of staining is required.

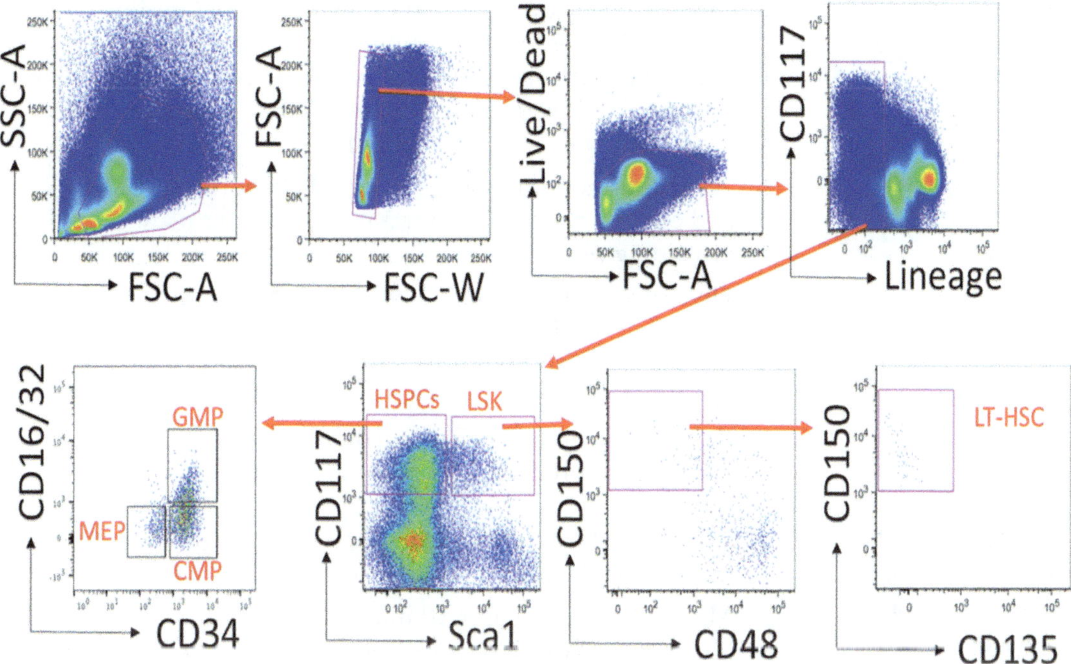

Fig. 1 Gating strategy for identification and isolation of 6–8-week-old B6/C57 mice bone marrow derived long term hematopoietic stem cells (LT-HSCs). Lin⁻Sca1⁺Kit⁺CD150⁺CD48⁻CD135⁻ cells are classified as LT-HSC

17. Centrifuge at $290 \times g$ 4 °C for 5 min and wash with staining buffer twice.

18. Suspend cells in staining buffer and store in 4 °C until FACS analyzed (few hours) or you can fix the cells with 1% PFA overnight. After fixing cells can be analyzed with little difference in staining profile for 2–3 days (*see* **Note 3**).

3.2 Purification of Human Mononuclear Cells by Density Gradient Centrifugation

1. Collect 4–5 mL of fresh bone marrow using aspiration needle into EDTA vials. Umbilical Cord blood and G-CSF or AMD3100 mobilized peripheral blood samples can alternately be used for HSC isolation. The viability of frozen samples is lower and entirely dependent on freezing method used and personal handling (*see* **Note 4**).

2. Dilute the sample with five times volume of DPBS (0.5%BSA) *or* IMDM media and pass it through 21G needle several times slowly at 4 °C to make homogenous suspension and break clumps of cells/tissue.

3. Remove bone fragments or large clumps by passing the sample through 70-µ cell strainer in a 50 mL falcon tube.

4. Layer the diluted bone marrow on 10–15 mL sterile filtered Ficoll Hypaque (1.077 g/mL) in 50 mL falcon tube slowly along the side of the tube and centrifuge at $1600 \times g$ 20 °C for 30 min in swinging bucket rotor without any brakes.

5. After centrifugation, store the upper phase plasma into another tube and transfer the mononuclear cells containing HSCs from the buffy coat colored interphase to another 15 mL falcon. The lower phase contains RBCs sediments, granulocytes and tube other stromal cells.

6. Wash the mononuclear cells with PBS containing 0.1% BSA and centrifuge at $340 \times g$ for 5 min to remove platelets and any remaining plasma.

7. Discard supernatant and WBC pellet should be clear of RBCs. Treat with ACK lysis buffer for 20–30 min at 4 °C and repeat the PBS wash and centrifugation step to get clear mononuclear cell pellet.

8. Resuspend the cell pellet in 100 µL PBS (0.5% BSA, 2 mM EDTA) and proceed for CD34$^+$ MACS enrichment step.

3.3 MACS Based Magnetic Positive Selection of Human CD34+ Cells

1. Add 100 µL of FcR blocking reagent and 100 µL of CD34$^+$ microbeads (Miltenyi Biotec) for up to 10^8 cells.

2. Mix well and incubator 30 min at 4 °C. Wash the cells by adding 5–10 mL of PBS (0.5%BSA and 2 mM EDTA) and centrifuge at $340 \times g$ for 5 min at 4 °C.

3. Aspirate the supernatant and resuspend the cells in 0.5 mL PBS (0.5%BSA and 2 mM EDTA) followed by proceeding for magnetic separation method.

4. Place the MS or LS enrichment column (LS column is usually used for larger number of cells) in the magnetic field and prepare the column by rinsing 0.5–1 mL of PBS (0.5% BSA and 2 mM EDTA) without the column to drying up completely.

5. Load the cells suspension on the column and collect the CD34 negative unlabeled cells flow through the column in a collection tube.

6. Add another 1 mL PBS (0.5% BSA and 2 mM EDTA) on the column to completely remove the unlabeled cells.

7. Remove the column from the separator and put on 15 mL falcon tube.

8. Add 1 mL of PBS (0.1% BSA) and flush out the labeled CD34$^+$ cells from column by pushing the plunger in to the column and repeat the procedure to completely remove any remaining cells.

9. Add another 5 mL PBS to the suspension and spin at $340 \times g$ for 10 min at 4 °C.

10. Discard the supernatant and resuspend the CD34$^+$ enriched cells in 100 µL in staining buffer or PBS (0.5% BSA).

11. Dilute fluorescent antibodies to their predetermined optimal concentrations in staining buffer and add small aliquots of

the diluted antibodies to the small FACS tubes or 1.5 mL microcentrifugation tube containing CD34$^+$ enriched cells suspensions.

12. Incubate for 45–60 min on ice protected from light. Mix the cell suspension every 15 min by pipetting gently.

13. Wash the cells by adding 1 mL of PBS (0.1% BSA) to remove unbound antibodies. Centrifuge cells as 340×g for 5 min at 4 °C. After centrifugation, carefully aspirate the supernatant from the cell pellet.

14. Resuspend the cells in 100–200 µL of live dead violet viability dye (final conc. 0.5–1 µg/mL) in staining solution and incubate for 20 min on ice or at 4 °C in dark.

15. Wash the cells by adding 1 mL of PBS (0.1% BSA) and centrifuge cells as 340×g for 5 min at 4 °C. After centrifugation, carefully aspirate the supernatant from the cell pellet.

3.4 Cells Staining Protocol for Human Cells

1. Dilute the different fluorescent antibodies for CD34, CD38, CD90, CD45, CD45RA, CD49f, CD135, CD123, CD10 and other lineage markers to their predetermined optimal concentrations in staining Buffer (dilution of antibodies is given in Table 3) and add the aliquots of the diluted antibodies to the FACS tubes containing CD34+ enriched cells suspensions (*see* **Note 5**).

Table 3
List of antibodies and their dilution used for isolation of mice and human hematopoietic stem cells

Antibody Used	Source	Cat. No.	Dilution
Anti-mouse CD16/32	BioLegend	101302	1:200
Anti-mouse CD3	BioLegend	100334	1:100
Anti-mouse CD4	BioLegend	100428	1:200
Anti-mouse CD8	BioLegend	100725	1:100
Anti-mouse B200	BioLegend	103227	1:200
Anti-mouse CD19	BioLegend	115523	1:200
Anti-mouse CD11b	BioLegend	101224	1:200
Anti-mouse Gr-1	BioLegend	108430	1:200
Anti-mouse Ter119	BioLegend	116232	1:100
Anti-mouse CD16/32	BioLegend	101324	1:50
Anti-mouse CD34	BD Biosciences	560238	1:25
Anti-mouse Sca1	BioLegend	108134	1:100

(continued)

Table 3
(continued)

Antibody Used	Source	Cat. No.	Dilution
Anti-mouse c-Kit	BioLegend	105826	1:100
Anti-mouse CD127	BioLegend	135014	1:50
Anti-mouse CD150	BioLegend	115912	1:100
Anti-mouse CD48	BioLegend	103412	1:100
Anti-mouse CD135	BioLegend	135306	1:50
Anti-human CD235a	BD Biosciences	559944	1:50
Anti-human CD41	BD Biosciences	559768	1:50
Anti-human CD14	eBioscience	15-0149-41	1:50
Anti-human CD11b	BD Biosciences	561686	1:50
Anti-human CD16	BD Biosciences	555408	1:50
Anti-human CD3	BD Biosciences	555341	1:50
Anti-human CD4	BD Biosciences	561004	1:50
Anti-human CD8	BD Biosciences	555368	1:50
Anti-human CD56	BD Biosciences	555517	1:50
Anti-human CD19	BD Biosciences	555414	1:50
Anti-human CD10	BioLegend	312216	1:50
Anti-human CD38	BioLegend	356624	1:50
Anti-human CD34	BioLegend	343610	1:100
Anti-human CD90	BD Biosciences	555595	1:25
Anti-human CD45RA	BioLegend	304146	1:100
Anti-human CD135	BioLegend	313306	1:50
Anti-human CD123	BioLegend	306026	1:100
Anti-human CD49f	BioLegend	313616	1:50
Anti-human CD45	BioLegend	368516	1:100
LIVE/DEAD™ Fixable Violet Dead Cell Stain	Thermo Scientific	L34955	1 µg/mL

2. Incubate for 45–60 min on ice protected from light.

3. Wash the cells with 1 mL of PBS (0.1% BSA) to remove unbound antibodies. Centrifuge cells as $453 \times g$ for 5 min at 4 °C.

4. After each centrifugation, carefully aspirate the supernatants from the cell pellet.

5. Resuspend the cells in 100–200 µL of live dead violet viability dye (final conc. = 0.5–1 µg/mL) in PBS (0.1% BSA) solution and incubate for 20 min in ice bucket or at 4 °C in dark.

6. Wash the cells with 1 mL of PBS (0.1% BSA) and Centrifuge cells as $453 \times g$ for 5 min at 4 °C.

7. Discard the supernatant and resuspend the cell pellet in 500–100 µL PBS (0.5% BSA) or IMDM media (1–2% FBS conc.) and proceed to Flow sorting.

3.5 High Speed Cell Sorting Using BD ARIAIII Sorter

1. Single stain compensation mice/human beads are prepared by mixing antibody with BD comp beads. First, vortex the mice or human specific compensation beads for 5 seconds and mix with equal volume of negative control beads in a tube and dispense 30-40µL in small FACS tubes for each florochrome. Add 0.25 µL of single stain antibody (mice/human comp beads prepared separately) for each color in each tube and keep in dark for 15 min.

2. Wash with 1 mL PBS, centrifuge at $290 \times g$ for 5 min at 4 °C, discard supernatant and resuspend in 500–1000 µL PBS. Keep the single stains in dark on ice until use in flow sorter.

3. Turn on power and fluidics of ARIAIII and fill the tank with sheath fluid.

4. Start FACSDIVA software (5.1 or higher) and replace closed loop nozzle with 70–85-µ nozzle and turn on stream.

5. Wait for the stream to stabilize and set up camera, amplitude, and drop frequencies and temp at 4 °C.

6. CS&T module for setup and tracking: Set up 70 or 85 micro configuration and run CST beads after lasers have been turned on for at least 20–30 min and make sure PMT voltages are in their desired ranges.

7. Select four-way holders for microcentrifuge tubes and make sure that the stream hits the center of the tube while collecting the waste stream (*see* **Notes 6**). Adjust the Plate Voltage until all streams are well separated and hit the tubes.

8. Determining the correct and robust drop delay is one of the most important steps in the machine setup. Turn on the sweet spot and run Accudrop beads in drop-delay experiment at the event rate of 800–2000 events/s without collecting them in collection tube, select the best drop-delay value.

9. Set up new experiment and select 10+ fluorochrome combinations and set up compensation beads for each channel. Make sure MFI is brightest for that fluorochrome than others. Calculate compensation and apply and save the settings. Clean flow cell line by running PBS for 2 min.

10. Place the sample collection microcentrifuge tube with 500–700 µL of 10% FBS IMDM media.

11. Load the sample, start data acquisition and recording at 2000–3000 events/s, set up gates on live cells and sort the desired

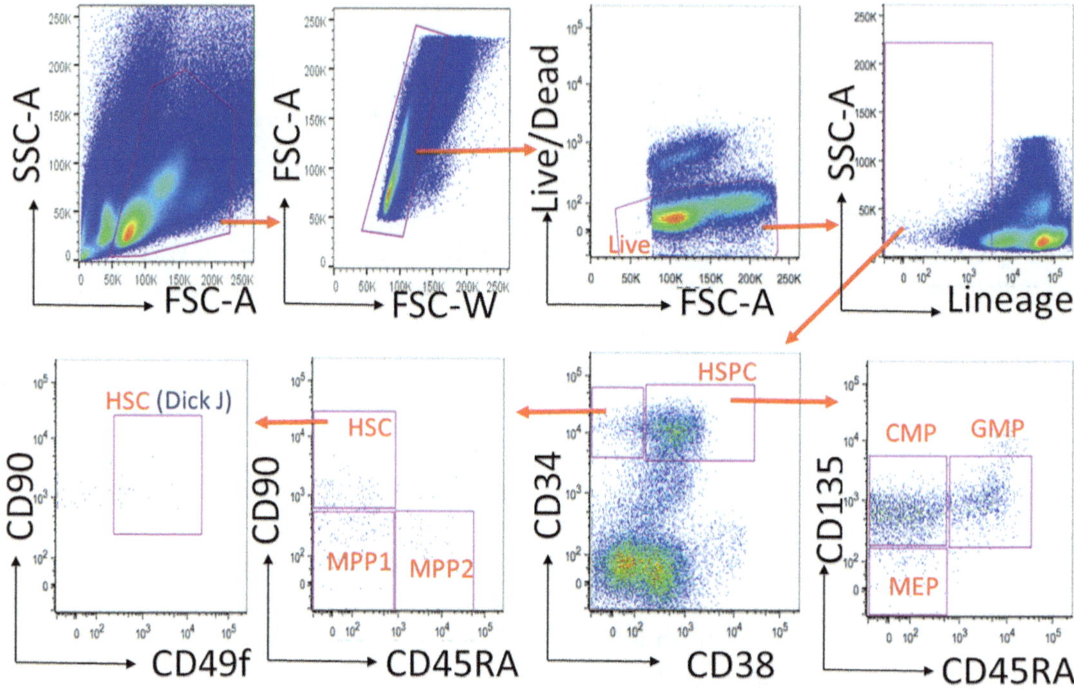

Fig. 2 Gating strategy for isolation of adult human bone marrow derived HSCs and other progenitors

HSCs population in two-way sorting and select the precision method to purity.

12. Double sort the HSCs to confirm their purity and resuspend them 1 mL IMDM media.

13. After double sorting, stop sorting and export the FCS file for the analysis later.

14. Centrifuge the cells at $453 \times g$ for 10 min at 4 °C.

15. Once sorted, HSC cells can be used for gene expression profile studies, cell signaling, in vitro culture, or transplantation studies, or can be cryopreserved. The HSCs can be cryopreserved in CryoStor CS5 (a serum-free freezing media) or filtered IMDM media containing 5–10% DMSO, 10% FBS, 1% penicillin–streptomycin, and 0.1% β-mercaptoethanol (β-ME) (Fig. 2).

4　Conclusion

Hematopoiesis has been extensively studied as a paradigm of stem cell biology and development. The most recent progress in hematopoietic stem cells (HSCs) identification and isolation methods has laid the groundwork to access the complexity of hematopoietic hierarchy at the HSC and progenitors' levels. Identification of different cell surface markers expression metabolic properties, and lately stage-specific transcriptomics analysis of stem cells and progenitors during emergence/development have paved way for their classification and isolation. Further, the advent of FACS-based transcriptomic techniques like Single-Cell RNAseq has helped in making possible the long-sought identification of age, stress-related, and hematological disorders associated with changes in stem cells and their therapeutic interventions.

5 Notes

1. Recommended volume of beads is 20 μL but higher percentage of beads will lower cells viability and yield difference is not much).

2. Ensure to make a homogenous suspension and that there are not any clumps of cells.

3. It is advised to run sorting on the same days based on downstream application of sorted cells without fixations, viz., if RNA is to be isolated from the cells or cells to be used in transplantation experiment.

4. It is advised to use fresh samples for HSC functional studies experiments.

5. It is important to setup negative controls as well to check the baseline staining of the antibodies. Negative controls can be unstained cells, isotype controls, or FMO depending on your application. For multicolor staining, single color compensation controls without viability dye and completely unstained sample sorting cells is also required to setup PMT voltages in flow cytometer.

6. The point of stream collection is very critical in sorting and would determine where sorted cells will fall while sorting.

References

1. Spangrude GJ, Heimfeld S, Weissman IL (1988) Purification and characterization of mouse hematopoietic stem cells. Science 241:58–62

2. Morrison SJ, Weissman IL (1994) The long-term repopulating subset of hematopoietic stem cells is deterministic and isolatable by phenotype. Immunity 1:661–673

3. Christensen JL, Weissman IL (2001) Flk-2 is a marker in hematopoietic stem cell differentiation: a simple method to isolate long-term stem cells. Proc Natl Acad Sci U S A 98:14541–14546

4. Boyer SW, Schroeder AV, Smith-Berdan S et al (2011) All hematopoietic cells develop from hematopoietic stem cells through Flk2/Flt3-positive progenitor cells. Cell Stem Cell 9:64–73

5. Morita Y, Ema H, Nakauchi H (2010) Heterogeneity and hierarchy within the most primitive hematopoietic stem cell compartment. J Exp Med 207:1173–1182

6. Yang L, Bryder D, Adolfsson J et al (2005) Identification of Lin(−)Sca1(+)kit(+)CD34(+)Flt3- short-term hematopoietic stem cells capable of rapidly reconstituting and rescuing myeloablated transplant recipients. Blood 105:2717–2723

7. Kiel MJ, Yilmaz OH, Iwashita T et al (2005) SLAM family receptors distinguish hematopoietic stem and progenitor cells and reveal endothelial niches for stem cells. Cell 121:1109–1121

8. Oguro H, Ding L, Morrison SJ (2013) SLAM family markers resolve functionally distinct subpopulations of hematopoietic stem cells and multipotent progenitors. Cell Stem Cell 13(1):102–116

9. McCune JM, Namikawa R, Kaneshima H et al (1988) The SCID-hu mouse: murine model for the analysis of human hematolymphoid differentiation and function. Science 241:1632–1639

10. Link H, Arseniev L, Bahre O et al (1996) Transplantation of allogeneic CD34+ blood cells. Blood 87(11):4903–4909

11. Michallet M, Philip T, Philip I et al (2000) Transplantation with selected autologous peripheral blood CD34+Thy1+ hematopoietic stem cells (HSCs) in multiple myeloma: impact of HSC dose on engraftment, safety, and immune reconstitution. Exp Hematol 28:858–870

12. Baum CM, Weissman IL, Tsukamoto AS, Buckle AM, Peault B (1992) Isolation of a candidate human hematopoietic stem-cell population. Proc Natl Acad Sci U S A 89:2804–2808

13. Peault B, Weissman I, Baum C (1993) Analysis of candidate human blood stem cells in "humanized" immune-deficiency SCID mice. Leukemia 7(Suppl 2):S98–S101

14. Murray L, Chen B, Galy A et al (1995) Enrichment of human hematopoietic stem cell activity in the CD34+Thy-1+Lin- subpopulation from mobilized peripheral blood. Blood 85:368–378

15. Majeti R, Park CY, Weissman IL (2007) Identification of a hierarchy of multipotent hematopoietic progenitors in human cord blood. Cell Stem Cell 1(6):635–645

16. Notta F, Doulatov S, Laurenti E et al (2011) Isolation of single human hematopoietic stem cells capable of long-term multilineage engraftment. Science 333(6039):218–221

17. Goodell MA, Rosenzweig M, Kim H et al (1997) Dye efflux studies suggest that hematopoietic stem cells expressing low or undetectable levels of CD34 antigen exist in multiple species. Nat Med 3(12):1337–1345

18. Ergen AV, Jeong M, Lin KK et al (2013) Isolation and characterization of mouse side population cells. Methods Mol Biol 946:151–162

19. Hirschmann-Jax C, Foster AE, Wulf GG et al (2004) A distinct "side population" of cells with high drug efflux capacity in human tumor cells. Proc Natl Acad Sci U S A 101(39):14228–14233

20. Kim M, Cooper DD, Hayes SF et al (1998) Rhodamine-123 staining in hematopoietic stem cells of young mice indicates mitochondrial activation rather than dye efflux. Blood 91(11):4106–4117

21. McKenzie JL, Takenaka K, Gan OI et al (2007) Low rhodamine 123 retention identifies long-term human hematopoietic stem cells within the Lin-CD34$^+$CD38$^-$ population. Blood 109: 543–545

Chapter 5

Identification and Characterization of Hair Follicle Stem Cells

Xiaoyang Wang, Yizhan Xing, and Yuhong Li

Abstract

Hair follicle stem cells (HFSCs) are epithelial cells that inhabit in the bulge region of hair follicles. They govern development of hair follicle as well as periodically regeneration of hair follicle. Under special condition, they also play roles in homeostasis of skin and other skin appendages. To characterize HFSCs in vitro, HFSCs must be isolated and cultured. In this chapter, we introduce a mechanical method to isolate HFSCs from mouse vibrissa hair follicle, and a modified method to culture isolated HFSCs. We also describe methods to characterize HFSCs, including clone formation assay and chamber graft assay.

Key words Hair follicle stem cell, Vibrissa hair follicle, Cell culture, Clone formation assay, Chamber graft assay

1 Introduction

The hair follicle is a dynamic structure that generates hair through a complicated and exquisitely regulated cycle of growth and regression controlled by numerous genes. Hair follicles undergo episodic cycles of growth (anagen), degeneration (catagen), and rest (telogen). In adult skin, each hair follicle contains a reservoir of hair follicle stem cells (HFSCs) located in a region called bulge. The activities of HFSCs control the cycling of the hair follicles. At the start of each cycle, the HFSCs proliferate and migrate downward to start growing phase and form mature hair follicle. At catagen and telogen phase, HFSCs pause proliferation. Under special condition, they can regenerate epidermis, blood vessels, sweat gland, sebaceous gland, and even neural tissue. With appropriate culture condition, HFSCs can maintain self-renewal and multipotency in vitro. To better understand the role of HFSCs, the identification and characterization of HFSCs is the first step.

Expression of specific molecules is required to maintain stem cell properties so the combination of these specific markers can help us identifying HFSCs from terminally differentiated cells.

Shree Ram Singh and Pranela Rameshwar (eds.), *Somatic Stem Cells: Methods and Protocols*, Methods in Molecular Biology, vol. 1842, https://doi.org/10.1007/978-1-4939-8697-2_5, © Springer Science+Business Media, LLC, part of Springer Nature 2018

A combination of CD34 and α6-integrin antibodies is commonly used. CD34 is a well-known stem cell marker. It is also used in the identification of HFSCs of mouse or rat. Alpha-6-integrin is widely expressed in epidermal cells, and is usually used together with CD34 to identify HFSCs. It is better to use a transgenic mouse line that has epithelium-specific marker to exclude potential contamination from other tissues. The most popular transgenic mouse is K15 promoter-driven reporter mouse, such as Krt1-15-EGFP (The Jackson Laboratory, stock No: 005244) [1]. Several techniques have been identified to help isolating HFSCs, including fluorescence-activated cell sorting (FACS), magnetic activated cell sorting (MACS), and mechanical isolation [2]. FACS based isolating methods have been widely used, but expensive equipment are needed [3]. In this chapter, we discuss how to isolate HFSCs with mechanical method. In 2009, Nowak, J.A. and E. Fuchs published a protocol about the isolation and culture of epithelial stem cells [4]. The culture of HFSCs is modified from that protocol. HFSCs have the characteristics of self-renewal and multipotency. We will also discuss several commonly used methods to characterize the identified HFSCs, including clone formation assay and chamber graft assay.

2 Materials

2.1 Cell Culture

1. Cholera toxin.
2. Chelex.
3. Collagen IV.
4. Compressed CO_2.
5. Dispase I.
6. Dulbecco's modified Eagle's medium/Ham's F-12 Nutrient Medium (3:1 Mix) without calcium.
7. Dulbecco's modified Eagle's medium.
8. F-12 Nutrient Mixture (Ham).
9. Fetal bovine serum.
10. Hydrocortisone.
11. Hank's Balanced Salt Solution (HBSS).
12. Insulin.
13. L-Glutamine.
14. Mitomycin C.
15. Phosphate-buffered saline.
16. Penicillin–streptomycin solution.

17. Sodium bicarbonate.
18. T3 (3,3′,5-triiodo-L-thyronine).
19. Transferrin.
20. Trypsin–EDTA 0.25%.

2.2 Characterization of HFSCs

1. BSA.
2. Crystal violet dye.
3. Collagenase.
4. DNase I.
5. Betadine.
6. Ethanol.
7. Methanol.
8. 2-Mercaptoethanol.
9. Normal goat serum.
10. Normal donkey serum.
11. Paraformaldehyde.
12. Trypan Blue solution.
13. Triton X-100.

2.3 Equipment

1. 4-1/2″ dissecting forceps.
2. 4-1/2" Iris Scissors, curved.
3. #21 blade steel scalpel.
4. 20 G Syringe needles.
5. 27 G Syringe needles.
6. 40 µM nylon cell strainers.
7. 70 µM nylon cell strainers.
8. 5 mL round-bottom cap tubes.
9. 50 mL conical tubes.
10. Electric shaver.
11. Nalgene MF75 Disposable Sterile Filtration Unit.
12. Reichert Brightline Hemocytometer.
13. Stereo microscope with fluorescence (Leica Microsystems).
14. Sterile grafting domes.
15. Tissue culture dish, 60-mm.
16. Tissue culture dish, 100-mm.
17. Ultra-pure H_2O equipment.

3 Methods

3.1 Identification of HFSCs

3.1.1 Preparation of F Media for 3T3 Fibroblast Culture

1. Add 3.5 L ultra-pure water to a 4 L conical flask.
2. Add three 1 L packets of Dulbecco's modified Eagle's medium and one 1 L packet of F-12 Nutrient Mixture (Ham's) to the flask and make sure to rinse any residue from the packets.
3. Add 12.28 g sodium bicarbonate.
4. Add 1.9 g L-glutamine.
5. Add 40 mL 100× Pen/Strep solution.
6. Stir for 20 min with foil covered on top of conical flask.
7. Check and adjust pH to 7.2 ± 0.05 with HCl; this process will take several hours.
8. Adjust volume to 4 L with ultrapure water.
9. Apply compressed CO_2 to the media for 10 min until the media changes to an amber color.
10. Filter-sterilize and prepare 225 mL aliquots in the hood.
11. Label the date prepared and store at −20 °C.
12. Add 25 mL fetal bovine serum to make a completed media.

3.1.2 Preparation of Chelated Fetal Bovine Serum (500 mL)

Day 1

1. Add 200 g dry Chelex resin for 500 mL Chelated fetal bovine serum preparation and place in a 2 L beaker (*see* **Note 1**).
2. Add 2 L ultrapure water, stir continuously and cover with aluminum foil all the time.
3. Check and adjust pH to 7.5 ± 0.05 with HCl every 20 min. This step will take several hours and when pH stabilizes at 7.5 ± 0.05 for more than 30 min, place aluminum foil-covered beaker at 4 °C overnight to let the Chelex resin settle down at the bottom of the flask.

Day 2

1. Carefully decant the water and keep the Chelex resin pellet in the beaker.
2. Add 2 L fresh ultra-pure water to the beaker, stir continuously and cover with aluminum foil all the time.
3. Check and adjust pH to 7.5 ± 0.05 every 20 min as in day 1.
4. Once pH has stabilized at 7.5 ± 0.05 for more than 30 min, place the aluminum foil- covered solution at 4 °C overnight.
5. Thaw 500 mL frozen fetal bovine serum at 4 °C overnight for the following day.

Day 3

1. If pH has not stabilized from the day 2, continue adjusting pH until it is stable at 7.5 ± 0.05. Leave the aluminum foil-covered

beaker at 4 °C for 1 h to allow Chelex to form a compact pellet.

2. Carefully decant the water as in day 1.

3. Add the thawed 500 mL fetal bovine serum to the Chelex slowly to minimize bubbles.

4. Stir slowly for no longer than 1 h at 4 °C to minimize bubbles.

5. Place aluminum foil-covered beaker at 4 °C for 1 h or overnight to allow Chelex to form compact pellet.
Day 4

1. Under sterile conditions, decant the serum into a Nalgene bottle top filter unit and filter through. Add 37.5 mL 100× cocktail, 375 μL 10^{-6} M cholera toxin, and 375 μL 4 mg/mL hydrocortisone to 500 mL chelated FBS (*see* **Notes 2–4**).

2. Filter-sterilize and prepare 38 mL aliquots in the hood.

3. Label the date prepared and store at −20 °C.

3.1.3 Preparation of E Media without Calcium

1. Add 1.5 L ultrapure water to a 4 L conical flask.

2. Add contents of 2× Gibco defined media packet (DMEM: F 12 = 3:1, calcium free) and make sure to rinse any residue from the packets.

3. Add 6.14 g sodium bicarbonate.

4. Add 0.95 g L-glutamine.

5. Add 20 mL of 100× Pen/Strep solution.

6. Stir for 20 min with aluminum foil covered on top of conical flask.

7. Check and adjust pH to 7.2 ± 0.05 with HCl. This process will take several hours.

8. Adjust volume to 2 L with ultrapure water.

9. Apply compressed CO_2 to the media for 10 min until the media changes to an amber color.

10. Filter-sterilize and prepare 212 mL aliquots.

11. Label and mark date prepared. Store at −20 °C.

12. Add 38 mL chelated fetal bovine serum with added components to make a completed media.

3.1.4 Preparation of Feeder Cells

1. Always feed 3T3 cells the day before mitomycin C treatment.

2. Dissolve mitomycin C in sterile PBS to make a 0.4 mg/mL stock solution. Mix well and store in brown bottle to avoid light at 4 °C.

3. Remove F media from 60 mm tissue culture dishes and wash once with 4 mL 1× PBS.

4. Add 4 mL fresh F-media and 100 μL of mitomycin C to make a final concentration 10 μg/mL into the 60 mm tissue culture dishes (*see* **Note 5**).

5. Incubate for 2 h in cell culture incubator with 5% CO_2.

6. Remove the media from 60 mm tissue culture dishes and wash with 1× PBS once.

7. Aspirate off the 1× PBS and add 4 mL fresh F-media. Mitomycin C treated cells will be viable for several weeks, if fresh F-media can be provided two to three times per week.

3.1.5 Preparation of Collagen IV-Coated Plates

1. Make 0.05 mol/L HCl solution and 1× PBS and autoclave.

2. Dilute Collagen: Place type IV collagen in a benchtop cooler in 4 °C to thaw slowly (this process will take approximate 48 h). Shake to mix collagen well after thaw. Add 0.05 mol/L HCl into thawed collagen to make an 80 μg/mL solution. Aliquot unused diluted collagen and store in −80 °C.

3. Coating: add 200 μL diluted collagen into each well of a 24-well plate to make an 8 μg/cm² final concentration.

4. Place the 24-well plate at room temperature for 1 h.

5. Aspirate off supernatant.

6. Wash the 24-well plate with 1× PBS for three times.

7. Seal up the coated 24-well plate and place at 4 °C (coated plates can be preserved at 4 °C for 1 week).

3.1.6 Isolation of HFSCs

1. Take a 9-day-old healthy C57 mouse (*see* **Note 6**).

2. Euthanize the mouse following standard protocol.

3. Disinfect vibrissa pads with alcohol pads and cut it off bilaterally with an iris scissor. Place vibrissa pads into a 60-mm plate containing 1× HBSS solution.

4. Clean the vibrissa pads with 1× HBSS solution. Aspirate off HBSS solution.

5. Dissect each hair follicle together with its connective tissue sheath using 27G syringe needles under dissecting microscope with 3.2× magnification. Transfer the dissected hair follicles into another 60-mm plate with 1× HBSS solution.

6. Aspirate off HBSS solution after collecting enough hair follicles. Add 2 mL 0.25% dispase and incubate for 20 min at room temperature.

7. Transfer dissected hair follicles into 100-mm plate. Add 1× HBSS solution to dilute dispase. To get a clean background during dissection, wash three times with HBSS solution. Aspirate off HBSS solution after thorough washing.

8. Dissect hair follicles out of connective tissue sheath under dissecting microscope. In order to get hair follicle bulge region,

make a horizontal cut directly above dermal papilla to remove dermal papilla region; Make another horizontal cut directly above bulge region to remove upper part of hair shaft (*see* **Note 7**).

9. Transfer dissected bulges into collagen IV coated 24-well plate, place only one dissected bulge into each well. Incubate the plate in 37 °C for 30 min.

10. Add fresh 0.3 mM calcium E media and incubate the plate. Change media and feed cells with fresh 0.3 mM calcium E media every 3 days. HFSCs can be visualized by day 3.

11. Cells reach confluence after 1–2 weeks and passage cells on mitomycin C treated 3T3 feeder cells (*see* **Note 8**).

3.1.7 Culturing of HFSCs on Feeder Layers

1. At 2 h before plating cells, plate mitomycin C treated feeder cells growing in F medium in the desired culture vessel, typically a 24-well plate, and allow them to adhere.

2. Remove the cultured bulge tissues from the 24-well plate with sterile forceps. Aspirate off 0.3 mM calcium E media and rinse with 1× HBSS solution three times (*see* **Note 9**).

3. Add 200 µL 0.25% trypsin to digest HFSCs. Usually, 10–20 min is needed.

4. Add 500 µL 0.3 mM calcium and 15% FBS-supplemented E medium to stop the reaction of trypsin. Spin the 0.25% trypsin digested HFSC cells for 5 min at 1000 rpm (150 g) to get cell pellet and carefully remove supernatant.

5. Resuspend HFSCs in a small volume of 0.3 mM calcium and 15% FBS-supplemented E medium. Count the cell density under microscopy.

6. Aspirate off F medium from feeder cells and replace with pre-warmed 0.3 mM calcium and 15% FBS-supplemented E medium before HFSCs are ready to plate.

7. Typically, 10^3–10^4 HFSCs are plated per well with feeder cells in a 24-well plate. Cultures grow with a doubling time of 24 h and are passaged while still in the exponential growth phase.

8. Feed cells with fresh 0.3 mM calcium and 15% FBS-supplemented E media every 2 days. Typically, colonies can be visualized by day 3 (Fig. 1)

3.2 Characterization of HFSCs

The cultured HFSCs can be further identified by FACS analysis with α6-integrin and CD34 antibodies. Regular experiments such as BrdU quantification, cell cycle analysis, RNA isolation, protein isolation, and immunofluorescence staining can be performed to characterize cultured HFSCs. Here we will discuss two special methods for the characterization of HFSCs, clone formation assay and chamber graft assay [5, 6].

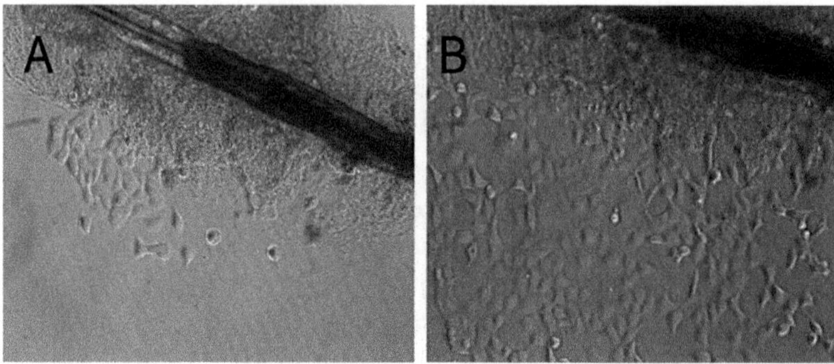

Fig. 1 Bulge culture on collagen IV coated plate (a) Six days after plating, HFSCs begin to grow in bulge area. (**b**) Eight days after plating, HFSCs enter exponential growth phase. Magnification 100×

3.2.1 Clone Formation Assay

1. An equal number of cells is plated onto mitomycin C-treated 3T3 fibroblasts (*see* **Note 10**).

2. Allow the cells to settle down in wells in cell culture incubator with 5% CO_2 and don't change the culture media until 48 h after plating.

3. Aspirating off old culture media and washing the culture carefully with prewarmed sterile 1× PBS once and add appropriate amount of 0.3 mM Ca^{2+} E media every other day.

4. Small colonies can be visualized under microscope starting from day 3 and if colonies grow to 30–50 cells, it reaches to the end (*see* **Note 11**).

5. Fix the colonies with 100% methanol for 10 min at room temperature.

6. Stain the cells with 100 mL 0.5% crystal violet dye in the dark. (To make 0.5% crystal violet, add 0.5 g crystal violet in 20 mL methanol and 80 mL distilled water, filter the solution with 0.45-µm strainer before use). Leave the dish at room temperature for 10–30 min until the colonies are easily visualized. Remove the crystal violet; it can be reused or disposed of as toxic waste.

7. Wash the wells carefully with distilled water for 10 min or until the background is clear enough to visualize colonies.

8. Inverse the dishes on a paper towel and leave there for a couple of hours until the wells are completely dry.

9. Count and analyze the colonies (Fig. 2).

3.2.2 Chamber Graft Assay

1. Place newborn (born within 24 h) C57BL/6J mice into culture dishes and transfer to a CO_2 euthanizing chamber for 20 min.

2. Immerse mice in 70% ethanol for 2 min immediately after euthanasia, and place dishes on ice until next step.

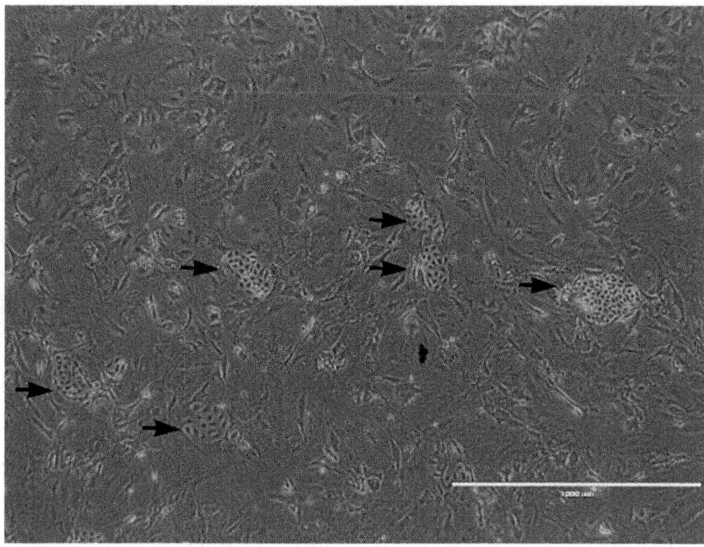

Fig. 2 Clonal growth of HFSCs on 3T3 feeder cells. Arrowheads depict HFSC colonies. Scale bar = 1000 μm

3. Grasp the body of mice firmly between a pair of sterile forceps and cut off the whole full-thickness back skin with sterile sharp scissors.

4. Place the released back skin in a clean 100-mm culture dish with dermis side down. Spread out the skin completely flat. Up to 6–8 mice back skin can be placed into a 100-mm culture dish.

5. Pour 10 mL freshly thawed 0.25% Trypsin without EDTA into the 100-mm dish. Make sure back skins float rather than attach to bottom of dish. Store the dish with floating skins at 4 °C overnight in a refrigerator.

6. The next day, aspirate trypsin and add 20 mL 10% FBS supplemented DMEM medium to the dish to neutralize reaction.

7. For each skin, lift the dermis up straight above the epidermis and transfer dermis to a new 100-mm culture dish for fibroblasts preparation.

8. Rinse dermises with 1× PBS and add 10 mL 0.2% collagenase I. Incubate dermises at 37 °C in a shaking incubator for 30 min.

9. Add 60 μL DNase and continue incubation for 10 min.

10. Aspirate the medium and neutralize reaction with 10% FBS supplemented DMEM medium.

11. Filter through 70-μm strainer. Then filter through 40-μm strainer.

12. Centrifuge the filtrate at 1000 rpm for 5 min and save the pellet which contains the bulk of the dermal fibroblasts.

13. Resuspend the fibroblasts pellet in 10 mL cold DMEM–F12 (1:1) medium and determine the total cell number by counting a suitable diluted aliquot.

14. Combine HFSCs with primary fibroblasts in cold DMEM–F12 (1:1) medium. For each chamber graft assay, mix 2 million freshly prepared HFSCs with 20 million primary fibroblasts.

15. Centrifuge the cell mixtures at 1000 rpm for 5 min. Save the pellet and resuspend in 50 µL DMEM–F12 (1:1). Store the resuspension in 1 mL syringe. Put the syringe on ice to be taken to animal room for grafting.

16. Place nude mice in the induction chamber under hood. Adjust the oxygen flowmeter to approximately 0.9 L/min and the isoflurane vaporizer to approximately 3.5% for induction and approximately 1.5% for maintenance.

17. Disinfect backs of the sedated mice first with Betadine, then with 70% ethanol.

18. Pull up middle back skin with forceps and remove a portion of lifted skin with curved scissors to create a 1-cm diameter hole.

19. Place the chamber into the hole and clip skin to the rim to prevent chamber from falling off.

20. Take the syringe and inject the cell suspension through the hole onto the chamber.

21. Transfer grafted mice to a clean cage and observe until recover from anesthesia.

22. Check mice daily.

23. Cut off upper part of chamber 7 days after grafting to expose wound. Remove chamber 14 days after grafting. Observing the wound until visible hair grow out above the skin surface and take pictures (Fig. 3).

4 Notes

1. Use only plastic beakers when making FBS because all of compounds added on the third day stick to glass.

2. 100× Cocktail: Mix 140 mL sterile 1× PBS with 20 mL of 5 mg/mL Insulin, 20 mL of 5 mg/mL Transferrin and 20 mL of 2×10^{-8} M T3. Sterilize through a filter and store in 37.5 mL aliquots in sterile tubes at -20 °C.

3. 4 mg/mL Hydrocortisone: Dissolve one vial in 95% EtOH to produce desired concentration. Sterilize through a filter and store in 1 mL aliquots at -20 °C.

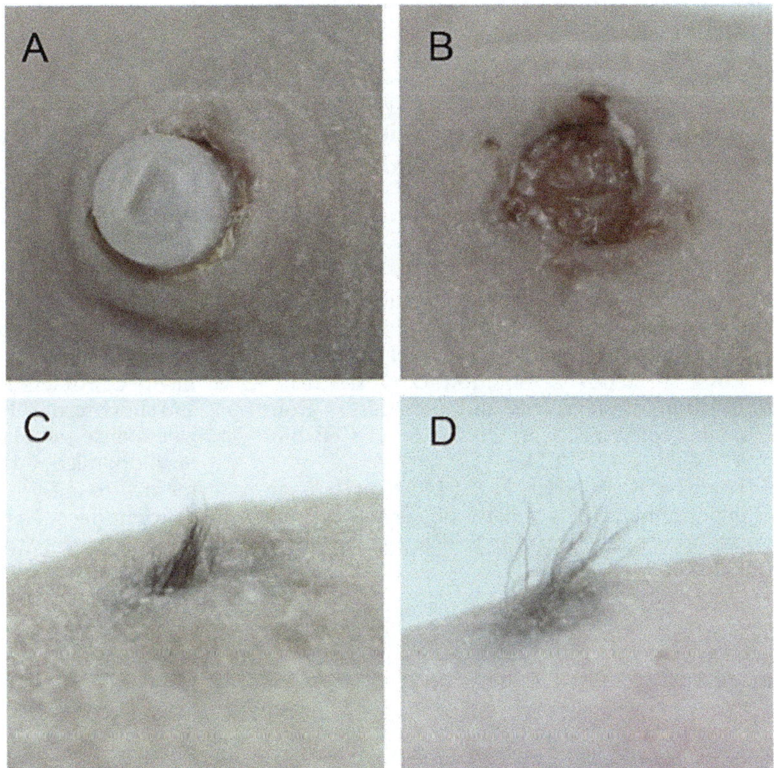

Fig. 3 Chamber graft assay. (**a**) Seven days after chamber graft. (**b**) Fourteen days after chamber graft (7 days after chamber removing). (**c**) Twenty-one days after chamber graft. Hair follicles grow outside of the skin. (**d**) Twenty-eight days after chamber graft

4. 10^{-6} M Cholera Toxin: Dissolve 1 mg vial of Cholera Toxin in 11.9 mL of ultra-pure water. Sterilize through a filter and store in 1 mL aliquots at 4 °C.

5. Mitomycin C working solution must be made freshly.

6. Better isolation can be done with a Krt1-15-EGFP mouse.

7. Dissection is more accurate under immunofluorescence microscope if a Krt1-15-EGFP mouse is applicable.

8. HFSCs can be monitored under immunofluorescence if a Krt1-15-EGFP mouse is applicable; GFP-positive HFSCs can be cultured and passaged on mitomycin C-treated 3T3 feeder cells.

9. HFSCs can adhere firmly to the plate very well.

10. Number of cells is determined by square centimeters of each well. A quantity of 1000 cells is plated in 1 cm².

11. Sterile environment is not necessary from **step 5**.

Acknowledgments

This work was supported by the National Natural Science Foundation of China (No. 81472895) and the Natural Science Foundation of Chongqing (No. cstc2015jcyjA1219).

References

1. Morris RJ, Liu Y, Marles L et al (2004) Capturing and profiling adult hair follicle stem cells. Nat Biotechnol 22(4):411–417
2. Kostic L, Sedov E, Soteriou D et al (2017) Isolation of stem cells and progenitors from mouse epidermis. Curr Protoc Stem Cell Biol 41:1C.20.1–1C.20.11
3. Kandyba E, Kobielak K (2014) Wnt7b is an important intrinsic regulator of hair follicle stem cell homeostasis and hair follicle cycling. Stem Cells 32(4):886–901
4. Nowak JA, Fuchs E (2009) Isolation and culture of epithelial stem cells. Methods Mol Biol 482:215–232
5. Lichti U, Anders J, Yuspa SH (2008) Isolation and short-term culture of primary keratinocytes, hair follicle populations and dermal cells from newborn mice and keratinocytes from adult mice for in vitro analysis and for grafting to immuno-deficient mice. Nat Protoc 3(5):799–810
6. Lee LF, Jiang TX, Garner W et al (2011) A simplified procedure to reconstitute hair-producing skin. Tissue Eng Part C Methods 17(4):391–400

Chapter 6

Methods of Mesenchymal Stem Cell Homing to the Blood–Brain Barrier

Peter Conaty, Lauren S. Sherman, Yahaira Naaldijk, Henning Ulrich, Alexandra Stolzing, and Pranela Rameshwar

Abstract

Mesenchymal stem/stromal cells (MSC) are multipotent cells that can be isolated from adult and fetal tissues. In vitro, MSCs show functional plasticity by differentiating into specialized cells of all germ layers. MSCs are of relevant to medicine and have been proposed for several disorders. MSCs can be transplanted across allogeneic barriers as "off the shelf" cells. This chapter focuses on methods to deliver MSCs to the brain because neurological pathology such as damage due to stroke can lead to debilitating mental and physical problems. In general, neurological diseases are difficult to treat, partly due to the challenge in getting drugs across the blood–brain barrier (BBB). MSCs as well as other stem cells can cross the BBB. The described method begins to develop procedures, leading to efficient delivery of drugs to the brain. Here describe how MSCs can be propagated from bone marrow aspirates and their utility in delivering small RNA to the brain. The chapter discusses the issue to enhance efficient delivery of MSCs to the brain.

Key words Mesenchymal stem cell, Bone marrow, Blood–brain barrier, Noncoding RNA, Drug delivery

1   Introduction

The application of stem cells (SC) for neurological damage caused by events such as a stroke or hypoxic-ischemia (HI) are well studied [1, 2]. Systematic research studies have been done to determine if SCs can form functional neurons. Among the studied SCs are mesenchymal SC (MSC), which can be induced to generate cells of all germ layers, including neural cells [3]. MSC are obtained from adult and fetal tissues. These include but are not limited to cord blood, fetal liver, bone marrow (BM), and adipose tissue [4]. Once removed from in vivo sources, the MSCs are expanded in vitro, and characterized to ensure consistency with the literature.

Shree Ram Singh and Pranela Rameshwar (eds.), *Somatic Stem Cells: Methods and Protocols*, Methods in Molecular Biology, vol. 1842, https://doi.org/10.1007/978-1-4939-8697-2_6, © Springer Science+Business Media, LLC, part of Springer Nature 2018

There is no clear answer on the method of using MSCs to reverse brain damage. The question that remains is whether the MSCs should be injected directly to the injured site versus other routes such as intravenous injection [5–7]. Direct engraftment of the MSC could be an advantage since the cells will be implanted directly to the damaged site. This approach might be more efficient if the process does not require an invasive procedure, which could lead to damage to the site of engraftment, including to the blood brain barrier (BBB) [5]. Systemic delivery of MSCs is safe but may be less efficient to reach the brain as compared to direct implantation of MSCs [8]. One of the problems of indirect delivery is the lung effect, indicating that intravenous injection of MSCs may be trapped in the lungs [6, 9]. This would limit the number of MSCs that could have the potential to reach the brain. As an alternative, intranasal application of cells provides a noninvasive method to deliver SC directly into the central nervous system (CNS) [10].

Experimental trials with MSCs have indicated improvement in neural outcomes, without evidence of cell replacement [10, 11]. This suggested that the MSCs may be acting through mediators to correct the pathological problems. The secretome of MSCs could be soluble factors as well as microvesicles such as exosomes [12, 13]. The identification of the MSC-derived secretome is a complex process since the particular factor could be different within the complex microenvironment in vivo. The identification of how MSCs secrete distinct secretome is thus crucial. Answers to these questions could continue to be sought as drug delivery of MSCs continue in experimental studies. An additional layer of complexity arises when autologous therapies are being designed as age and disease state might alter the secretome: it is well known that age alters the secretome of MSC and that it can even become harmful, making it necessary to think about cell quality [14].

The BBB is a major physical barrier that MSC must overcome [8]. This highly selective, semipermeable membrane is made up of interacting CNS endothelial cells which are held together by tight junctions [9]. Under normal conditions, the BBB prevents most cell types from entering into the brain, however there have been no detailed studies conducted regarding how much the BBB is an active barrier for MSC in treating the brain. This limits interaction with the CNS that might benefit neural functions. The efficient protection by the BBB could be damaged by injury such as stroke, brain tumor, and aging [15–17]. During pathological conditions, the tight connection of the BBB is disrupted and would allow cells to cross into the neural tissues. Among the cells that may get into the brain are stem cells such as MSCs. However, it is important to be able to control the modulation of the BBB in order to allow a greater number of MSCs or their secreted secretome to reach the desired site in the brain. Mannitol, which has been used in clinical practice to reverse intracranial pressure, could be used to temporarily get MSCs into the brain [18].

The method by which mannitol has been used to release excess fluids from the brain may apply to allow MSCs to enter the brain. In fact, there is support to test the effectiveness of mannitol as a potential modulator of the BBB for the effective use of systemic MSC therapy [18] (*see* **Note 1**). The clinical application of mannitol has been used to ease the intracranial pressure, which could occur by the blockage of veins that drain blood from the brain, or an increase in volume of blood or brain tissue [19]. The osmolarity of mannitol releases the pressure through a temporary increase in the size of the tight junctions in the BBB via contraction of the endothelial cells [20]. This property of mannitol has been proposed to allow the MSCs to get into the brain after the junctions have been temporarily opened. This opening into the brain is believed to be able to allow for the efficient crossing of MSCs and/ or their secretome to the desired site in the brain [19, 20]]. Experimental studies are needed to determine how the treatment may get to the desired site within the brain for targeted outcome. Research studies include methods to identify how inflammatory mediators and their receptors could aid in the precision method to get injected cells into the intended target. The dosing and timing of mannitol will need to be further studied to ensure that there is no damage caused by the opening of the tight junctions as prolonged use of mannitol can lead to swelling of the brain.

The method to isolate large numbers of MSCs has been established in the majority of academic/clinical centers. However, methods to get MSCs in the brain remain an unmet need. More importantly, if methods could be established to get MSCs into the brain, these stem cells could be used as drug delivery vehicle. The instruction part of this chapter will describe how a representative drug, noncoding RNA, can be delivered within MSCs for target in the brain. Additionally, the chapter will include instruction on experimental methods to label MSCs for tracking both in vitro and in vivo.

Several studies have attempted to track MSCs in animal models, some yielding confounding results [2, 6–8]. In all cases, intravenously injected MSCs are initially trapped in the lungs where they can secrete various factors for a systemic response [7]. Over time, many of the MSCs migrate from the lung to various tissues, including the brain, bone marrow, liver, spleen, kidney, and pancreas [7]. However, different groups have reported varying degrees and efficiencies of engraftment of transplanted MSCs. These differences can be explained, perhaps, by varied efficiency of MSC from distinct sources (bone marrow, adipose, placenta) to migrate, the cell culture conditions, and the number of times the MSCs were passaged [9, 10]. Further, the delivery method itself can influence the MSCs' homing ability, with direct injection permitting—or increasing efficiency of—homing to target organs not efficiently reached by intravenous injection [11]. As such, cells administered via intranasal administration are able to migrate and to reside in the CNS [21–23]. Migration initiates in the nasal mucosa through the cribriform plate following the olfactory neural pathway and other nasal routes (e.g., trigeminal and perivascular routes) [23]. Thus, the described methods can be used to improve the in vivo tracking of transplanted MSCs in a time-dependent manner.

2 Methods

2.1 Culturing Bone Marrow-Derived MSCs

1. Add 10 mL of MSC media to vacuum-gas plasma treated tissue culture Falcon 3003 petri dishes.

2. Add 1–2 mL of bone marrow aspirate to the dish. Gently swirl the plates mix the aspirate within the tissue culture media.

3. Incubate at 37 °C in 5% CO_2 for 3 days.

4. Remove nonadherent cells and slowly transfer into a 50-mL tissue culture tube containing 25 mL of Ficoll Hypaque 1.077 g/mL (*see* **Notes 2–4**).

5. Immediately after removing the nonadherent cells, add 5 mL of MSC media to the petri dish. The media should be replaced as quickly as possible to avoid the dish to become overly dry.

6. Separate the red blood cells and neutrophils placed on top of the Ficoll Hypaque by centrifuging for 30 min at 500 × *g* at room temperature. To aid in a clear separation, place the brake on the centrifuge in the "off" position.

   After centrifugation if you notice a cloudy upper layer, this could be due to remaining mononuclear cells in the upper layer. To ensure recovery of all mononuclear cells, continue to centrifuge for ~5 min. If needed, you may prolong the centrifugation time. You should be cautious at this point since you may centrifuge the desired middle layer containing the mononuclear fraction within the Ficoll Hypaque layer.

7. Immediately after centrifugation, aspirate the top layer, which contain media without interrupting the middle layer, which contains the desired mononuclear cells.

8. Immediately use a Pasteur pipette to carefully remove the mononuclear layer and transfer to a tube containing any tissue media with 2% FCS (*see* **Note 5**).

9. Wash the cells by centrifuging at 500 × *g*, room temperature for 20 min. If you notice a cloudy upper layer, this will likely indicate that the mononuclear cells have not been pelleted. To recover all of your cells, recentrifuge for a longer time, ~5 min,

but this could require a longer time until the upper layer is clear.

10. Immediately aspirate the media and disassociate the pellet by forcefully tapping the pellet with your fingers. This should not be delayed since the pellet, if remained in the tube for a prolonged period, will compromise the cell viability. Add fresh media with 2% FCS and then centrifuge for 15 min (Wash step).

11. Add MSC media to the pellet and then resuspend.
    Rule of thumb: Add 5 mL of the suspension cells/original petri dish. Specifically, if you began with ten petri dishes and the contents of the ten dishes were added to one 50 mL tube then you should add 50 of media.

12. Mix the cells and distribute the cell suspension equally into the dishes. In this described method, you should add 5 mL cell suspension/dish.

13. Reincubate the culture dishes. At weekly intervals, replace 50% of the media (*see* **Note 6**).

14. At 80–90% confluence, trypsinize the adherent cells and split at a ratio of 1:3 to 1:6 (*see* **Note 7**).

2.2 Preparation of MSCs

This step is needed if further research is conducted with a particular population of MSCs. Also, if the experiment needs to track the MSCs such as for in vivo transplantation, the MSCs should be labeled with a fluorescence dye (see below) or the use of a vector containing a fluorescence reporter gene.

2.2.1 Cell Sorting

1. Wash MSCs with 1× PBS.

2. Add 2 mL of 0.05% trypsin to cells in T75 Flask and incubate at 37 °C for 4 min (*see* **Note 8**).

3. Transfer trypsinized cells to media containing 5–10% FCS (*see* **Notes 9–11**).

4. Pellet the cells by centrifuging at $500 \times g$ for 10 min. Aspirate the supernatant and resuspend in 10 mL of 1× PBS.

5. Repeat **step 4** (*see* **Note 12**).

6. Label MSCs with the desired fluorescence-tagged antibody. For example, if you need the population expressing leptin receptor, able accordingly. Incubate with the antibody for 30 min; wash once by filling the tube with PBS and then centrifuge.

7. Resuspend cells in 10 mL of sorting buffer.

8. Centrifuge at $500 \times g$ for 5 min.

9. Aspirate the supernatant and resuspend the pellet in 2 mL of sorting buffer.

10. Determine cell concentration and dilute with sorting buffer as needed (*see* **Note 13**).

11. Filter cells through a strainer cap to remove any clumps that may clog the instrument (*see* **Note 14**).

12. Use the tube consistent with the institutional cell sorter for collection. Add 2 mL of complete media to the collection tubes (*see* **Note 15**).

2.3 Cell Transfection

This section describes the method to take the MSCs and load noncoding RNA for delivery to the brain.

1. On the first day, seed MSCs to achieve 60–80% confluence for transfection (*see* **Notes 16** and **17**).

2. On day of transfection, dilute 9 μL of Lipofectamine RANiMAX Reagent in 150 μL of Opti-MEM medium per well.

3. Dilute 3 μL of miRNA (10 μM) in 150 μL of Opti-MEM medium per well.

4. Add 150 μL of diluted miRNA to 150 μL of diluted Lipofectamine RNAiMAX Reagent per well and incubate for 5 min.

5. Add 250 μL of miRNA–lipid complex into each well. In each well, this will contain 25 pmol of the RNA and 7.5 μl of Lipofectamine RNAiMAX.

6. Incubate the cells for 1–3 days at 37 °C. Take an aliquot of cells to ensure you have transfected your miRNA. This can be done by PCR for the miRNA. Alternatively, if you tag the miRNA with a fluorochrome, you can test your MSCs by flow cytometry.

2.4 In Vivo Injection

1. Resuspend ~10^6 anti-miRNA or fluorochrome-labeled transfected MSCs in 0.2 mL PBS.

2. Inject the entire volume in a mouse, intraorbitally or intravenously. If you are using human cells, you could use immune deficient mice. However, this is not necessary since human MSCs can survive in xenografts.

Track the homing of MSCs after sacrificing the mice, and then label for human protein using anti-human NUMA [4]. If you opted to use fluorescence-labeled MSCs, sections of tissue can be analyzed for cell fluorescence. While traditional fluorescent labeling may be beneficial for short-term experiments (e.g., DiR), long-term tracking requires fluorochromes that will not degrade during the experiment period. Qdots overcome this difficulty, with an ability to identify as few as 1000 labeled transplanted MSCs after greater than 30 days [1, 2]. Like DiR and Qdots, labeling with firefly luciferase allows for tracking transplanted MSCs in vivo, by imaging systems [3]. In addition to these staining methods, MSCs can be labeled with contrast or trackers for identification by magnetic resonance imaging (MRI), positron emission tomography (PET), and single-photon emission CT (SPECT). However, the effect(s) of these labeling mechanisms on MSC multipotency and function have largely yet to be evaluated [5].

2.5 Intranasal Application

1. Hold the mouse via hand constrain and recline the mouse onto its back while the skull is immobilized.

2. Apply 6 µL of 100 U hyaluronidase solution per nostril twice to a final volume of 24 µL. The hyaluronidase solution is applied by nose drops placing 10 µL pipette tip carefully in the nostril allowing the drop to be inhaled. After the first application wait 2 min before administering the 2nd application [24] (*see* **Note 18**).

3. Place the mouse back into the cage and wait 30 min prior to administering the cells.

4. Resuspend 10^6 anti-miRNA or fluorochrome-labeled transfected MSC in 24 µL PBS.

5. Apply transfected MSC as described in **steps 1** and **2** without the hyaluronidase.

2.6 Summary

The method to expand MSCs seems to be established within the community of stem cell scientists. However, it is unclear if all sources of MSCs will show similar efficiency to home to the brain. This chapter provides the basis for more in-depth research studies. Figure 1 shows the initiating culture with BM aspirate following by phenotypic characterization. The expanded MSCs are sorted if needed and then injected into mice. The MSCs if loaded with a tracking dye can be imaged. Alternatively, sections of the brain can be used to identify the homing of human cells.

3   Notes

1. Studies are needed to optimize dosing and administration times, but the following was found in a study done by Gonzales-Portillo et al. and can be used as a reference for further study [10]. Mannitol (1.1 mol/L at 4 °C) is administered at the same injection site as the umbilical cord blood with injections taking place over 10 min. For adults, 1.5–2.0 g/kg of mannitol was administered while the dose for pediatrics was 0.25–1.0 g/kg.

2. This assumes that you have multiple culture plates from the same donor. If not, use a 15-mL conical polystyrene tube. In cases where there are multiple plates, pool the nonadherent cells from the same donor into the same tube of Ficoll

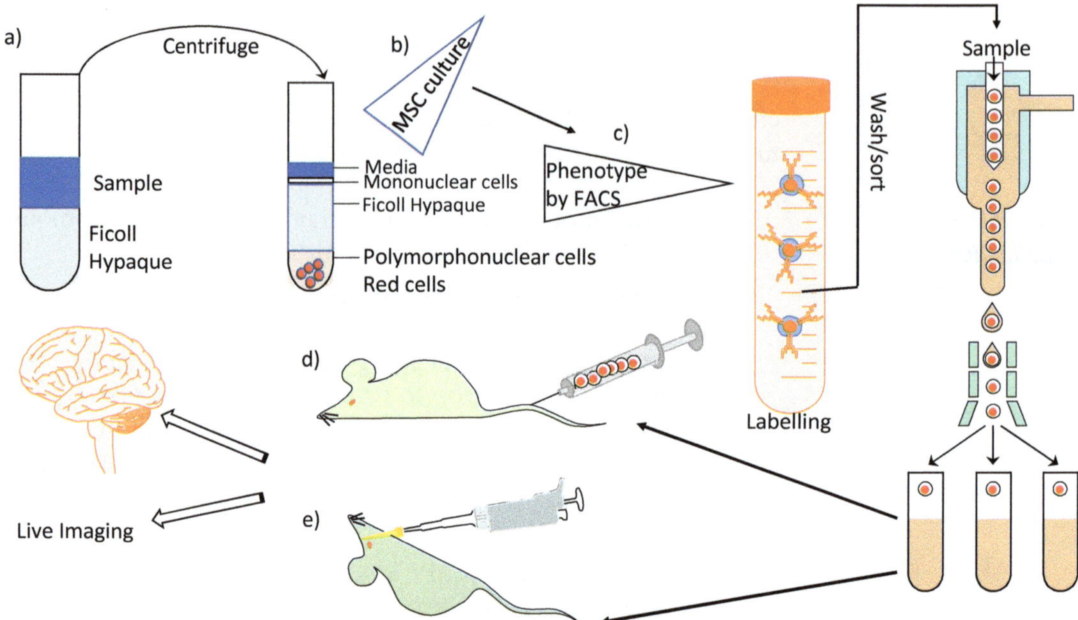

Fig. 1 Work flow with MSC to studies on brain homing. (**a**) 3-day nonadherent media from MSC cultures are subjected to Ficoll Hypaque gradient separation. (**b**) The MSCs are expanded and then characterized phenotypically. (**c**) MSCs are subjected to phenotypic characterization by FACs and if needed by sorting. (**d**) The expanded MSCs are injected intravenously into mice. (**e**) Alternatively, the MSCs are injected intra-nasally to maximize cell delivery to the brain. Tracking of the MSCs can be done by live imaging or by sectioning of the brain

Hypaque. As you transfer the nonadherent cells onto the Ficoll, try to remove most of the red blood cells from the culture dish. This is important because as the red cells begin to lyse this could become toxic to the MSCs.

3. Pipet the nonadherent cells with a 2-mL pipette or a glass 9" Pasteur pipette. Slowly add to the Ficoll Hypaque gradient by placing the tip of the pipette to the corner of the tube. Rapid addition to the Ficoll Hypaque could prevent a clean separation of the different fractions and this would result of the mononuclear cells.

4. Failure to remove the red blood cells will lead to the subsequent lysis during the period of expansion. If this occurs, the lysed material will be toxic to the adherent cells. To avoid this, use a 5 or 10 mL pipette to mix the nonadherent cells before removing them from the culture dish.

5. Ficoll Hypaque is toxic to the cells. Thus, immediately after centrifugation, remove the mononuclear cells. Since you are likely to pipette some of the Ficoll Hypaque, try to dilute this with sera-free media; otherwise it will be difficult to pellet the cells. The latter could remain within the Ficoll. As a rule of

thumb, you can use conical tubes: E.g., 15 mL tubes could contain ~7 mL of Ficoll and ~5–7 mL of cell suspension.

6. Avoid confluency of the MSC cultures as cell contact could lead to spontaneous differentiation.

7. After four passages the adherent cells should be asymmetrical and spindle shaped, and display the following phenotypes: CD14$^-$, CD29$^+$, CD44$^+$, CD34$^-$, CD45$^-$, SH2$^+$, prolyl-4-hydroxylase$^-$.

8. Deadhered cells should be floating with trypsin. If not, hit sides of flasks to dislocate the cells. If your cells are in petri dishes you can dislocate the cells by pipetting up and down.

9. If you are planning to study surface markers or if the cells are sensitive to trypsin, use Accutase or TrypLE to deadhere.

10. The volume of the media needed to transfer the trypsinized cells will depend on the volume of trypsin. The sera within the media served as a substrate for residual trypsin. However, if the trypsin is diluted sufficiently, it will become inactive.

11. Media containing diluted FCS is necessary to stop the trypsin activity. FCS will act as a substrate for any active diluted trypsin.

12. If residual FCS is believed to remain, another wash may be performed. However, avoid excess washing as each wash would result in cell loss.

13. If the cells form clumps, incubate with 5 mM EDTA for 10 min and then resuspend in sorting buffer.

14. Cell concentration used for sorting will depend on the number of final cells required. Also, the starting number of cells will depend on the expected frequency in your population. As a general rule, the cell concentration is 1–2 × 10^6 cells/mL, otherwise sorting time will be long. It is always better to start with a higher concentration of cells and then dilute as needed.

15. A small amount of bovine sera albumin or serum can result in healthier cells after sorting.

16. The following procedure is scaled for the use of 6-well plates. If using 24- or 96-well plates, adjust accordingly. Also, you may look online at the Invitrogen Lipofectamine RNAiMAX Reagent protocol.

17. If the reduction in viable cells is not an issue, you may also transfect by electroporation.

18. Hyaluronidase allows for invasion of the cells by breaking up the barrier of the nasopharyngeal mucosa. Diluted hyaluronidase should be aliquoted in small volumes to avoid freeze–thaw.

References

1. van Velthoven CT, Kavelaars A, van Bel F, Heijnen CJ (2010) Mesenchymal stem cell treatment after neonatal hypoxic-ischemic brain injury improves behavioral outcome and induces neuronal and oligodendrocyte regeneration. Brain Behav Immun 24:387–393

2. Lee JS, Hong JM, Moon GJ, Lee PH, Ahn YH, Bang OY, collaborators S (2010) A long-term follow-up study of intravenous autologous mesenchymal stem cell transplantation in patients with ischemic stroke. Stem Cells 28:1099–1106

3. Mareschi K, Novara M, Rustichelli D, Ferrero I, Guido D, Carbone E, Medico E, Madon E, Vercelli A, Fagioli F (2006) Neural differentiation of human mesenchymal stem cells: evidence for expression of neural markers and eag K+ channel types. Exp Hematol 34:1563–1572

4. Wei X, Yang X, Han ZP, Qu FF, Shao L, Shi YF (2013) Mesenchymal stem cells: a new trend for cell therapy. Acta Pharmacol Sin 34:747–754

5. Cunningham MG, Bolay H, Scouten CW, Moore C, Jacoby D, Moskowitz M, Sorensen JC (2004) Preclinical evaluation of a novel intracerebral microinjection instrument permitting electrophysiologically guided delivery of therapeutics. Neurosurgery 54:1497–1507

6. Harting MT, Jimenez F, Xue H, Fischer UM, Baumgartner J, Dash PK, Cox CS (2009) Intravenous mesenchymal stem cell therapy for traumatic brain injury. J Neurosurg 110:1189–1197

7. Kean TJ, Lin P, Caplan AI, Dennis JE (2013) MSCs: delivery routes and engraftment, cell-targeting strategies, and immune modulation. Stem Cells Int 2013:732742

8. Kim SM, Jeong CH, Woo JS, Ryu CH, Lee JH, Jeun SS (2016) In vivo near-infrared imaging for the tracking of systemically delivered mesenchymal stem cells: tropism for brain tumors and biodistribution. Int J Nanomedicine 11:13–23

9. Liu L, Eckert MA, Riazifar H, Kang DK, Agalliu D, Zhao W (2013) From blood to the brain: can systemically transplanted mesenchymal stem cells cross the blood-brain barrier? Stem Cells Int 2013:435093

10. Sharp J, Keirstead HS (2009) Stem cell-based cell replacement strategies for the central nervous system. Neurosci Lett 456:107–111

11. Maumus M, Jorgensen C, Noel D (2013) Mesenchymal stem cells in regenerative medicine applied to rheumatic diseases: role of secretome and exosomes. Biochimie 95:2229–2234

12. Drago D, Cossetti C, Iraci N, Gaude E, Musco G, Bachi A, Pluchino S (2013) The stem cell secretome and its role in brain repair. Biochimie 95:2271–2285

13. Li Y, Chopp M (2009) Marrow stromal cell transplantation in stroke and traumatic brain injury. Neurosci Lett 456:120–123

14. Sarkar P, Redondo J, Kemp K, Ginty M, Wilkins A, Scolding NJ, Rice CM (2018) Reduced neuroprotective potential of the mesenchymal stromal cell secretome with ex vivo expansion, age and progressive multiple sclerosis. Cytotherapy 20:21–28

15. Yonemori K, Tsuta K, Ono M, Shimizu C, Hirakawa A, Hasegawa T, Hatanaka Y, Narita Y, Shibui S, Fujiwara Y (2010) Disruption of the blood brain barrier by brain metastases of triple-negative and basal-type breast cancer but not HER2/neu-positive breast cancer. Cancer 116:302–308

16. Prakash R, Carmichael ST (2015) Blood-brain barrier breakdown and neovascularization processes after stroke and traumatic brain injury. Curr Opin Neurol 28:556–564

17. Stone LA, Smith ME, Albert PS, Bash CN, Maloni H, Frank JA, McFarland HF (1995) Blood-brain barrier disruption on contrast-enhanced MRI in patients with mild relapsing-remitting multiple sclerosis: relationship to course, gender, and age. Neurology 45:1122–1126

18. Rangel-Castilla L, Gopinath S, Robertson CS (2008) Management of intracranial hypertension. Neurol Clin 26:521–541

19. Borlongan CV, Hadman M, Sanberg CD, Sanberg PR (2004) Central nervous system entry of peripherally injected umbilical cord blood cells is not required for neuroprotection in stroke. Stroke 35:2385–2389

20. Gonzales-Portillo GS, Sanberg PR, Franzblau M, Gonzales-Portillo C, Diamandis T, Staples M, Sanberg CD, Borlongan CV (2014) Mannitol-enhanced delivery of stem cells and their growth factors across the blood-brain barrier. Cell Transplant 23:531–539

21. Leovsky C, Fabian C, Naaldijk Y, Jager C, Jang HJ, Bohme J, Rudolph L, Stolzing A (2015) Biodistribution of in vitro-derived microglia applied intranasally and intravenously to mice: effects of aging. Cytotherapy 17:1617–1626

22. Danielyan L, Beer-Hammer S, Stolzing A, Schafer R, Siegel G, Fabian C, Kahle P,

Biedermann T, Lourhmati A, Buadze M, Novakovic A, Proksch B, Gleiter CH, Frey WH, Schwab M (2014) Intranasal delivery of bone marrow-derived mesenchymal stem cells, macrophages, and microglia to the brain in mouse models of Alzheimer's and Parkinson's disease. Cell Transplant 23(Suppl 1):S123–S139

23. Danielyan L, Schafer R, von Ameln-Mayerhofer A, Buadze M, Geisler J, Klopfer T, Burkhardt U, Proksch B, Verleysdonk S, Ayturan M, Buniatian GH, Gleiter CH, Frey WH 2nd (2009) Intranasal delivery of cells to the brain. Eur J Cell Biol 88:315–324

24. Zwijnenburg PJ, van der Poll T, Florquin S, van Deventer SJ, Roord JJ, van Furth AM (2001) Experimental pneumococcal meningitis in mice: a model of intranasal infection. J Infect Dis 183:1143–1146

Chapter 7

3D Bioprinting and Stem Cells

**Caitlyn A. Moore, Niloy N. Shah, Caroline P. Smith,
and Pranela Rameshwar**

Abstract

Three-dimensional (3D) in vitro modeling is increasingly relevant as two-dimensional (2D) cultures have
been recognized with limits to recapitulate the complex endogenous conditions in the body. Additionally,
fabrication technology is more accessible than ever. Bioprinting, in particular, is an additive manufacturing
technique that expands the capabilities of in vitro studies by precisely depositing cells embedded within a
3D biomaterial scaffold that acts as temporary extracellular matrix (ECM). More importantly, bioprinting
has vast potential for customization. This allows users to manipulate parameters such as scaffold design,
biomaterial selection, and cell types, to create specialized biomimetic 3D systems.

The development of a 3D system is important to recapitulate the bone marrow (BM) microenvironment since this particular organ cannot be mimicked with other methods such as organoids. The 3D system
can be used to study the interactions between native BM cells and metastatic breast cancer cells (BCCs).
Although not perfect, such a system can recapitulate the BM microenvironment. Mesenchymal stem cells
(MSCs), a key population within the BM, are known to communicate with BCCs invading the BM and to
aid in their transition into dormancy. Dormant BCCs are cycling quiescent and resistant to chemotherapy,
which allows them to survive in the BM to resurge even after decades. These persisting BCCs have been
identified as the stem cell subset. These BCCs exhibit self-renewal and can be induced to differentiate.
More importantly, this BCC subset can initiate tumor formation, exert chemoresistance, and form gap
junction with endogenous BM stroma, including MSCs. The bioprinted model detailed in this chapter
creates a MSC-BC stem cell coculture system to study intercellular interactions in a model that is more
representative of the endogenous 3D microenvironment than conventional 2D cultures. The method can
reliably seed primary BM MSCs and BC stem cells within a bioprinted scaffold fabricated from CELLINK
Bioink. Since bioprinting is a highly customizable technique, parameters described in this method
(i.e., cell–cell ratio, scaffold dimensions) can easily be altered to serve other applications, including studies
on hematopoietic regulation.

Key words Alginate, Bone marrow, Breast cancer, Stem cells, Dormancy, 3D bioprinting, Extracellular
matrix

1 Introduction

The microenvironment in which stem cells reside dictates cell
morphology, polarity, growth, motility, signal transduction, gene and
protein expression, and biochemical activities [1–8]. As a result,

Shree Ram Singh and Pranela Rameshwar (eds.), *Somatic Stem Cells: Methods and Protocols*, Methods in Molecular Biology, vol. 1842,
https://doi.org/10.1007/978-1-4939-8697-2_7, © Springer Science+Business Media, LLC, part of Springer Nature 2018

routinely used 2D cell culture systems are limited in terms of therapeutic in vitro testing and system modeling as the basis for in vivo studies. On the other hand, 3D cultures allow scientists to more closely mimic natural tissues, promoting cell interactions between adjacent cells and extracellular matrix (ECM) in all directions [9]. Due to advances in additive manufacturing, bioprinting has become a widely utilized technique for creating 3D in vitro systems [10]. Critically, 3D bioprinting applications require an interdisciplinary collaboration of cell biology and biomedical medical engineering, as outlined in this chapter.

1.1 Central Approaches to Bioprinting

There are three central approaches to bioprinting, depending on the needs of the tissue, system complexity, or experimental needs: (1) biomimicry; (2) autonomous self-assembly; and (3) mini-tissues. Hence, the approach selected ultimately depends on the end application.

Biomimicry seeks to reproduce the specific cellular functional components of tissues through "mimicking" the cellular microenvironment [11]. This is achieved by printing a suitable microenvironment, including relevant cell types, ECM, and gradients of soluble and insoluble factors.

Autonomous self-assembly utilizes the primitive properties of stem cells and embryonic organ development as a guide for creating more complex 3D biostructures by directly printing spheroids of primitive cells that undergo fusion and cellular organization to mimic organogenesis [12]. In doing so, the spheroid can self-produce the ECM and cell signaling gradient to direct the composition, localization, functional and structural properties for proper histogenesis and biological microarchitecture [13, 14]. This methodology requires an intimate knowledge of organogenesis to manipulate the process. It also requires a unique cell source of primitive cells, often derived from adult stem cell (ASCs) donors or induced pluripotent stem cells (iPSCs), bringing about issues regarding differentiation patterns, immunocompatibility, and engraftment [15].

Mini-tissues overcome the developmental issues of autonomous self-assembly by printing smaller, functional building blocks on scaffolds and integrating them into a larger, macrostructure via rational design or self-assembly [16, 17]. Thus, this method draws upon a strong knowledge of biomimicry and autonomous self-assembly, as mini-tissues integrate the two approaches [18].

1.2 Considerations in Selecting the Appropriate Bioprinter

Selecting a bioprinter requires one to consider the parameters that are most important to the end application. These parameters include, but are not limited to, surface resolution, cell viability, and bioink material consistency. This section reviews the various types of bioprinters used in contemporary research.

Thermal inkjet bioprinting electrically heats the printer head to 200–300 °C to produce air-pressure pulses that ejects consistent bioink droplets [19, 20]. Importantly, the high temperature of the printer head does not have a substantial impact on the cell macromolecules or viability, as the ~2 μs contact time only raises the temperature of the droplet about 4–10 °C [21, 22]. The advantages of thermal inkjet bioprinters are their high print speed, low cost, and versatility of application [23]. On the other hand, disadvantages of thermal inkjet bioprinting include exposure to thermal and mechanical stress, poor nonuniform droplet directionality and size, frequent nozzle clogging, and unreliable cell encapsulation.

Acoustic inkjet bioprinters contain a voltage-mediated piezoelectric crystal that creates a 15–25 kHz acoustic wave frequency to break bioink droplets at regular intervals and eject the droplets from the nozzle [24, 25]. The advantages of an acoustic inkjet bioprinter include its ability to control uniform droplet size and ejection directionality without exposure to thermal or mechanical stressors [26–28]. Additionally, an open-pool nozzle-less ejection system can be used to avoid sheer stress, clogging, and loss of cell viability from the nozzle tip [29]. Disadvantages include potential for cell lysis and membrane damage due to the 15–25 kHz pulse and the limitation of a 10 centipoise maximum viscosity for bioink [30, 31].

Laser-assisted bioprinting (LAB) utilizes laser-induced forward transfer by focusing laser pulses in an absorbing layer of ribbon [10]. This generates a high-pressure bubble that can propel the ejection of cellular substrates onto a collector scaffold or plate. Resolution of LAB printing can be optimized to each print by adjusting laser fluency (energy/area), surface tension, substrate hydrophilicity, and the air gap between the ribbon and substrate layers [29, 32, 33]. LAB is compatible with viscosities of 1–300 mPa/s and can deposit densities up to 10^8 cells with negligible changes in cell viability and function at a laser pulse rate of 5 kHz and speeds of 1600 mm/s [34–36]. However, LAB requires rapid gelatin kinetics for shape fidelity and a ribbon layer for each cell type used, leading to slow print times and high costs [37, 38].

Microextrusion bioprinters provide a temperature-controlled, continuous extrusion of material via a pneumatic or mechanical dispensing system. This provides the greatest versatility in compatibility with various biomaterials (viscosity range: 30 mPa/s to 6×10^7 mPa/s) [39, 40]. Mechanical (piston or screw) dispensing systems provide greater control over material flow and spatial control than pneumatic dispensing systems, which is advantageous for higher viscosity materials [41]. Although microextrusion bioprinters can deposit high cell densities relative to other modes of boiprinting, they tend to show lower cell viabilities due to shear stresses inflicted on the cells during extrusion, a property that is independent of nozzle diameters [31, 42, 43].

Additionally, microextrusion printers tend to have slower print times compared to inkjet bioprinters. The method described in this chapter utilizes a commercially available pneumatic extrusion-based bioprinting system.

1.3 Considerations for Bioink and Cell Selection: Recapitulating the Microenvironment

In general, ECM forms a structural framework to stabilize tissues and provide mechanical support for cell attachment and motility [1]. It also plays a crucial role in cell functionality and differentiation through receptor-mediated signaling and regulation of gene [12, 44, 45]. In terms of bioprinting, the ECM is initially provided in the form of "bioink," or the biomaterial in which cells are embedded [46]. Bioink-based ECM must have seven major properties: (1) cross-linking to prevent the scaffold from falling apart in various medias or during transplantation; (2) porous to allow for cells to intercalate and adhere; (3) biodegradable in a time-sensitive fashion to allow for sufficient cellular engraftment prior to complete scaffold resolution; (4) proper surface topography and chemistry; (5) biocompatibility to prevent immunorejection upon transplantation; (6) mimics the physiological microenvironment in terms of high surface area to volume ratio, proper ECM component ratios, and hydrophobic; and (7) commercial accessibility and feasibility of all ECM scaffold materials [47].

It is important to note that variability in composition, potential cytotoxic breakdown metabolites, and other aspects of the bioink must be considered. Hence, scaffold biomaterials must be optimized for each specific cell type or experimental model. Major aspects regarding cell function to be considered when selecting the optimal bioink include the following.

1.3.1 Viscosity

Higher-viscosity materials often provide structural support for the printed construct and lower-viscosity materials provide a suitable environment for maintaining cell viability and function [30].

1.3.2 Proliferation Rate

The cell line of choice should be able to expand sufficiently for printing and engraftment, while capable of mimicking physiological ratios [48]. Too low proliferation rate may lead to a loss of viability of the transplanted construct, whereas too high of a proliferation rate may lead to hyperplasia or apoptosis. However, one must note that the proliferation rate might be irrelevant, depending on the model. As an example, it might be desired for the supporting BM microenvironment to remain in a nonproliferative state but to remain viable.

1.3.3 Cell Durability

The cells used for bioprinting must be durable enough to survive the stresses of the bioprinting process, including shear stress during extrusion, presence of toxins and enzymes, and nonphysiological pH [49, 50]. Also, if the scaffold for transplantation, it is important that the embedded cells can withstand physiological stresses in addition to the stresses mentioned above.

1.3.4 Potency

Though many primary cells offer therapeutic options, they are limited in lifespan and long-term engraftment. Stem cells offer a promising source for bioprinting due to their undifferentiated, yet multilineage differentiation potential and self-renewal ability [51]. When selecting stem cells to incorporate into the 3D bioprinted model, consider the experimental and therapeutic boundaries, as well as the source, immunocompetence, and long-term engraftment potential dictated by the potency of the cell [52].

1.4 Selection of Elements for Proposed Model

The proposed BM-mimicking model described in the sections below is bioprinted from a hydrogel bioink (CELLINK Bioink, CELLINK) using a pneumatic extrusion-based bioprinter (Inkredible Standard 3D Bioprinter, CELLINK) as depicted in Fig. 1. Primary MSCs isolated from BM aspirate are incorporated into the model to most accurately represent MSCs from the BM niche. MDA-MB-231 BCCs stably transfected with an Oct4a-GFP plasmid are used to allow to isolate specific subsets of BCCs based on the expression of *Oct4a*, a stem cell-associated gene [53].

2 Materials

1. Dulbecco's Modified Eagle Medium (DMEM); high glucose, L-glutamine, penicillin–streptomycin, 10% fetal calf serum (FCS).

2. Ficoll-Hypaque solution.

3. PBS without Ca^{2+} and Mg^{2+}.

4. Accutase.

5. Trypan Blue.

6. CELLINK Bioink; alginate, nanocellulose.

7. Cross-linking solution; 100 mM $CaCl_2$, deionized water.

3 Methods

3.1 MSC Culture from BM Aspirates

1. Unfractionated aspirates from healthy volunteers ranging from 18 to 30 years of age are diluted in DMEM containing 10% FCS, seeded in vacuum gas plasma-treated plates, and cultured as described [54, 55].

2. After 3 days, red blood cells and granulocytes are removed by Ficoll-Hypaque density gradient centrifugation. The remaining mononuclear fraction is seeded in the plates. At weekly intervals, 50% of media is replaced with fresh media. Adherent cells are serially passages five times after the growth attained ~80% confluence.

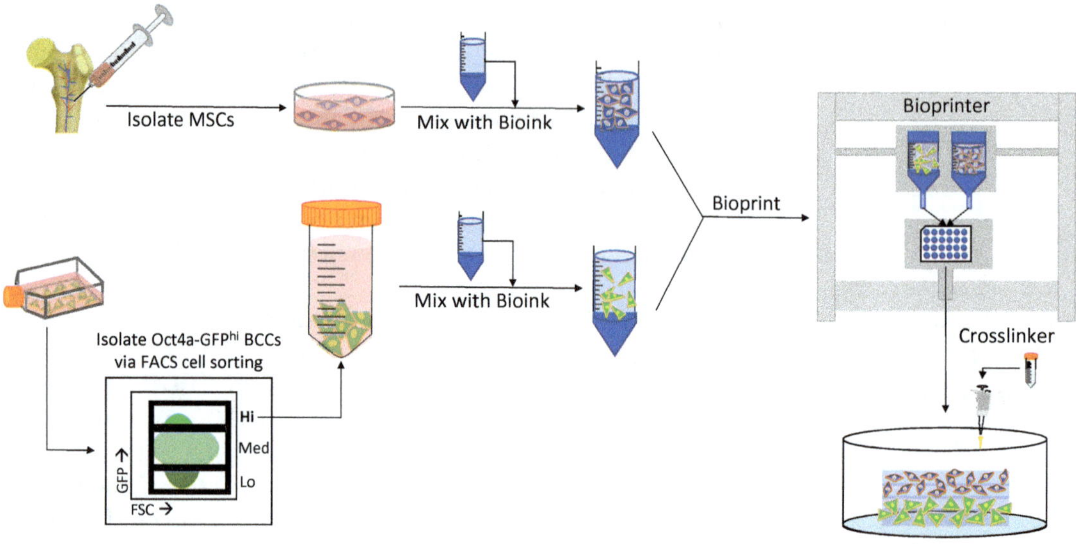

Fig. 1 Pneumatic extrusion-based bioprinting of mesenchymal stem cells (MSCs) and breast cancer stem cells. MSCs are isolated from bone marrow aspirates from healthy volunteers aged 18–35 years and mixed into CELLINK Bioink. Breast cancer stem cells are isolated from the heterogeneous population of breast cancer cells (BCCs) by Oct4a-GFP expression using FACS cell sorting. The Oct4^{hi} subset, which is the breast cancer stem cells, is isolated and mixed into CELLINK Bioink. The two printer cartridges containing cell-laden bioink are inserted into the bioprinter and printed into scaffolds in a 24-well plate. The scaffolds are subsequently cross-linked using 100 mM CaCl$_2$, washed with media, and then incubated at 37 °C and 5% CO$_2$ for final application

3. After four cell passages, the adherent cells are symmetric, show multilineage differentiation, and are CD14$^-$, CD29$^+$, CD44$^+$, CD34$^-$, CD45$^-$, CD105$^+$, and prolyl-4-hydroxylase$^-$.

3.2 Isolation of BC Stem Cells

1. BCCs such as MDA-MB-231 are stably transfected with pOct4a-GFP which acts as a fluorescent reporter for the relative expression of Oct4a. The method used by this reporter gene system to isolate BC stem cell is described [53, 55].

2. Oct4a-GFP-MDA-MB-231 are cultured in DMEM containing 10% FCS. Media is replaced twice weekly and cells are grown to ~80% confluence. High GFP levels in these cells (highest 5% expression, Oct4a-GFP^{hi}) have been identified as the BC stem cell subpopulation and are sorted with the FACS cell sorter [53].

3.3 Prepare Cells and Equipment for Bioprinting

3.3.1 Calibrate the Bioprinter (See **Note 1**)

1. Attach air compressor to bioprinter. Attach each to a separate, grounded power-source. Turn on air compressor. Then, turn on bioprinter by pressing the circular button on the front-right of the machine.

2. Insert empty printing cartridge with printing tip attached into the printer head

3. Calibrate the bioprinter axes: Prepare Bioprint → Home Axes.

4. Place 24-well plate onto platform and secure.

5. Align printer tip with the center of the 1st well (front left-most well): Utilities → Move Axes → 0.1 mm → X-Axis. Then, Utilities → Move Axes → 0.1 mm → Y-Axis.

6. Lower the printer tip until there is only enough space between the tip and the bottom of the well to fit a single sheet of paper: Utilities → Move Axes → 0.1 mm → Z-Axis.

7. Calibrate Z-axis position: Prepare Bioprint → Calibrate Z.

8. Set printer head pressures to 30 kPa using dials on the right side of the bioprinter (*see* **Note 2**).

9. Remove empty printing cartridge and set aside.

3.3.2 Prepare MSCs for Bioprinting

1. After validating the identity of the MSCs in culture, use Accutase to detach MSCs from plates.

2. Pellet cells by centrifuging at $500 \times g$ at RT for 10 min. Remove accutase and replace with desired volume of media.

3. Count and determine viability of MSCs using Trypan Blue.

4. Remove 72×10^4 MSCs and place in sterile tissue culture-grade centrifuge tube. Pellet cells by centrifuging at $500 \times g$ at RT for 10 min.

5. Resuspend pellet in 72 μL of media (*see* **Note 3**).

3.3.3 Prepare BC Stem Cells (See Above) for Bioprinting

1. Count and calculate the number of viable cells. The latter is assessed by with Trypan Blue.

2. Remove 72×10^4 BC stem cells. Pellet cells by centrifuging at $500 \times g$ at RT for 10 min.

3. Resuspend pellet in 72 μL of media (*see* **Note 3**).

3.4 Bioprint and Culture Scaffolds

1. Load 72 μL MSC suspension into Luer-lock syringe. Load 720 μL bioink into Luer-lock syringe. Connect syringes using Luer-lock adapter.

2. Gently mix the contents of each syringe until the cell suspension is indistinguishable and the mixture is homogeneous. At which point, move mixture into one syringe and remove empty syringe and adaptor.

3. Dispense mixture into bioprinter cartridge and insert plunger. Using tweezers, push down on the plunger to move the mixture to the base of the cartridge without losing any through the nozzle opening.

4. Repeat **steps 1–3** with 72 μL BCSC suspension and 720 μL bioink. When completed, there will be two bioprinter cartridges containing 720 μL of homogenized cell–bioink mixture.

5. Screw printer tips into nozzle openings (where Luer-lock adapter previously fit).

6. Attach air pressure line to the top of the cartridge and secure the cartridges into the printer heads.

7. Select the 24-well plate program designated for the brand of 24-well plate utilized in the study (*see* **Note 4**).

8. As scaffolds are printed, cover each completely with 100 mM $CaCl_2$ cross-linking solution. Allow scaffolds to cross-link for 15–20 min (*see* **Note 5**).

9. After cross-linking, remove cross-linker and replace with media and incubate scaffolds at 37 °C for 20 min.

10. After 20 min, replace media with fresh media. Incubate scaffolds until desired time point, refreshing 50% of the media twice per week (*see* **Note 6**).

4 Notes

1. Commands given here are specific to the CELLINK Inkredible 3D Bioprinter (Standard Model). Commands will vary between models and brands of bioprinter but workflow/logic will be relatively similar.

2. 30 kPa is optimized for CELLINK Bioink using a 0.4 mm diameter nozzle tip. This value can be optimized (for validation or for use with other bioinks) by turning the pressure for the printer head to 0 kPa and slowly increasing the pressure until a continuous, steady stream of bioink leaves the nozzle. The pressure at which this is observed is the "optimal" printing pressure.

3. All volumes are optimized for 24-well plate program that deposits 24 scaffolds of dimension 6 mm × 6 mm × 1.2 mm. Each scaffold contains 60 µL of bioink; therefore, 1.44 mL of bioink is required to bioprint 24 scaffolds of the prescribed dimensions. In addition, media resuspension volume is 1/10th the bioink volume. For 1 MSC: 1 BCSC coculture within scaffolds of prescribed dimensions, the total volume of cell suspension will be 144 µL (or 72 µL of suspension for each cell type). Hence, bioink volume and corresponding cell suspension volumes will differ based on scaffold dimensions and the cell–cell ratio utilized in the 3D model. Also, in this model, the MSCs and BC stem cells are deposited separately in the scaffold. Cell types can be homogenized and added simultaneously if cell and bioink volumes are adjusted as required.

4. Plate dimensions vary by brand. For a seamless printing process, it is critical that the printing program is coded specifically for the brand of plate used.

5. Cross-linking is sufficient when the scaffolds can be gently moved with minimal deformation.

6. If scaffolds lose structure during culture, it is acceptable to add small volumes of cross-linking solution into the culture media to sustain cross-links in the scaffold.

Acknowledgment

In partial fulfillment for a Ph.D. by CAM. This work is supported by a grant awarded by F.M Kirby Foundation.

References

1. Birgersdotter A, Sandberg R, Ernberg I (2005) Gene expression perturbation in vitro- a growing case for three-dimensional (3D) culture systems. Semin Cancer Biol 15:405–412

2. Cukierman E, Pankov R, Yamada KM (2002) Cell interactions with three-dimensional matrices. Curr Opin Cell Biol 14:633–639

3. Even-Ram S, Yamada KM (2005) Cell migration in 3D matrix. Curr Opin Cell Biol 17:524–532

4. Yamada KM, Cukierman E (2007) Modeling tissue morphogenesis and cancer in 3D. Cell 130:601–610

5. Dai X, Ma C, Lan Q, Xu T (2016) 3D bioprinted glioma stem cells for brain tumor model and applications of drug susceptibility. Biofabrication 8:045005

6. Hakkinen KM, Harunaga JS, Doyle AD, Yamada KM (2011) Direct comparisons of the morphology, migration, cell adhesions, and actin cytoskeleton of fibroblasts in four different three-dimensional extracellular matrices. Tissue Eng Part A 17:713–724

7. Yu Y, Zhang Y, Martin JA, Ozbolat IT (2013) Evaluation of cell viability and functionality in vessel-like bioprintable cell-laden tubular channels. J Biomech Eng 135:91001

8. Zhou X, Zhu W, Nowicki M, Miao S, Cui H, Holmes B, Glazer RI, Zhang LG (2016) 3D bioprinting a cell-laden bone matrix for breast cancer metastasis study. ACS Appl Mater Interfaces 8:30017–30026

9. Guiro K, Arinzeh TL (2015) Bioengineering models for breast cancer research. Breast Cancer 9:57–70

10. Dababneh AB, Ozbolat IT (2014) Bioprinting technology: a current state-of-the-art review. J Manuf Sci Eng 136:061016

11. Ingber DE, Mow VC, Butler D, Niklason L, Huard J, Mao J, Yannas I, Kaplan D, Vunjak-Novakovic G (2006) Tissue engineering and developmental biology: going biomimetic. Tissue Eng 12(12):3265–3283

12. Kasza KE, Rowat AC, Liu J, Angelini TE, Brangwynne CP, Koenderink GH, Weitz DA (2007) The cell as a material. Curr Opin Cell Biol 19:101–107

13. Steer DL, Nigam SK (2004) Developmental approaches to kidney tissue engineering. Am J Physiol Renal Physiol 286:F1–F7

14. Derby B (2012) Printing and prototyping of tissues and scaffolds. Science 338(6109): 921–926

15. Marga F, Neagu A, Kosztin I, Forgacs G (2007) Developmental biology and tissue engineering. Birth Defects Res C Embryo Today 81:320–328

16. Mironov V, Visconti RP, Kasyanov V, Forgacs G, Drake CJ, Markwald RR (2009) Organ printing: tissue spheroids as building blocks. Biomaterials 30:2164–2174

17. Kelm JM, Lorber V, Snedeker JG, Schmidt D, Broggini-Tenzer A, Weisstanner M, Odermatt B, Mol A, Zünd G, Hoerstrup SP (2010) A novel concept for scaffold-free vessel tissue engineering: self-assembly of microtissue building blocks. J Biotechnol 148:46–55

18. Skardal A, Zhang J, Prestwich GD (2010) Bioprinting vessel-like constructs using hyaluronan hydrogels crosslinked with tetrahedral

polyethylene glycol tetracrylates. Biomaterials 31:6173–6181

19. Okamoto T, Suzuki T, Yamamoto N (2000) Microarray fabrication with covalent attachment of DNA using bubble jet technology. Nat Biotechnol 18:438–841

20. Goldmann T, Gonzalez JS (2000) DNA-printing: utilization of a standard inkjet printer for the transfer of nucleic acids to solid supports. J Biochem Biophys Methods 42(3):105–110

21. Xu T, Jin J, Gregory C, Hickman JJ, Boland T (2005) Inkjet printing of viable mammalian cells. Biomaterials 26(1):93–99

22. Xu T, Gregory CA, Molnar P, Cui X, Jalota S, Bhaduri SB, Boland T (2006) Viability and electrophysiology of neural cell structures generated by the inkjet printing method. Biomaterials 27(19):3580–3588

23. Cui X, Boland T, D D'Lima D, K Lotz M (2012) Thermal inkjet printing in tissue engineering and regenerative medicine. Recent Pat Drug Deliv Formul 6(2):149–155

24. Tekin E, Smith PJ, Schubert US (2008) Inkjet printing as a deposition and patterning tool for polymers and inorganic particles. Soft Matter 4(4):703–713

25. Fang Y, Frampton JP, Raghavan S, Sabahi-Kaviani R, Luker G, Deng CX, Takayama S (2012) Rapid generation of multiplexed cell cocultures using acoustic droplet ejection followed by aqueous two-phase exclusion patterning. Tissue Eng Part C Methods 18:647–657

26. Saunders R, Bosworth L, Gough J, Derby B, Reis N (2004) Selective cell delivery for 3D tissue culture and engineering. Eur Cells Mater 7:84–85

27. Saunders RE, Gough JE, Derby B (2008) Delivery of human fibroblast cells by piezoelectric drop-on-demand inkjet printing. Biomaterials 29:193–203

28. Nakamura M, Kobayashi A, Takagi F, Watanabe A, Hiruma Y, Ohuchi K, Iwasaki Y, Horie M, Morita I, Takatani S (2005) Biocompatible inkjet printing technique for designed seeding of individual living cells. Tissue Eng 11:1658–1666

29. Colina M, Serra P, Fernández-Pradas JM, Sevilla L, Morenza JL (2005) DNA deposition through laser induced forward transfer. Biosens Bioelectron 20:1638–1642

30. Kim JD, Choi JS, Kim BS, Choi YC, Cho YW (2010) Piezoelectric inkjet printing of polymers: stem cell patterning on polymer substrates. Polymer 51:2147–2154

31. Murphy SV, Skardal A, Atala A (2013) Evaluation of hydrogels for bio-printing applications. J Biomed Mater Res A 101:272–284

32. Chrisey DB (2000) The power of direct writing. Science 289:879–881

33. Guillemot F, Souquet A, Catros S, Guillotin B (2010) Laser-assisted cell printing: principle, physical parameters versus cell fate and perspectives in tissue engineering. Nanomedicine 5:507–515

34. Guillotin B, Souquet A, Catros S, Duocastella M, Pippenger B, Bellance S, Bareille R, Rémy M, Bordenave L, Amédée J (2010) Laser assisted bioprinting of engineered tissue with high cell density and microscale organization. Biomaterials 31:7250–7256

35. Hopp B, Smausz T, Kresz N, Barna N, Bor Z, Kolozsvári L, Chrisey DB, Szabó A, Nógrádi A (2005) Survival and proliferative ability of various living cell types after laser-induced forward transfer. Tissue Eng 11:1817–1823

36. Koch L, Kuhn S, Sorg H, Gruene M, Schlie S, Gaebel R, Polchow B, Reimers K, Stoelting S, Ma N (2009) Laser printing of skin cells and human stem cells. Tissue Eng 16:847–854

37. Kattamis NT, Purnick PE, Weiss R, Arnold CB (2007) Thick film laser induced forward transfer for deposition of thermally and mechanically sensitive materials. Appl Phys Lett 91:171120

38. Duocastella M, Fernández-Pradas J, Morenza J, Zafra D, Serra P (2010) Novel laser printing technique for miniaturized biosensors preparation. Sensors Actuators B Chem 145:596–600

39. Chang CC, Boland ED, Williams SK, Hoying JB (2011) Direct-write bioprinting three-dimensional biohybrid systems for future regenerative therapies. J Biomed Mater Res B Appl Biomater 98:160–170

40. Fedorovich NE, Swennen I, Girones J, Moroni L, Van Blitterswijk CA, Schacht E, Alblas J, Dhert WJ (2009) Evaluation of photocrosslinked lutrol hydrogel for tissue printing applications. Biomacromolecules 10:1689–1696

41. Chang R, Nam J, Sun W (2008) Effects of dispensing pressure and nozzle diameter on cell survival from solid freeform fabrication-based direct cell writing. Tissue Eng 14:41–48

42. Smith CM, Stone AL, Parkhill RL, Stewart RL, Simpkins MW, Kachurin AM, Warren WL, Williams SK (2004) Three-dimensional bioassembly tool for generating viable tissue-engineered constructs. Tissue Eng 10:1566–1576

43. Nair K, Gandhi M, Khalil S, Yan KC, Marcolongo M, Barbee K, Sun W (2009) Characterization of cell viability during bioprinting processes. Biotechnol J 4:1168–1177

44. Reed EJ, Klumb L, Koobatian M, Viney C (2009) Biomimicry as a route to new materials: what kinds of lessons are useful? Math Phys Eng Sci 367:1571–1585

45. Huh D, Torisawa YS, Hamilton GA, Kim HJ, Ingber DE (2012) Microengineered physiological biomimicry: organs-on-chips. Lab Chip 12:2156–2164

46. Hospodiuk M, Dey M, Sosnoski D, Ozbolat IT (2017) The bioink: a comprehensive review on bioprintable materials. Biotechnol Adv 35:217–239

47. Haycock JW (2011) 3D cell culture: a review of current approaches and techniques. Methods Mol Biol 695:1–15

48. Ambesi-Impiombato F, Parks L, Coon H (1980) Culture of hormone-dependent functional epithelial cells from rat thyroids. Proc Natl Acad Sci 77:3455–3459

49. Blaeser A, Duarte Campos DF, Puster U, Richtering W, Stevens MM, Fischer H (2016) Controlling shear stress in 3D bioprinting is a key factor to balance printing resolution and stem cell integrity. Adv Healthc Mater 5:326–333

50. Ozbolat IT, Yu Y (2013) Bioprinting toward organ fabrication: challenges and future trends. IEEE Trans Biomed Eng 60:691–699

51. Pittenger MF, Mackay AM, Beck SC, Jaiswal RK, Douglas R, Mosca JD, Moorman MA, Simonetti DW, Craig S, Marshak DR (1999) Multilineage potential of adult human mesenchymal stem cells. Science 284:143–147

52. Li J, Chi J, Liu J, Gao C, Wang K, Shan T, Li Y, Shang W, Gu F (2017) 3D printed gelatin-alginate bioactive scaffolds combined with mice bone marrow mesenchymal stem cells: a biocompatibility study. Int J Clin Exp Pathol 10:6299–6307

53. Patel SA, Ramkissoon SH, Bryan M, Pliner LF, Dontu G, Patel PS, Amiri S, Pine SR, Rameshwar P (2012) Delineation of breast cancer cell hierarchy identifies the subset responsible for dormancy. Sci Rep 2:906

54. Corcoran KE, Trzaska KA, Fernandes H, Bryan M, Taborga M, Srinivas V, Packman K, Patel PS, Rameshwar P (2008) Mesenchymal stem cells in early entry of breast cancer into bone marrow. PLoS One 3:e2563

55. Bliss SA, Sinha G, Sandiford OA, Williams L, Engelberth DJ, Guiro K, IL L, Greco SJ, Ayer S, Bryan M, Kumar R, Ponzio NM, Rameshwar P (2016) Mesenchymal stem cell-derived exosomes stimulates cycling quiescence and early breast cancer dormancy in bone marrow. Cancer Res 76:1–13

Chapter 8

Characterization of Gastrospheres Using 3D Coculture System

Carlos Antônio do Nascimento Santos, Radovan Borojevic, Luiz Eurico Nasciutti, and Christina M. Maedatakiya

Abstract

To understand the molecular mechanisms involved in gastric disorders and regeneration, we need an in vitro tridimensional (3D) culture model, which can mimic the in vivo gastric microenvironment. A 3D coculture system named gastrosphere is proposed herein, composed of primary human gastric epithelial and stromal cells. The primary cultures were obtained from endoscopic gastric biopsies, and after mechanical and enzymatic dispersion, epithelial (HGE3) and stromal (HGS12) cells were expanded. After extensive immunocytochemical characterization, cells were seeded onto 96-well round bottom plates previously covered with 1% agarose. Cells were cultured in KM-F12 culture medium with 10% fetal bovine serum (FBS), antibiotics, and antimycotics, in humidified air at 37 °C and atmosphere containing 5% CO_2 for 72 h or until spheres formation. Then gastrospheres were carefully transferred to a rotary cell culture system (RCCS-4), and maintained for 07, 14, 21, and 28 days. Gastrospheres were morphologically characterized by immunocytochemistry [cytokeratins (CK), vimentin, α-smooth muscle actin (α-SMA), laminin (LN), fibronectin (FN), and type IV collagen (CIV), proliferating cell nuclear antigen (PCNA)], and electron microscopy. In gastrospheres, the cytokeratin-positive epithelial cells were found in the outer layer, while vimentin-positive stromal cells were localized in the center of the gastrospheres. PCNA+ cells were mainly seen at the peripheral and in the intermediary region while nestin+ cells were also depicted in the latter zone. Scanning electron microscopy revealed groups of cohesive gastric cells at the periphery, while transmission electron microscopy demonstrated some differentiated mucous-like or zymogenic-like cells in the periphery and stromal structures located at the center of the 3D structures. Extracellular matrix was deposed between cells. Our data suggest that in vitro gastrospheres recapitulate the in vivo gastric microenvironment.

Key words Gastrospheres, Human gastric stromal and epithelial cells, Nestin-positive epithelial cells

1 Introduction

The human gastric mucosa is constituted by a self-renewing epithelium maintained by adult stem cells located in spatially restricted sites of the tissue. Their activity and proliferation is strongly dependent upon complex signaling pathways involving other epithelial cells as well as surrounding stromal cells. The human gastric

Shree Ram Singh and Pranela Rameshwar (eds.), *Somatic Stem Cells: Methods and Protocols*, Methods in Molecular Biology, vol. 1842, https://doi.org/10.1007/978-1-4939-8697-2_8, © Springer Science+Business Media, LLC, part of Springer Nature 2018

mucosa forms flask-like invaginations (gastric pits), which communicate with several tubular glands [1]. These gastric glands in turn are organized into three regions: isthmus, neck and base, and contain different cell populations, such as mucous neck cells, parietal cells, principal or zymogenic cells, and enteroendocrine cells. Surrounding the glands, a supportive stroma contains stromal cells [2]. The gastric mucosa is located in a harsh environment, in constant contact with mucosal secretion, nutrients, metabolites, and resident bacteria, which can lead to a loss of mucosa homeostasis. This depends upon a variety of mechanisms including its continuous self-renewal. This is provided by a population of adult stem cells, residing in specific regions such as the isthmus [3–6], and the bottom of gastric glands. Moreover, a reserve stem cell population has also been described at the gland base of the antrum/pyloric region [3]. The ability of long-term growth in vitro of primary nontransformed tissues derived from the gastrointestinal tissues allowed to maintain and form tridimensional (3D) tissue that can represent the complexity of in vivo gastrointestinal tissues [7–9]. Herein, we describe the formation and characterization of composite 3D gastrointestinal organoids, constituted of primary human gastric epithelia with the supporting stromal cells, the structure named "gastrospheres" [8]. Collectively, organoids represent a complex system that maintains some properties of the original tissue [10, 11], and the potentiality for long-term cultivation of primary human cells. Therefore, this model seems to be an excellent tool to study the endogenous processes of epithelial regeneration [3, 8, 12] as well as the interactions of the gastric mucosa with parasites/microbes like H. pylori [13, 14].

The de novo generation of the 3D human gastric mucosa in vitro (gastrosphere) was achieved through the spontaneous gastric epithelial and stromal cells interaction by cultivating cells in a nonadhesive culture system with routine culture medium (DMEM and DMEM-F12). Therefore, the gastric mucosa-like microenvironment formed in the gastrospheres may have allowed the establishment of the gastric epithelial stem cells niche that enabled the maintenance of undifferentiated cells along with some differentiated cells. Moreover, the approach utilized in this study for isolation and purification of both cell populations has great potential for future investigations, especially as a robust 3D culture system for investigating aspects of the gastric mucosa cell–cell and cell–parasite interactions in vitro.

2 Materials

2.1 Gastric Cell Isolation and Cell Culture

1. Cryogenic tubes 1.8 mL.
2. 15 mL centrifuge tubes.
3. 96-well cell culture cluster round bottom.
4. 96-well cell culture plates flat bottom sterile.
5. 24-well cell culture plates flat bottom Sterile.
6. Pasteur pipettes and pipettes as needed.
7. 35 mm culture dish.
8. Dulbecco's modified Eagle's medium/Nutrient mixture F-12 Ham.
9. Fetal bovine serum (FBS).
10. Penicillin–streptomycin solution, x100.
11. Amphotericin B.
12. Collagenase type I.
13. Dispase I protease.
14. Trypsin–EDTA solution 0.25%.
15. Agarose.
16. Dimethyl sulfoxide (DMSO).
17. Iris Scissors 110 mm Spring Action Blunt (micro scissors), straight or curved, Micro/Neuro Surgery Instruments.
18. Stainless steel curved scalpel.
19. Magnetic stirrer, Sargent Welch Scientific company.
20. Magnetic Stirring Bar (Spinbar Teflon Micro).
21. Corning Pyrex Glass Erlenmeyer flask, Measuring, 25 mL.
22. Rotary cell culture system (RCCS-4).
23. Freezing Container.
24. NIH/3T3 (ATCC® CRL1658™).
25. Cobalt-60 irradiator (type Gamma cell).
26. Prepare Dulbecco's Modified Eagle's Medium (DMEM) containing 1% of each penicillin and streptomycin (PS), and 1 µg/mL amphotericin.
27. Prepare 4 mL 0.1% collagenase type I solution (0.1% collagenase/EDTA w/v in DMEM without FBS) for three fragments (*see* **Note 1**).
28. Prepare 4 mL 0.125% dispase solution (0.125% dispase/EDTA w/v in DMEM without FBS).

29. Prepare feeder layer dishes (25 cm^2) with DMEM, and seed 3T3 irradiated cells, 24 h before use, and place the dishes in an incubator (37 °C, 5% CO$_2$).

30. Prepare 20 mL 1% agarose in PBS in a culture bottle.

2.2 Characterization of Cells Population by Immunostaining and Lectins Staining

1. Paraformaldehyde PA, EM grade.

2. Ammonium chloride PA (NH4Cl).

3. Gelatin from bovine skin type B.

4. Saponin.

5. Normal goat serum.

6. Triton X-100.

7. Tween 20.

8. Keratinocyte (KM-F12) culture medium

9. Albumin, from bovine serum (BSA), fraction V.

10. Optimum cutting temperature (OCT) medium, (Sakura (Saku-4583).

11. Polylysine microscope adhesion glass slides.

12. Mouse anti-PCNA (clone PC10).

13. Mouse anti-Nestin antibody (BD Transduction Laboratories).

14. Mouse anti-Cytokeratin 7 (CK-7) (Dako).

15. Mouse anti-Cytokeratin 8 (CK-8) (Dako)

16. Mouse anti-type IV Collagen (COL IV) (Sigma).

17. Mouse anti-laminin (LN) (Sigma).

18. Mouse anti-fibronectin (FN) (Sigma).

19. Mouse anti-chondroitin sulfate (CS) (clone CS-56; Sigma).

20. Mouse anti-vimentin (VIM) (Sigma).

21. Mouse anti-α-Smooth Muscle Actin (α-SMA) (Sigma).

22. Streptavidin/biotin blocking kit.

23. Biotinylated anti-mouse IgG (H+L), F(ab′)$_2$ fragment (Sigma).

24. Biotinylated anti-rabbit IgG (H+L), F(ab′)$_2$ fragment (Sigma).

25. Streptavidin–Alexa 546 (Invitrogen) or Streptavidin–Alexa 488 (Invitrogen).

26. 4′,6-diamidino-2-phenylindole dihydrochloride (DAPI).

27. Vectashield mounting Medium (Vector Laboratories).

28. Biotinylated lectin *Dolichos biflorus* agglutinin (DBA) (αGalNAc) (Vectors) for parietal [14].

29. Biotinylated *Ulex europaeus* Agglutinin I (UEA I) (D-GalNAc) (Vector Laboratories) for mucous pit cells [15].

30. Biotinylated *Griffonia (Bandeiraea) simplicifolia* Lectin II (GSL II, BSL II) (α- or β-GlucNAc) (Vector Laboratories) for mucous neck cells [15].

31. Leica CM 1850 cryostat (Leica Biosystems).

2.3 Solutions

1. 4% paraformaldehyde (PFA) solution, pH 7.4–4% sucrose.

2. For a 4% paraformaldehyde solution, add 4 g of EM grade paraformaldehyde to 50 mL of H_2O. Add 1 mL of 1 M NaOH and stir gently on a heating block at ~60 °C until the dissolution of paraformaldehyde. Add 10 mL of 10× PBS, allow the mixture to cool to room temperature. Mix 4% sucrose solution v/v. Adjust the pH to 7.4 with 1 M HCl (~1 mL), then adjust the final volume to 100 mL with H_2O.

3. Phosphate saline buffer (PBS).

4. 0.8 g NaCl, 0.2 g KCl, 1.44 g Na_2HPO_4 and 0.24 g KH_2PO_4. Adjust the pH to 7.4 with HCl. Add distilled water to a total volume of 1 L.

5. Sorensen's phosphate buffer (0.133 M, pH 7.4).

6. 0.133 M Na_2HPO_4, 0.133 M KH_2PO_4. Mix 80.4 mL of Na_2HPO_4 and 19.6 mL of KH_2PO_4 pH 7.4.

7. 50 mM ammonium chloride (NH4Cl) in PBS pH 8.

8. 0.2% gelatin in saponin–PBS solution.

9. 1% normal goat serum solution.

10. 0.3% Triton X-100 in PBS.

2.4 Electron Microscopy Sample Preparation

1. 2.5% glutaraldehyde solution in 0.1 M/L Sorensen's phosphate buffer pH 7.4.

2. 0.1 M cacodylate buffer (pH 7.4).

3. 1% osmium tetroxide solution, 0.8% potassium ferrocyanide solution, 5 mM calcium chloride solution in 0.1 M cacodylate buffer.

4. Epoxy Resin for transmission electron microscopy (TEM) (Epoxy embedding kit, Fluka).

5. Solution A: 5 mL epoxy embedding medium, 8 mL 2-docecenylsuccinic anhydride (DDSA).

6. Solution B: 8 mL epoxy embedding medium, 7 mL methylnadic anhydride (NMA).

7. Epoxy A+B+1.5–2% [2,4,6-tris(dimethylaminomethyl)phenol] DMP-30.

8. 1% uranyl acetate in water.

9. 5% uranyl acetate in water.

10. Lead citrate (Reynolds).

11. 1.33 g lead nitrate (Pb $(NO_3)_2$), 1.76 g sodium citrate (Na3C6H5O7.2H$_2$O).

12. Ultramicrotome RMC-MT 6000 XL.

13. Electron microscopy (JEOL 5310).

3 Methods

All procedures are performed using sterile techniques in a biological safety cabinet (BSC), and all the solutions used in cell culture are sterile.

3.1 Harvest Biopsies and Gastric Cells Isolation

3.1.1 Epithelial Cell Isolation

1. After obtaining human normal gastric mucosa biopsies (Fig. 1a), place them in flow tubes with 4 mL DMEM culture medium with 5% FBS, antibiotics (PS), and antimycotics (1 µg/mL)

2. In a 30 mm petri dish containing 500 µL DMEM culture medium (Fig. 1b) cut all biopsies into small fragments (about 1 mm) with a curve scissors or forceps in the BSC (*see* **Note 2**).

3. Add 2 mL fresh DMEM to the petri dish, and carefully rinse the gastric fragments briefly by pipetting up and down twice and discard the supernatant with a 2-mL glass Pasteur pipette, leaving the fragments settled on the bottom of the dish (*see* **Note 3**).

4. Transfer the fragments into a sterile Erlenmeyer flask with 4 mL of cold dispase solution (0.125%) and incubate at 4 °C for 16 h (Fig. 1c). This step is important to detach epithelial cells from the extracellular matrix of the basal membrane. (*see* **Note 4**)

5. After the incubation, put a sterile small magnetic bar into Erlenmeyer flask and agitate for 1 h at room temperature, using a magnetic stirrer (Fig. 1d). Aspirate the supernatant and transfer to a 15-mL conical tube with 4 mL of fresh DMEM, then centrifuge the 15-mL tube at $400 \times g$ for 5 min (Fig. 1e–g).

6. Resuspend the pellet in DMEM supplemented with 10% FBS, penicillin–streptomycin, and plate it in a 25 cm^2 tissue culture dish (Fig. 1h) for 2–6 h in humidified air with 5% CO_2 at 37 °C, collect the nonadherent cells and centrifuge. This step is done to withdraw the possible adherent stromal cells.

7. Resuspend the cells in KM-F12 medium and seed $0.5–1 \times 10^6$ cells into a 25 cm^2 tissue culture dish, previously plated with irradiated 3T3 cells (Fig. 1i), and cultivated at 37 °C in humidified air with 5% CO_2 for 7–12 days or until the gastric epithelial cells cover the bottom and form a feeder layer. Change the culture medium twice a week.

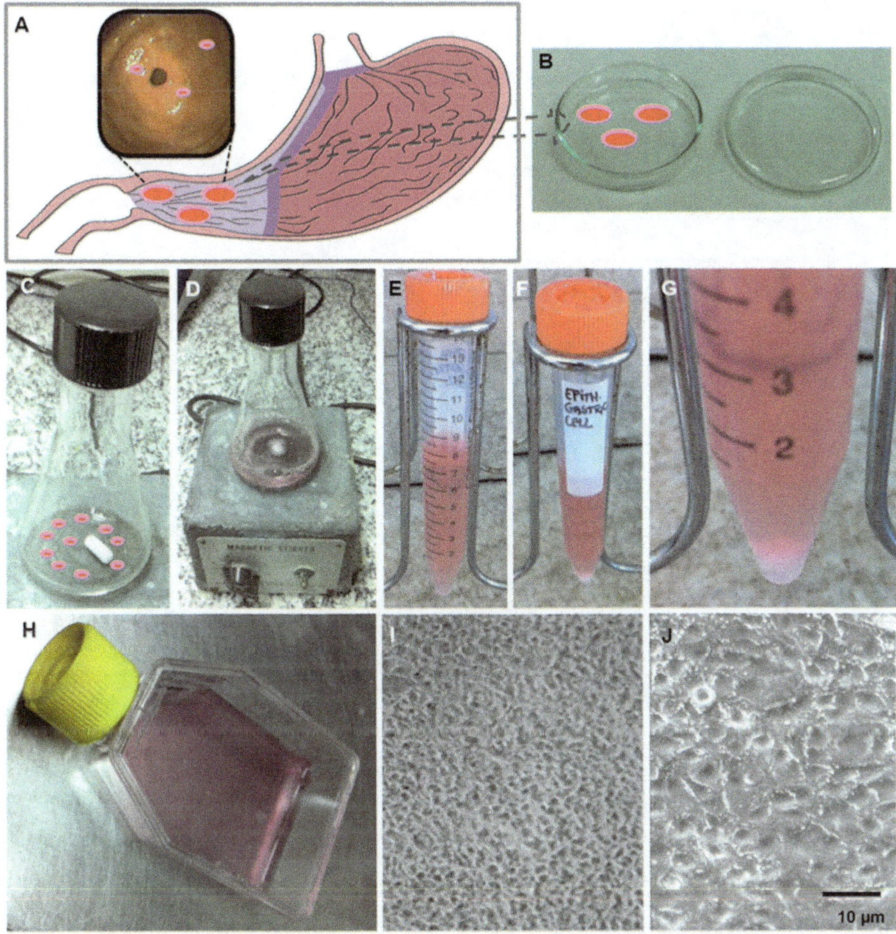

Fig. 1 Isolation of primary human gastric epithelial cells. (**a** and **b**) Schematic representation of the human gastric fragments harvested to get epithelial cells population; (**c**) Fragments of the gastric mucosa were transferred to a sterile Erlenmeyer flask with cold dispase solution (0.125%) and incubated at 4 °C for 16 h to detach epithelial cells from the ECM; (**d**) A sterile magnetic bar was introduced into the flask and agitated for 1 h at room temperature, in a magnetic stirrer. (**e**) The supernatant was transferred to a 15-mL conical tube with fresh DMEM, then centrifuged at $400 \times g$ for 5 min to obtain a pellet (**f** and **g**); (**h**) The pellet was resuspended in DMEM culture medium supplemented with 10% FBS, penicillin–streptomycin, and put in a 25 cm² tissue culture dish for 2–6 h in humidified air with 5% CO_2 at 37 °C; (**i**). The supernatant was collected and transferred to a culture dish containing irradiated 3T3 cells (feed layer) (**i**); (**j**) Epithelial cells grown onto irradiated 3T3 cells after 7 days of culture

8. After day 7 or 12, detach the epithelial cells (Fig. 1j) and centrifuge the cells in 15 mL tube at $400 \times g$ for 5 min, cultivate in culture dishes without 3T3 cells in DMEM/F12 with 10% FBS.

3.1.2 Stromal Cell Isolation

1. Repeat the **steps 1** and **2** of Subheading 3.1.1 described above (Fig. 2a, b).

2. In a 30 mm petri dish with 500 µL collagenase solution, cut all biopsies into small fragments (about 1 mm) with a curve scissors or forceps, and in the BSC.

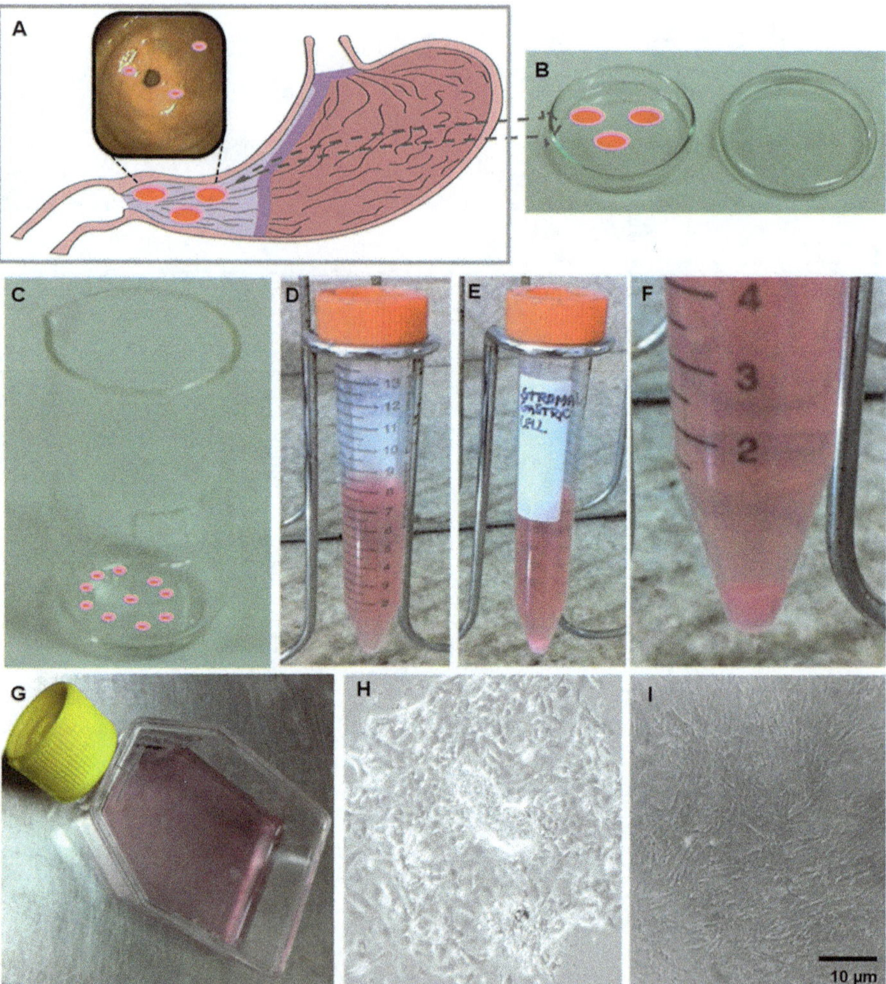

Fig. 2 Isolation of primary human gastric stromal cells. (**a** and **b**) Schematic representation of the human gastric fragments harvested to get stromal cells population; (**c**) To obtain the stromal cells population fragments were transferred to a sterile Becker containing collagenase solution (0.125%) and biopsies were cut into small fragments (about 1 mm and incubated at 37 °C for 1 h; (**d**) The solution was subsequently transferred to a 15 mL conical tube with fresh DMEM, then centrifuged at 400 × g for 5 min to obtain a pellet (**e** and **f**); (**g**) The pellet was resuspended in DMEM culture medium supplemented with 10% FBS, penicillin/streptomycin, and put in a 25 cm^2 tissue culture dish for 5–6 days or until achieving confluent cultures in humidified air with 5% CO_2 at 37 °C; (**h**) Images obtained at phase contrast microscope of cells after 24 h of culture and 60 days of culture (**i**)

3. Repeat the **step 3** of Subheading 3.1.1 described above.

4. Transfer the small fragments from 35 mm culture dish into a sterile 20 mL backer flask with 4 ml of collagenase solution (0.1%) and incubated at 37 °C for two cycles of 30 min (Fig. 2b, c). Carefully rinse the gastric fragments briefly by pipetting up and down 4 times every 15 min into a 15-mL sterile conical flask (Fig. 2d).

5. After the first 30 min of incubation, aspirate the supernatant and transfer to a sterile 15 mL conical tube with 4 mL of fresh DMEM and keep on ice.

6. Complete the tubes with the remaining tissues with 4 mL of collagenase solution (0.1%) and incubated at 37 °C for more 30 min, as in the **step 1**.

7. After the first 30 min of incubation, aspirate the supernatant and transfer to a sterile 15 mL conical tube with 4 mL of fresh DMEM and centrifuge the tubes at 400 × g for 5 min (Fig. 2e, f).

8. Resuspend the pellet in DMEM supplemented with 10% FBS, penicillin–streptomycin and plate 0.5–$1 × 10^6$ these cells into the 25 cm^2 tissue culture dishes and incubate in a humidified atmosphere of 95% air-5% CO_2 at 37 °C, for 5–7 days or up until the gastric stromal cells achieve 80% of the confluence (Fig. 2g–i). Change the DMEM every 3 days.

9. After the day 7, detach and plate the cells for about 12 times as in the **step 5**. In each passage, observe the cell morphology under phase contrast and epifluorescence microscopy. After the twelfth passage, connective tissue cells cultures are homogeneous without epithelial cells.

10. Cryopreserve the gastric cells in cryotubes with 10% DMSO in SFB freezing solution. Freeze cells with a specific cooling system which decrease slowly the temperature, one degree by minute, and put at −80 °C freezer at least 12 h. Aliquots of 10^6 cells per cryotubes can be stored at liquid nitrogen for longer term or until future use.

3.2 Immunostaining

1. Indirect immunofluorescence staining of the gastric cells was performed using antibodies against cytoskeleton filaments, extracellular matrix components, and a marker for progenitor cells, nestin (Fig. 3), and proliferation, the PCNA, after 5–7 days of culture.

2. To visualize the immune complexes, dilute the primary antibody in dilution buffer containing 0.2% gelatin in 0.05% saponin–PBS or 0.25% Tween 20–PBS.

3. To fix the plated gastric cells to the 24-well glass coverslips, remove the medium and add 300 μL of 4% paraformaldehyde (PFA) plus 4% sucrose in Sorensen's phosphate buffer (0.1 M, pH 7.4) at room temperature for 30 min. Gently wash the cells at least three times with 500-μL PBS to remove PFA.

4. After washing, block nonspecific binding of the antibodies using 50 nM NH4Cl in PBS pH 8.0 for 30 min. To permeabilize the cells, remove the block solution, and add 300 μL of 0.05% saponin in PBS for 30 min. Remove the solution and

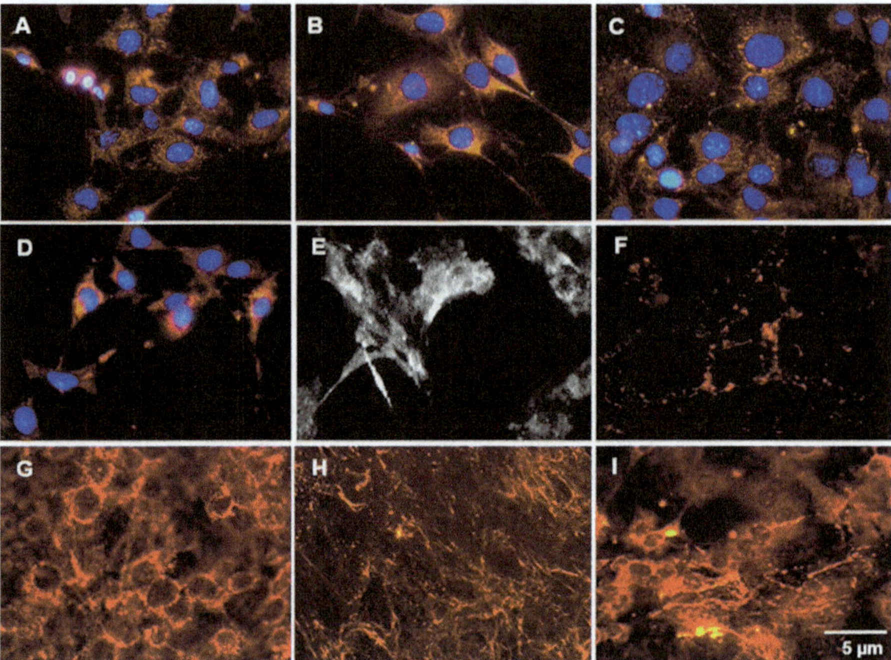

Fig. 3 Immunocytochemical staining of primary human antral gastric epithelial cells (5 days of culture). Cells are strongly positive to CK-7 (**a**); CK-8 (**b**); and pan-CK (**c**); faintly or strongly stained for nestin (**d**); some gastrin+ cells (**e**); cells form an extensive extracellular col IV network (**f**); strongly reactive for fibronectin (**g**); small deposition of laminin (**h**); cells are strongly reactive for CS (**i**)

incubate the cells in permeabilize solution containing 0.2% gelatin and 0.05% saponin–PBS for 30 min.

5. Incubate the cells with the primary antibodies (Figs. 3 and 4) (cytokeratin 7 (CK-7), cytokeratin 8 (CK-8), pan cytokeratin (pan CK), type IV collagen (CIV) fibronectin (FN), chondroitin sulfate (CS), α-smooth muscle actin (α-SMA), vimentin (VIM), nestin, Oct-4, and the PCNA, all the antibodies were made in mouse and rabbit, and all diluted 1:100 at 4 °C overnight.

6. After antibody incubation, wash the cells three times with 300 μL of gelatin/saponin–PBS solution, and detect the primary antibodies using biotinylated anti-mouse and anti-rabbit secondary antibodies (Sigma) for 30 min at room temperature, wash, and incubate with streptavidin Cy3 (1:1000) for 90 min. After washing, stain the nuclei of cells with DAPI diluted in PBS and mount the coverslips with Vectashield.

7. Observe and record the cells using a Nikon Eclipse E 800 epifluorescence microscope connected to a digital camera. Capture Images of high quality in 204,861,536-pixel buffer.

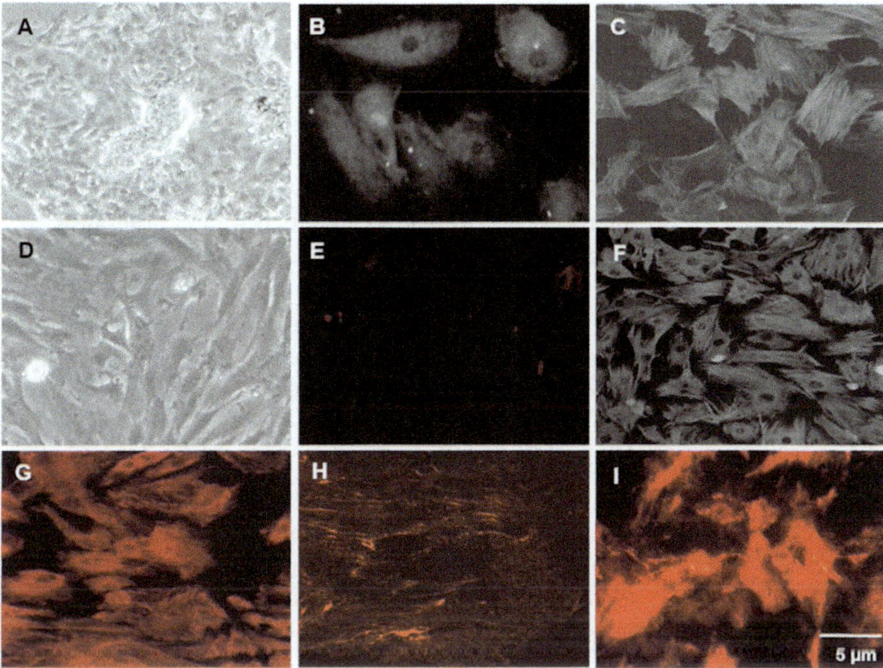

Fig. 4 Phase contrast microscopy (**a**) image of first passage cells of the primary stromal cells population isolated from human antral gastric mucosa. Immunofluorescence. (**b**) cells are CK7-, and (**c**) highly stained for smooth muscle α-actin; (**d**). Phase-contrast image of primary gastric cells after 60 days of culture; Immunofluorescence. Cells are negative for pan cytokeratin (**e**); strongly stained with vimentin (**f**); laminin (**g**); fibronectin (**h**); and COL IV (**i**)

3.3 Characterization of HGE3 Cell Glycoconjugates with Lectins

The staining of the gastric epithelial cells was performed using a series of biotinylated lectins: *Dolichos biflorus* agglutinin (DBA), *Ulex europaeus* Agglutinin I (UEA I), and *Griffonia* (*Bandeiraea*) *simplicifolia* Lectin II (GSL II) with strong affinity to carbohydrate residues in the parietal, and mucous cells of the gastric epithelia [15, 16].

Both the biotinylated lectins and the Streptavidin–Alexa conjugates are diluted in PBS.

1. Remove the medium of the plated gastric epithelial cells to the 24-well glass coverslips, wash with PBS three times for 10 min, fix with 300 µL of cold acetone for 5 min at 4 °C (or 4% PFA), and wash with phosphate-saline buffer (PBS). Then incubate with PBS containing 5% bovine serum albumin (BSA, pH 7.4), for 1 h at room temperature followed by a step for inhibiting endogenous biotin with the streptavidin/biotin blocking kit (incubate the streptavidin solution for 15 min, wash with PBS, incubate with the biotin solution for 15 min).

2. Subsequently, incubate the epithelial cells with biotinylated lectins (1:100) at 25 °C for 1 h. After then, wash the coverslips

in 0.25% Tween 20 in PBS, and incubated for 30 min at 25 °C with Streptavidin–Alexa 546 or 488 conjugate (1:1000).

3. After lectins incubation, wash the cells once in 300 μL of PBS and thrice with deionized water. Stain the nucleus of epithelial cells with DAPI, and mount the coverslips. Observe and record the cells using a Nikon Eclipse E 800 epifluorescence microscope connected to a digital camera as previously described above.

3.4   3D Coculture Assays of the Epithelial and Stromal Cells

1. Before 3D coculture formation, prepare 20 mL 1% agarose gel solution in PBS and pre-recover the 96-well round bottom plates, using an automatic pipette (1000 μL) to coat the inside of the each well that plate with thawed agarose (about at 40–50 °C) (*see* **Note 5**). Agarose gel solution can be sterilized in a microwave oven (cycle 3 of 10 s, potency 100).

2. To obtain the three-dimensional structures (spheres) cocultivate the stromal (HGS12) and epithelial (HGE3) cells in the same proportion of 1:1 (1.5×10^4 cells), in 96-well tissue culture plates (Fig. 5), previously covered with 1% agarose gel (Fig. 5a), in KM-F12 culture medium with 10% FBS, antibiotics (PS) and antimycotics (1 μg/mL) and maintain about 72 h or until spheres completely formation (Fig. 5b), in humidified air at 37 °C and atmosphere of 5% of CO_2.

3. Under a sterile condition, transfer carefully the spheres into the rotary cell culture system (RCCS-4) (Fig. 5c), using automatic pipette (1000 μL) with a pipette tip with cut off tip to avoid damaging spheres (*see* **Note 6**). Introduce 60 spheres per 12 mL of KM-F12 (about 5 spheres per mL) and maintain in humidified air at 37 °C and atmosphere of 5% of CO_2.

4. To maintain individual sphere aggregates in suspension, fix rotation speed in 35 rpm. Monitor spheres daily by observations of pH and of dissolved CO_2 and O_2 and bubble formation.

5. At each 48 h, change approximately 90% of the total culture medium volume by fresh medium. Maintain spheres cultured during 21 days or up to 75 days as shown in the Fig. 5d, under these conditions in the RCCS-4 prior their characterizations.

3.5   Scanning and Transmission Electron Microscopy

1. Perform the ultrastructural morphology of the gastric cells in 3D cell culture using scanning and transmission electron microscopy (Fig. 5e–h).

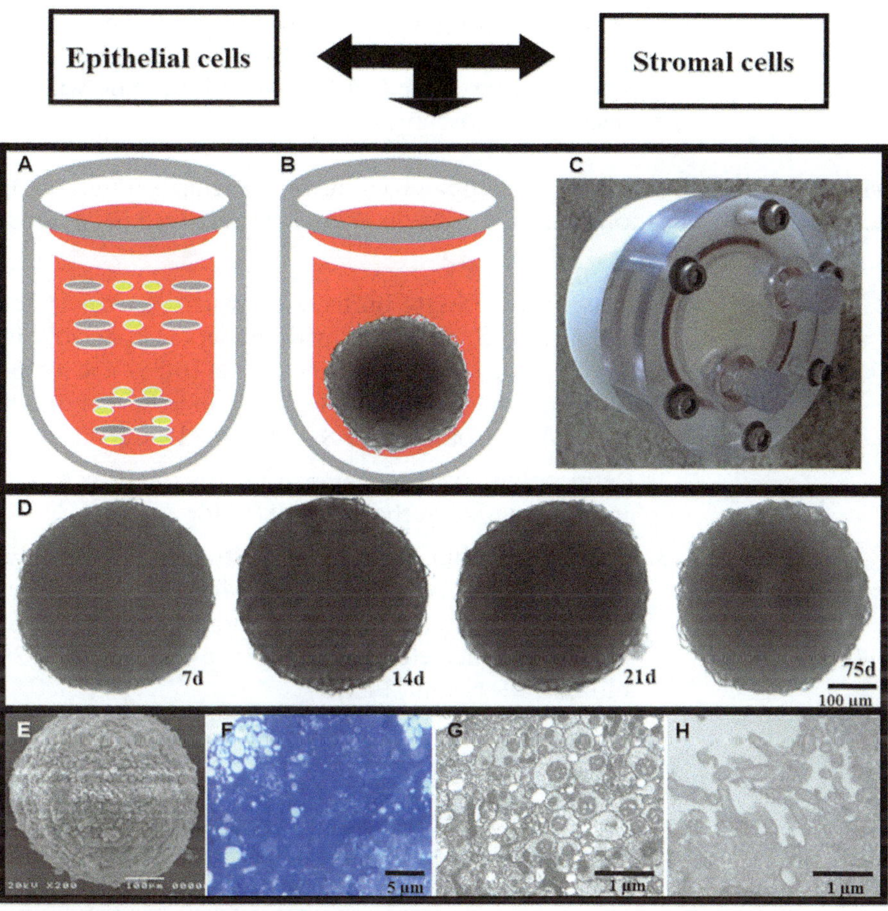

Fig. 5 Schematic illustration of the gastrospheres formation using primary human gastric epithelial and stromal cells in suspension in a 96-well plate pretreated with agarose (**a**); after 24 h, cells organize themselves in an single sphere per well (**b**); Image of the bioreactor device to culturing spheres in short-term and long-term culture (**c**); Size of the gastrospheres during timeline from 07 to 75 days (**d**); Scanning electron microscope image of the gastrosphere (**e**); Semithin section of a gastrosphere stained with Toluidine blue, showing some superficial cells with huge empty vacuoles reminding a mucous cell (**f**); Ultrathin section of a gastrosphere cell which looks like a zymogenic cell (**g**); Transmission electron microscope image of the surface of the gastrosphere showing the apical domain of a cell with thin, short cytoplasmic projections (h)

3.5.1 Scanning Electron Microscopy

1. To fix the plated gastric cells to the 24-well glass coverslips and spheres (96-well), remove the medium and add 200 μL of 2.5% glutaraldehyde solution in 0.1 M/L Sorensen's phosphate buffer (pH 7.4) for 1–2 h and rinse thrice in 0.1 M cacodylate buffer (pH 7.4).

2. Postfix the cells in a 1% osmium tetroxide solution containing 0.8% potassium ferrocyanide solution, and 5 mM calcium chloride solution in 0.1 M cacodylate buffer for 30 min.

3. Dehydrated in graded alcohol solutions (15, 30, 50, 70, 80, 90, and 100%) and critical-point drying (Baltec CPD 030). Then, sputter-coated with gold and examine with a scanning electron microscopy.

3.5.2 Transmission Electron Microscopy

To process the samples to transmission electron microscopy, repeat fixation and post fixation as in Subheading 3.5.1, **steps 1** and **2** above.

1. Rinse in cacodylate buffer, stain with 1% uranyl acetate overnight at 4 °C, dehydrate with graded acetone solutions (once 15, 30, 50, 70, 80, 90, and twice 100%) and embed in 1:1 epon–acetone solution overnight in a shaker, then in pure epon (A+B) solution for 8 h (Epon 812 substitute).

2. Resin polymerization—prepare epon solution A+B+DMP-30 and polymerize in the oven 60 °C for 3 days.

3. Prepare semithin (Fig. 5f) and ultrathin sections (Fig. 5g, h) with an Ultra microtome RMC-MT 6000 – XL, double-stain 5 min 5% uranyl acetate and 1 min in lead citrate (Reynolds), and view under electron microscopy at 80 kV (1200 JEOL and 268 Morgagni, FEI).

3.5.3 Immunohisto-chemical Characterization of Gastrospheres (See Note 8)

1. To fix the gastrospheres, transfer carefully the gastrospheres into the 15-mL sterile conical tube with 1 mL of 4% paraformaldehyde (PFA)–4% sucrose in Sorensen's phosphate buffer (0.1 M, pH 7.4) at room temperature for 30 min, using automatic pipette (1000 μL) with a pipette tip with cut off tip to avoid damaging spheres.

2. Remove the PFA solution, wash the spheres thrice with 1 mL of PBS, and freeze gastrospheres, embedding them in OCT medium. Put into a −80 °C freezer until use.

3. Cut in a cryostat (6 μm) and place frozen sections onto a polylysine-coated glass slides. Put slides inside the −80 °C freezer until use.

4. Perform indirect immunofluorescence staining using specific antibodies to Cytokeratin 7, Laminin, Type IV collagen, Fibronectin, Nestin and Proliferating cell nuclear antigen showed positive cells in this structure (Fig. 6a–f).

5. Remove slides from the freezer. Let for about 20 min at room temperature. Refix in 4% PFA–4% sucrose solution for about 10 min. Wash gastrospheres sections twice with PBS to remove PFA.

6. After washing, incubate in a 50 mM ammonium chloride solution in PBS to block free aldehyde residues for 30 min. To permeabilize gastrospheres remove the blocking solution, and incubate 300 μL of 0.05% saponin in PBS for 30 min. Remove

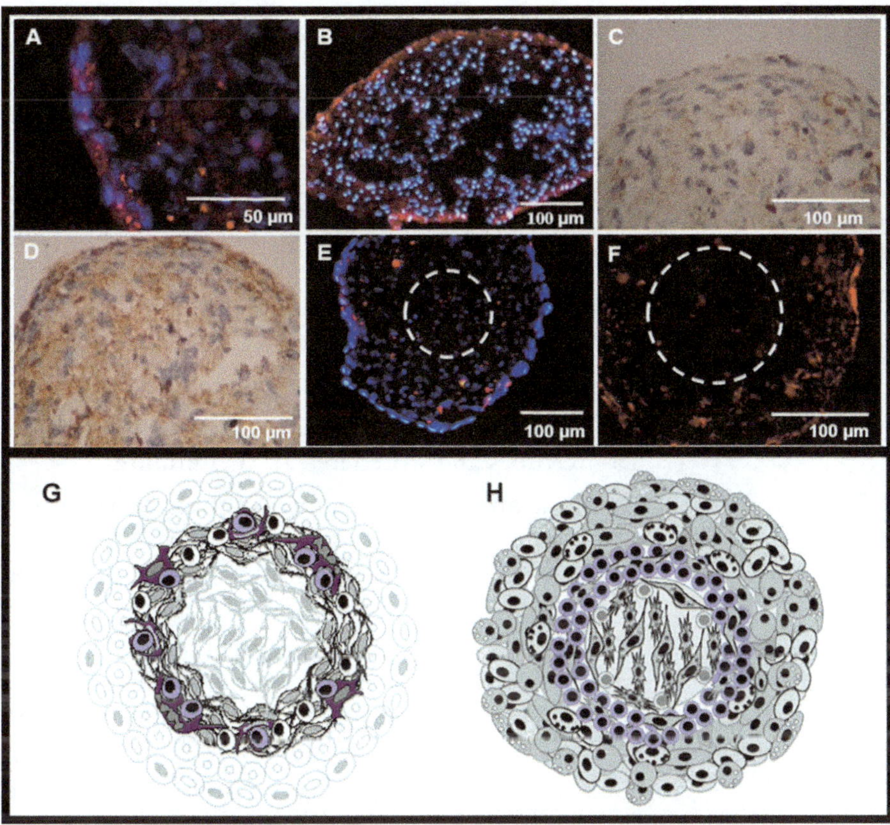

Fig. 6 (a) Three-dimensional images of isolated human primary gastric stromal and epithelial cells 21 days after being placed in bioreactor culture. Immunohistochemical staining of 6 μm frozen sections through gastrosphere using specific antibodies to Cytokeratin 7 (**a**), Laminin (**b**), Type IV collagen (**c**), Fibronectin (**d**), Nestin (**e**), and Proliferating cell nuclear antigen (**f**) showed positive cells in this structure. Schematic illustration of a proposed model for stem cell localization in gastrosphere in the short-term (**g**) and in the long-term culture (**h**)

the solution and incubate the gastrospheres sections in a blocking solution containing 5% BSA, 0.2% gelatin, and 0.05% saponin–PBS containing 20% normal goat serum for 1 h (*see* **Note 8**). Then, inhibit endogenous biotin using the biotin block in kit. Wash gastrospheres sections with PBS for 5 min. Incubate the streptavidin solution for 15 min. Wash sections in PBS for 5 min, then incubate in the biotin solution for 15 min. Then wash in PBS once.

7. Incubate sections with the diluted primary antibodies in a humid chamber, at 4 °C overnight.

8. Take out the gastrospheres sections from the refrigerator, wash twice in PBS solution, and incubate the biotinylated anti-mouse or anti-rabbit secondary antibodies for 60 min at room temperature.

9. Wash twice with PBS solution. Incubate with streptavidin Cy3 (1:1000) for 60 min at room temperature, protecting slides from the light. After washing twice with PBS stain the nuclei of cells with DAPI diluted in PBS, for 5 min. Wash with distilled water, twice 5 min, and mount with Vectashield.

10. Observe the sections using an epifluorescence microscope connected to a digital camera. Capture Images of high quality in 204,861,536-pixel buffer.

4 Notes

1. The dispase and collagenase preparations can be aliquots in 15 mL tubes and stored in -20 °C until use.

2. Dissociation of gastric fragments should be as soon as possible after biopsies uptake the from patient's stomach. This will ensure that the cells are healthy and will improve cell viability.

3. The biopsies can be cutting in small fragments into the small backer (5 mL) using microsurgical scissors, and in the BSC (Iris Scissors 110 mm Spring Action Blunt/Blunt, Straight or Curved, Micro/Neuro Surgery Instruments, Simrix).

4. After the fragment incubation in dispase solution as in Subheading 3.1.1, **steps 4** and **5**, reminiscent fragments can be used to obtain stromal cells. In the same way, stromal cells adhered in 25 cm^2 tissue culture dish for 2–6 h as in Subheading 3.1.1, **step 6** can be used after characterized as stromal cell.

5. Agarose gel solution can be sterilized by elevating the temperature until ebullition in microwave (cycle 3 of 10 s, potency 100) or in an autoclave (at 127 °C for 20 min)

6. Handle the spheres using a 1000 µL tip with point cut off to avoid damaging and shape alteration after its transference into the rotary cell culture system (RCCS-4). Note that the number of cells per gastrospheres is critical to determine spheres size.

7. The size of gastrospheres did not vary during the temporal evolution. For instance, spheres formed from 3×10^4 cells seeded measure about 300 to 350 µm diameter. Images of gastrospheres were obtained in a phase contrast microscope and were measured at 3, 7, 14, 21, and 75 days, using the ImageJ software (Fig. 5).

8. The gelatin/saponin–PBS solution can be used onto coverslips as negative controls of the immunocytochemical reaction instead of the primary antibodies.

Acknowledgments

The authors would like to express their gratitude to Prof. Leonardo P. Andrade and Mrs. Sonia Oliveira Souza for their technical help with transmission electron microscopy, and Prof. Heitor Siffert Souza for the gastric biopsy obtention. This study was supported by the Fundação de Amparo à Pesquisa Carlos Chagas Filho do Estado do Rio de Janeiro (FAPERJ), and Conselho Nacional de Desenvolvimento Científico e Tecnológico (CNPq).

References

1. Karam SM (1995) New insights into the stem cells and the precursors of the gastric epithelium. Nutrition 11:607–613

2. Karam SM, Straiton T, Hassan WM, Leblond CP (2003) Defining epithelial cell progenitors in the human oxyntic mucosa. Stem Cells 21:322–336

3. Barker N, Huch M, Kujala P, van de Wetering M, Snippert HJ, van Es JH et al (2010) Lgr5(+ve) stem cells drive self-renewal in the stomach and build long-lived gastric units in vitro. Cell Stem Cell 6:25–36

4. Bjerknes M, Cheng H (2002) Multipotential stem cells in adult mouse gastric epithelium. Am J Physiol Gastrointest Liver Physiol 283:G767–G777

5. Karam SM, Leblond CP (1993) Dynamics of epithelial cells in the corpus of the mouse stomach. I. Identification of proliferative cell types and pinpointing of the stem cell. Anat Rec 236:259–279

6. McDonald SA, Greaves LC, Gutierrez-Gonzalez L, Rodriguez-Justo M, Deheragoda M, Leedham SJ et al (2008) Mechanisms of field cancerization in the human stomach: the expansion and spread of mutated gastric stem cells. Gastroenterology 134:500–510

7. Ootani A, Li X, Sangiorgi E, Ho QT, Ueno H, Toda S et al (2009) Sustained in vitro intestinal epithelial culture within a Wnt-dependent stem cell niche. Nat Med 15:701–706

8. Santos CA, Andrade LR, Costa MH, Souza HS, Granjeiro JM, Takiya CM et al (2016) Gastrospheres of human gastric mucosa cells: an in vitro model of stromal and epithelial stem cell niche reconstruction. Histol Histopathol 31:879–895

9. Sato T, Vries RG, Snippert HJ, van de Wetering M, Barker N, Stange DE et al (2009) Single Lgr5 stem cells build crypt-villus structures in vitro without a mesenchymal niche. Nature 459:262–265

10. Kale S, Biermann S, Edwards C, Tarnowski C, Morris M, Long MW (2000) Three-dimensional cellular development is essential for ex vivo formation of human bone. Nat Biotechnol 18:954–958

11. Tong JZ, Sarrazin S, Cassio D, Gauthier F, Alvarez F (1994) Application of spheroid culture to human hepatocytes and maintenance of their differentiation. Biol Cell 81:77–81

12. Slack JM (2000) Stem cells in epithelial tissues. Science 287:1431–1433

13. Bartfeld S, Bayram T, van de Wetering M, Huch M, Begthel H, Kujala P et al (2015) In vitro expansion of human gastric epithelial stem cells and their responses to bacterial infection. Gastroenterology 148:126–36.e6

14. Schlaermann P, Toelle B, Berger H, Schmidt SC, Glanemann M, Ordemann J et al (2016) A novel human gastric primary cell culture system for modelling Helicobacter pylori infection in vitro. Gut 65:202–213

15. Fischer J, Klein PJ, Vierbuchen M, Skutta B, Uhlenbruck G, Fischer R (1984) Characterization of glycoconjugates of human gastrointestinal mucosa by lectins. I. Histochemical distribution of lectin binding sites in normal alimentary tract as well as in benign and malignant gastric neoplasms. J Histochem Cytochem 32:681–689

16. Kessimian N, Langner BJ, McMillan PN, Jauregui HO (1986) Lectin binding to parietal cells of human gastric mucosa. J Histochem Cytochem 34:237–243

Chapter 9

Markers and Methods to Study Adult Midgut Stem Cells

Nathan Pinto, Beyoncé Carrington, Catharine Dietrich, Rachit Sinha, Cristopher Aguilar, Tiffany Chen, Poonam Aggarwal, Madhuri Kango-Singh, and Shree Ram Singh

Abstract

Stem cells have emerged as a promising cell source to heal, replace or regenerate tissue and organs damaged by aging, injury or diseases. The intestinal epithelium is the most rapidly renewing tissue in our body, which is maintained by intestinal stem cells (ISCs), located at the bottom of the crypts. ISCs continuously replace lost or injured intestinal epithelial cells in organisms ranging from *Drosophila* to humans. The adult *Drosophila* midgut provides an excellent in vivo model system to study ISC behavior during stress, regeneration, aging and infection. There are several signaling pathways/genes have been identified to regulate ISCs self-renewal and differentiation during normal and pathological conditions. A significant number of genetic tools and markers have been developed in the last one decade to study *Drosophila* ISCs behavior. Here, we describe some of the markers and methods used to study ISCs behavior in adult midgut of *Drosophila*.

Key words *Drosophila*, Midgut, Intestinal stem cell, Immunostaining

1   Introduction

Adult stem cells have been identified in many tissues and organs. They are immature cells that maintain tissue homeostasis and have ability to replace cells that loss because of injury or diseases. Both intrinsic and extrinsic factors affect the self-renewal and differentiation ability of stem cells, and therefore, a fine balance between self-renewal and differentiation of stem cell is crucial. Disruption of this balance leads to organ degeneration, premature aging, and cancer [1–7]. Understanding the mechanisms of this balance is crucial in adult stem cell biology and regenerative medicine.

The adult mammalian intestinal epithelium is a highly self-renewing tissue, which is maintained by intestinal ISCs, located at the base of the crypts [3, 8]. ISCs proliferate and differentiate toward the lumen to produce absorptive and secretory cells types to maintain normal homeostasis of the gut. ISCs are labeled by

Shree Ram Singh and Pranela Rameshwar (eds.), *Somatic Stem Cells: Methods and Protocols*, Methods in Molecular Biology, vol. 1842, https://doi.org/10.1007/978-1-4939-8697-2_9, © Springer Science+Business Media, LLC, part of Springer Nature 2018

several markers including leucine-rich repeat-containing G-protein coupled receptor 5 (Lgr5)+ and B lymphoma Mo-MLV insertion region 1 homolog (Bmi1)+ cells [3, 8]. There are various genes/ signaling pathways have been identified to regulate ISCs behavior in the mammalian gut [3, 8–10].

Drosophila is a highly sophisticated genetic model organism to study ISCs behavior. Like mammalian intestine, the *Drosophila* adult intestinal epithelium (Fig. 1a) are highly proliferative because it contains quiescent and active stem cells [11–13]. *Drosophila* ISCs are distributed evenly throughout the gut but with variable proliferation capacity in different zones. ISCs self-renew and form a transient committed progenitor enteroblasts (EBs). The EBs differentiate into absorptive enterocyte (ECs) and the secretory enteroendocrine (ee) cells (Fig. 1b). Some recent studies suggest that ISCs can directly differentiate into ee or EC cells [14, 15] and a subset of progenitor cells among ISCs can produce ee cells upon mechanical stress [16]. In contrast, Chen et al. [17] reported that an ISC divides first to generate a new EE progenitor cell (EEP), which undergoes on round of division before generating EE cells (Fig. 1b). They found that 71% of EEPs undergo one round of mitosis to generate a pair of EEs; the remaining EEPs directly differentiate into a single EE [17]. ISCs are marked by expression of high levels of Delta (Dl), a ligand for the Notch receptor that activates Notch signaling in neighboring EBs [13]. Escargot (esg), a Snail family transcription factor is expressed in ISCs and EBs. EBs are also marked by suppressor of Hairless $S(u(H))GBE\text{-}lacZ$, a transcriptional reporter of Notch signaling [11–13]. The ee cells express the homeodomain transcription factor Prospero (Pros) and the ECs can be judged by their polyploid nuclei and the expression of nubbin/POU domain protein 1 (Pdm1).

Over the last 10 years, several signaling pathways/genes have been identified to regulate the ISCs behavior, tumorigenesis and survival in *Drosophila* midgut [11–51]. Accumulative studies also suggest that ISCs in the *Drosophila* midgut are activated in response to tissue stress, damage, aging, cell death, feeding chemicals or microbial pathogens, metabolism and by mechanical stress [16, 18–42]. Some of the proteins also regulate symmetric and asymmetric division of ISCs [52]. Recently, it was demonstrated that loss of ECs slows when ISC division is suppressed and accelerates when ISC division increases [53]. Aberrant division of ISCs and or EB differentiation results in tumor formation [53]. Recently, it has been demonstrated that after severe depletion because of starvation, tissue damage and aging, ISCs in the midgut can be replaced by spindle-independent ploidy reduction of cells in the enterocyte lineage through amitosis [54]. Males and females also differ in ISCs division and differentiation [55]. Further, a feedback signaling mechanism between differentiated cells (EC cells) and ISCs has been reported to play a crucial role in maintaining organ size in the intestine [56].

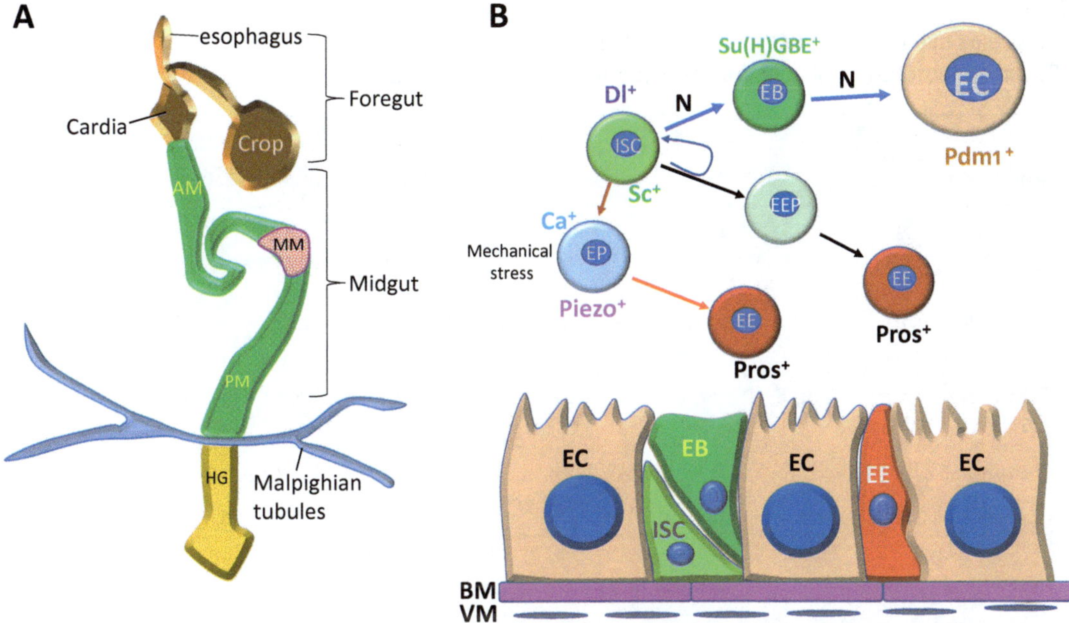

Fig. 1 The adult *Drosophila* midgut is maintained by intestinal stem cells (ISCs). (**a**) A schematic 3D model of adult *Drosophila* gastrointestinal system. (**b**) Model diagram of the adult midgut in cross section. Bottom panel, ISCs (light green) occupy a basal position in a niche near basement membrane (BM, pink) and the visceral muscle (VM, gray). ISCs self-renew and give rise to enteroblast and differentiated into enteroendocrine (ee) cells (red) and enterocytes (Ecs; orange). Upper panel, a diagram of ISCs lineage. Abbreviations: *AM* anterior midgut, *PM* posterior midgut, *HG* hindgut, *Dl* delta, *N* Notch, *EP* enteroendrocrine precursor, *EEP* EE progenitor cell, *Ca* calcium, *Sc* Scute

Drosophila serves as an excellent genetic model system to study adult stem cells. In this chapter, we have presented the markers and methods used to study ISCs in adult *Drosophila*. The antibodies and transgenes expressed in the different cell types in the adult midgut are presented in Tables 1 and 2, respectively.

2   Materials

For preparing all solutions, use ultrapure water and analytical grade reagents. Prepare and store all reagents at room temperature (unless otherwise indicated).

2.1   Drosophila Culture

1. Control and transgenic flies.

2. Fly culture incubator (18 °C, 25 °C and 29 °C).

3. *Drosophila* plastic vials and bottles.

4. Foam or cotton plugs.

5. *Drosophila* food: Cornmeal, agar, sucrose, yeast, and acid medium. Detail recipes are described in our previous protocols [57].

Table 1
Antibodies used as a marker in *Drosophila* ISCs and differentiated cells (Based on literatures published in ISC system from 2006–early 2018)

Antibody	Cell type expression in midgut	Species	Dilution	Source
Delta (Dl)	ISC	Mouse	1:100	DSHB
Notch (N)	ISC, EB	Mouse	1:10	DSHB
Armadillo (Arm)	Outer layer of ISC, EB and EC	Mouse	1:20	DSHB
Transient receptor potential ankyrin 1 (TRPA1)	ISCs	Rat	1:100	Paul Garrity
Phosphor-4EBP1 (Thr37/46)	Low in ISC, high in EB	Rabbit	1:200	Cell Signaling
Fragile X mental retardation 1 (FMR1)	ISC and EB	Mouse	–	DSHB
LIN28	Higher in Ebs than ISCs	Rat	–	Nicholas Sokol
Fork head (Fkh)	High in ISC and EB and weak in EE	Guinea pig	1:100	Ralf Pflanz
SRY-box 21 (Sox21)	ISC and EB	Rabbit	1:1000	Ringwen Xi
Piwi	ISCs and Ebs	Mouse	1:50	Brennecke lab
Anilin	Mitotic ISC	Rabbit	1:1000	Chris Field
Daughterless	ISC, EB, ee, EC	Rabbit	1:1000	Y.-N. Jan
Cyclin-dependent kinase 1(Cdc2)	ISC, EB	Rabbit	1:500	Santa Cruz
Acetylated histone (AcH3)	ISC, EB	Rabbit	1:500	Millipore
Phospho-histone 3 (pH3)	Dividing ISC	Rabbit	1:1000	Upstate
Hairless	ISC and EB, some differentiated progeny	Guinea pig	1:500	A. Preiss
Epidermal Growth Factor Receptor (EGFR)	ISC and/or EB	Goat	1:200	Santa Cruz
Yorkie (Yki)	ISC/EB	Rat Rabbit	1:1000 1:400	Helen McNeill Kenneth Irvine
Fat	ISC, EB	Rat	1:2000	Michael Simon
Asense	EE cells	Rabbit	1:5000	Y.-N. Jan
scute RNA	ISC, EB, and/or ee	–	–	
Diphospho-ERK (dpERK)	ISC, EB	Rabbit	1:1000	Cell Signaling
Signal-transducer and activator of transcription protein at 92E (STAT92E)	ISC, EB	Rabbit	1:500	S. Hou
Roundabout 2 (Robo2)	ISCs and EBs	Mouse	1:100	Barry Dickson

(continued)

Table 1
(continued)

Antibody	Cell type expression in midgut	Species	Dilution	Source
Brahma (Brm)	ISCs/EBs and some ee	Rabbit		Zhang lab
Windpipe (Wdp)	Ubiquitously expressed in intestines	Mouse	1:1000	Xinhua Lin
Connectin (Con)	Localized to the membrane junctions between ISC and EB	Mouse	1:4	DSHB
Phospho-moesin (pMoe)	Interface membranes between the ISC and EB cells and their corners	Rabbit	1:500	Cell Signaling
E-cadherin (E-cad)	Interface membranes between the ISC and EB cells	Rat	1:20	DSHB
Atypical protein kinase C (aPKC)	Apical areas of the ISC/EB	Rabbit	1:100	Santa Cruz
Wingless (wg)	Circular muscles, basement membrane, and ISCs	Mouse	1:200	DSHB
Slit	EE cells and weakly expressed ISCs and EBs	Mouse	1:20	DSHB
Caudal (Cad)	All epithelial cells of gut	Guinea Pig	1:400	East Asian Distribution Centre, JAPAN
Prospero (Pros)	ee cells	Mouse	1:100	DSHB
Allatostatin (Ast)	ee cells	Mouse	1:10	DSHB
Tachykinin (Tk)	ee cells	Rabbit	1:2000	D. Nässel
Bruchpilot (Brp)	ee cells	Mouse	1:20	DSHB
Peptidylglycine α-hydroxylating monoxygenase (Phm)	ee cells	Rabbit	1:250	Paul Taghert
Bursicon (Burs)	ee cells	Rabbit	1:250	Aaron J. W. Hsueh
POU domain protein 1 (Pdm1)	Mature EC	Rabbit	1:100	Xiaohang Yang
Dachsous (Ds)	Borders of EC cells	Rat	1:5000	Michael Simon
Phospho-MAD3 (pMAD3)	EC cells	Rabbit	1:300	Epitomics
Spectrin	Apical side of enterocytes	Mouse	1:100	DSHB
Alpha-tubulin (α-Tub)	Microtubule and spindle	Rat	1:50	Immunologicals direct
Gamma-tubulin (γ-Tub)	Centrosomes	Mouse	1:500	Sigma
CyclinT (CycT)	Ubiquitously expressed in posterior midgut	Rabbit	1:1000	Xinhua Lin

Table 2
Transgenes expressed in *Drosophila* ISCs and differentiated cells (Based on literatures published in ISC system from 2006–early 2018)

Transgenes	Expression in midgut
esg-Gal4-UAS-GFP	ISC, EB
PiezoP-GAL4	EE precursor cells
Sc-GFP	ISC
Kr-Gal4-UAS-GFP	ISC, EB
Su(H)Gbe-lacZ	EB
Pvf2-lacZ	ISC/EB
Dl-LacZ	ISC
delta-Gal4	ISC
phyl$^{3.4}$-GFP	ISC
Su(H)Gbe-Gal4	EB
polo-GFP	ISCs metaphase plate
vn-lacZ	Midgut visceral muscle cells
spi (spi-Gal4^{NP0261})	ISC, EB and low in EC
rho^{X81}-lacZ	Midgut visceral muscle cells
upd-lacZ	ISC and EB
upd-3-lacZ	EC
10XSTAT-DGFP	ISC, EB
myo1A-Gal4-UAS-GFP	ECs cells
24B-Gal4 (howGal4)	Visceral muscle-specific
ds-LacZ	EC and some EE cells
E(spl)mβ-CD2	EB cells
mira-promoter-GFP	ISC
vkg-GFP	Basement membrane
cad-Gal4-UAS-GFP	Posterior midgut cells
wg-lacz	Circular muscle cells of midgut
collagenIV-GFP	Basement membrane structure
gstD1-lacZ	ISC and EE
D-p38b-lacZ	ISC and EB
Slit-lacz	EE cells and weakly expressed ISCs
28E03-GAL4	Subset of EBs
Sox21a-GFP	ISC and EB
Dad-lacZ	EC cells

6. Active yeast granules.

7. Autoclave for preparing fly food.

8. Morgue for discarding dead flies.

2.2 Lineage Tracing, Gene Manipulation and RNAi Knockdown

1. Stocks for generating tubulin lacZ, positively marked mosaic lineage (PMML) and mosaic analysis with a repressible cell marker (MARCM) are easily available from the Bloomington Stock Center (https://flystocks.bio.indiana.edu).

2. RNAi lines can be obtained from Vienna Drosophila Resource Center (VDRC) (https://stockcenter.vdrc.at), Bloomington (https://flystocks.bio.indiana.edu) DRSC/TriP Functional Genomics Resources https://fgr.hms.harvard.edu/fly-in-vivo-rnai), National Institute of Genetics (NIG) (https://shigen.nig.ac.jp/fly/nigfly/) and TsingHua Fly Center (http://fly.redbux.cn/rnai.php?lang=en).

3. Circulating water bath for heat shock.

4. Incubator to maintain fly crosses (18 °C, 25 °C and 29 °C)

5. Some of important stocks and transgenic lines needed in these experiments are presented in Table 2.

2.3 Oil Red O Staining

1. Oil Red O.

2. Isopropanol solution.

3. Double-distilled water.

2.4 Immunofluorescence Staining of Gut

2.4.1 Gut Dissection

1. *Drosophila* adult flies (3–5 days old).

2. Carbon dioxide (CO_2) source for anesthetizing the flies (*see* **Note 1**).

3. Dissecting forceps or needles.

4. Glass microslides or petri dish.

5. Dissecting microscope.

6. *Drosophila* Ringer's solution for dissection. Store the dissecting solution at 4 °C.

2.4.2 Gut Fixation

1. 1× phosphate-buffered saline (PBS) with pH 7.4. Store at room temperature.

2. Triton X-100 (*see* **Note 2**).

3. Nitrile gloves.

4. Washing solution: Dissolve the 0.1% Triton X-100 in 1× PBS plus 0.5% BSA. Store at room temperature.

5. Paraformaldehyde or formaldehyde (37%) solution (*see* **Note 3**).

6. Fixation solution: (4% formaldehyde). This solution should be prepared fresh every time.

7. Parafilm to seal the Eppendorf tubes.

2.4.3 Immunostaining

1. Normal goat serum (NGS). Store at 4 °C.
2. Bovine serum albumin (BSA). Use 0.5%.
3. Blocking solution (2% NGS). Store at 4 °C.
4. Minivortex.
5. Tube shaker.
6. Aluminum foil.
7. Microcentrifuge tube rack.
8. Primary antibodies: The primary antibodies are listed in Table 1. Store all primary antibodies at 4 °C. For longer stability, store at −20 °C or −80 °C (*see* **Note 4**).
9. Secondary antibodies: Secondary antibodies can be purchased from several vendors. Storage methods, dilution and species of secondary antibodies are described in our previous protocols [57].
10. 4,6-Diamidino-2-phenylindole dihydrochloride (DAPI) to stain nuclear DNA. Store in the dark at 4 °C.

2.4.4 Mounting, Imaging, and Data Analysis

1. Glass micro coverslips.
2. Microscope slides.
3. Quick-dry nail polish.
4. Dissecting microscope.
5. Mounting medium: VECTASHIELD Antifade Mounting Medium with DAPI.
6. Permanent marker to label the slides.
7. Slide holder
8. Confocal microscope.
9. Computer and appropriate software (image browser, ImageJ, and Adobe Photoshop) for image processing.

3 Methods

Perform all procedures at room temperature unless otherwise mentioned.

3.1 Lineage Tracing and Gene Manipulation

Important stocks in the lineage tracing and gene manipulation are described in Singh et al. [57].

3.1.1 Generation of β-galactosidase (lacZ)-Marked PMML and MARCM Clones

1. The lineage marking system to generate clones of *lacZ* expressing cells has been developed by Harrison and Perrimon [58]. Detail methods are described in our previous protocols [57].

2. PMML: This system utilized the heat shock-inducible flippage (FLP) to reconstitute a functional *actin5C-gal4* gene from two complementary inactive alleles, *actin5C FRT52B* and *FRT52B gal4*. The *actin5C-gal4* gene drives green fluorescent protein (*GFP*) expression to mark cells and at the same time, activate or knock down the gene function by having upstream activation sequence (UAS) constructs in the marked cells [59]. Detail protocols are described in our previous protocols [57].

3. MARCM technique: It is used to create individually labeled homozygous cells in an otherwise heterozygous background. MARCM relies on recombination during mitosis mediated by FLP-flippase recognition target (FRT) recombination [60]. Detail MARCM protocols for gut are described our previous protocols [57].

4. Follow fixation, staining and examination, and confocal microscopy and analyses as described in Subheading 3.3 [57].

3.1.2 RNAi-Mediated Gene Depletion

1. Cross male *UAS-RNAi* transgene flies with virgin females of different ISC, EB, EE, and EC specific Gal4s, *UAS-GFP; tub-Gal80^s*.

2. Culture the flies at 18 °C.

3. Collect appropriate genotype of adult flies.

4. Place 3- to 5-day-old flies (preferably females) with the appropriate genotype in a new vial at 29 °C for 5 days, 7 days, or 10 days before dissection. Please place 10-15 flies in each vial.

5. Follow fixation, staining, confocal microscopy and analyses as described in Subheading 3.3 [57].

3.2 Oil Red O Staining

Oil Red O staining is used to detect neutral fat (lipid droplets) in the guts. Oil Red O staining was performed as described [33, 61].

1. Dissect the fresh gut tissues in 1× PBS or Ringer's solution.

2. Fix the gut tissues in 4% formaldehyde for 30 min.

3. Rinse the gut tissues three times in 1× PBS, in double-distilled water, and in 60% isopropanol solution.

4. Prepare working solution from stock solution by mixing 6 mL of 0.1% Oil Red O in isopropanol and 4 mL of double-distilled water.

5. Incubate gut tissues with this working solution for 20 min.

6. Wash the gut tissues in 60% isopropanol and water. Avoid washing with Triton X-100 based washing solution.

7. Mount in VECTASHIELD with DAPI.

8. Analyze the staining using a confocal microscope.

**3.3 Immunofluores-
cence Staining
of Midgut**

The detail staining protocols have been described [57].

3.3.1 Dissection

1. Anesthetize the flies with CO_2.

2. Use glass microslide or petri dish and place a few drops of Ringer's solution or 1× PBS.

3. Using fine forceps place flies on a slide or in a petridish.

4. Using fine forceps or fine needles, turn the body upside down, hold the top of the abdomen, and pull out the external genitalia with the other pair of forceps.

5. Pull the gut out from abdomen and remove the other debris.

6. Transfer the gut (which is cream color) into cold Ringer's solution or 1× PBS.

3.3.2 Fixation

1. Prepare fixing solution (4% formaldehyde in 1× PBS).

2. Fix the gut in 4% formaldehyde solution at room temperature for 20–40 min or overnight at 4 °C (*see* **Note 5**).

3. Remove the fixing solution and rinse and wash the gut tissues for 2 min each in 1× PBX.

*3.3.3 Blocking
and Immunostaining*

1. Prepare the blocking solution in 1× PBX and incubate the gut tissues for 30 min at room temperature or overnight at 4 °C (*see* **Note 6**).

2. Remove the blocking solution.

3. Dilute the primary antibody in washing solution and mix well on minivortex.

4. Incubate gut tissues overnight at 4 °C in 50–100 μL of diluted primary antibody. Make sure to wrap tube with parafilm. (*see* **Note 7**).

5. Remove the primary antibodies and rinse the gut tissues three times with washing buffer.

6. Wash the gut tissues on a shaker at room temperature for 15 min (three times) in washing buffer.

7. Prepare secondary antibody to desired concentration in washing buffer.

8. Add 200–400 μL of diluted secondary antibody to the gut tissues.

9. Incubate the gut tissues with secondary antibodies on a shaker at room temperature for 2 h or overnight at 4 °C (*see* **Note 8**).

10. Remove the secondary antibody and rinse the gut tissues three times with washing buffer.

11. Then wash the gut tissues on a shaker at room temperature for 15 min (three times) with washing buffer.

12. Finally, rinse the gut three times with washing buffer.

3.3.4 Mounting, Microscopy, and Data Analysis

1. Add few drops of VECTASHIELD Antifade Mounting Medium with DAPI to the gut tissues.

2. Store gut tissues overnight at 4 °C to allow tissues to equilibrate.

3. Using pipette, transfer the gut tissues to a glass microslides.

4. Arrange the gut tissues with forceps or needle in low light.

5. Carefully place a coverslip to the slide.

6. Remove excess mounting medium and seal the edges of the micro coverslip with nail polish (*see* **Note 9**).

7. Using a permanent marker label the slide for the specific genotype.

8. Store the prepared slides in the dark at 4 °C until observation (*see* **Note 10**).

9. Use confocal microscopy to capture images.

10. Use appropriate confocal image browser or ImageJ software to process the images.

11. Adobe Photoshop can be used to further processing/analysis of images.

12. Representative examples using the above techniques and staining protocols are presented in Fig. 2. However, for best results, optimize the condition as per requirements of specific experiments. The above protocols could be useful to study other stem cells in adult *Drosophi la* and other tissues as well.

4 Notes

1. Carbon dioxide is nontoxic; however, it can induce hypoxia and cause headaches or dizziness due to long exposure.

2. Triton X-100 allows the cell membrane permeable to the antibody. However, higher concentration can disrupt the epithelial membrane of the gut and poor staining.

3. Formaldehyde is a fixative and carcinogenic. Use extra caution. Use personal protective equipment (PPE).

4. Store the primary antibodies at 4 °C with 0.02% sodium azide. To maintain primary antibodies high quality store at −20 °C with 50% glycerol and at −80 °C for long-term storage. Do not frequently freeze and thaw because it can degrade proteins.

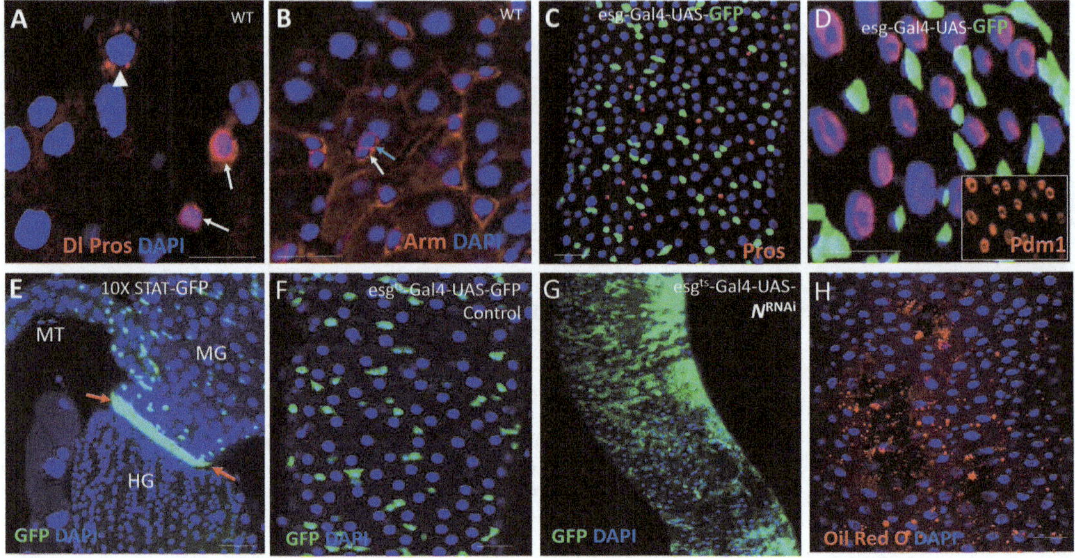

Fig. 2 ISCs markers, RNAi knockdown phenotype, transgenes expression, and lipid droplet staining in the adult midgut. (**a**) wild-type gut stained with anti-Dl (red—punctate expression in ISCs, anti-prospero (pros) (red-nuclear staining in ee cells), DAPI (blue). (**b**) wild-type gut stained with Armadillo (Arm, red) and DAPI, blue. (**c**) *esg-Gal4-UAS-GFP* flies gut stained with anti-Pros (red), GFP (green), and DAPI (blue) GFP mark the ISCs and Ebs (green) and anti-Pros (red, ee cells). (**d**) *esg-Gal4-UAS-GFP* flies gut stained with anti-Pdm1 (red), GFP (green), and DAPI (blue). GFP mark the ISCs and Ebs (green), Pdm1-mark the EC cells (red), DAPI (blue). Modified after Singh [57]. (**e**) *10X-STAT-GFP* gut stained for GFP (green-labeled the stem cells in MG-HG junction, red arrows; midgut (MG) and Malphigian tubules (MT), DAPI (blue). (**f**) *esg^{ts}-Gal4-UAS-GFP* control flies gut stained with GFP (green), DAPI (blue). (**g, f**) *esg^{ts}-Gal4-UAS-N^{RNAi}* flies gut stained with GFP (green), DAPI (blue). Knockdown of *N* resulted in ISC and ee tumors. (**h**) wild-type flies gut stained with Oil Red O (red), and DAPI (blue). Scale bars: **a, b, d** (5 μm) **c, e–h** (10 μm)

5. When fixing on room temperature limit the fixation time to 25–40 min because longer fixation results in poor staining. Alternatively, staining can be performed overnight at 4 °C for some antibodies such as delta.

6. The blocking can be performed overnight at 4 °C.

7. Primary antibodies can be incubated at room temperature for 2–3 h. But for better results, incubate overnight at 4 °C.

8. Store the secondary antibodies in the dark at 4 °C. Wrap the tube with aluminum foil to avoid the exposure of light.

9. To avoid coverslip to slip away during imaging, it is better to seal the edges with nail polish.

10. Store the slides in the dark at 4 °C and perform imaging within few days.

References

1. Singh SR (2012) Stem cell niche in tissue homeostasis, aging and cancer. Curr Med Chem 19(35):5965–5974

2. Soteriou D, Fuchs Y (2018) A matter of life and death: stem cell survival in tissue regeneration and tumour formation. Nat Rev Cancer 18(3):187–201

3. Simons BD, Clevers H (2011) Strategies for homeostatic stem cell self-renewal in adult tissues. Cell 145:851–862

4. Amcheslavsky A, Jiang J, Ito N, Ip YT (2011) Tuberous sclerosis complex and Myc coordinate intestinal stem cell growth and division in Drosophila. J Cell Biol 193:695–710

5. Xu N, Wang SQ, Tan D, Gao Y, Lin G, Xi R (2011) EGFR, Wingless and JAK/STAT signaling cooperatively maintain Drosophila intestinal stem cells. Dev Biol 354:31–43

6. Singh SR, Liu Y, Zhao J, Zeng X, Hou SX (2016) The novel tumour suppressor Madm regulates stem cell competition in the Drosophila testis. Nat Commun 7:10473

7. Hou SX, Singh SR (2017) Stem-cell-based tumorigenesis in adult Drosophila. Curr Top Dev Biol 121:311–337

8. Barker N, van Es JH, Kuipers J, Kujala P, van den Born M, Cozijnsen M, Haegebarth A, Korving J, Begthel H, Peters PJ, Clevers H (2007) Identification of stem cells in small intestine and colon by marker gene Lgr5. Nature 449:1003–1007

9. Zhu L, Gibson P, Currle DS, Tong Y, Richardson RJ, Bayazitov IT, Poppleton H, Zakharenko S, Ellison DW, Gilbertson RJ (2009) Prominin 1 marks intestinal stem cells that are susceptible to neoplastic transformation. Nature 457:603–607

10. Sangiorgi E, Capecchi MR (2008) Bmi1 is expressed in vivo in intestinal stem cells. Nat Genet 40:915–920

11. Ohlstein B, Spradling A (2006) The adult Drosophila posterior midgut is maintained by pluripotent stem cells. Nature 439:470–474

12. Micchelli CA, Perrimon N (2006) Evidence that stem cells reside in the adult Drosophila midgut epithelium. Nature 439:475–479

13. Ohlstein B, Spradling A (2007) Multipotent Drosophila intestinal stem cells specify daughter cell fates by differential notch signaling. Science 315:988–992

14. Guo Z, Ohlstein B (2015) Bidirectional notch signaling regulates Drosophila intestinal stem cell multipotency. Science 350(6263): aab0988

15. Zeng X, Hou SX (2015) Enteroendocrine cells are generated from stem cells through a distinct progenitor in the adult Drosophila posterior midgut. Development 142(4):644–653

16. He L, Si G, Huang J, Samuel ADT, Perrimon N (2018) Mechanical regulation of stem-cell differentiation by the stretch-activated Piezo channel. Nature 555(7694):103–106

17. Chen J, Xu N, Wang C, Huang P, Huang H, Jin Z, Yu Z, Cai T, Jiao R, Xi R (2018) Transient Scute activation via a self-stimulatory loop directs enteroendocrine cell pair specification from self-renewing intestinal stem cells. Nat Cell Biol 20(2):152–161

18. Biteau B, Hochmuth CE, Jasper H (2008) JNK activity in somatic stem cells causes loss of tissue homeostasis in the aging Drosophila gut. Cell Stem Cell 3:442–455

19. Choi NH, Kim JG, Yang DJ, Kim YS, Yoo MA (2008) Age-related changes in Drosophila midgut are associated with PVF2, a PDGF/VEGF-like growth factor. Aging Cell 7:318–334

20. Amcheslavsky A, Jiang J, Y. T. I (2009) Tissue damage-induced intestinal stem cell division in Drosophila. Cell Stem Cell 4:49–61

21. Apidianakis Y, Pitsouli C, Perrimon N, Rahme L (2009) Synergy between bacterial infection and genetic predisposition in intestinal dysplasia. Proc Natl Acad Sci U S A 106:20883–20888

22. Buchon N, Broderick NA, Chakrabarti S, Lemaitre B (2009) Invasive and indigenous microbiota impact intestinal stem cell activity through multiple pathways in Drosophila. Genes Dev 23:2333–2344

23. Buszczak M, Paterno S, Spradling AC (2009) Drosophila stem cells share a common requirement for the histone H2B ubiquitin protease scrawny. Science 323:248–251

24. Bardin AJ, Perdigoto CN, Southall TD, Brand AH, Schweisguth F (2010) Transcriptional control of stem cell maintenance in the Drosophila intestine. Development 137:705–714

25. Jiang H, Patel PH, Kohlmaier A, Grenley MO, McEwen DG, Edgar BA (2009) Cytokine/Jak/Stat signaling mediates regeneration and homeostasis in the Drosophila midgut. Cell 137:1343–1355

26. Jiang H, Grenley MO, Bravo MJ, Blumhagen RZ, Edgar BA (2011) EGFR/Ras/MAPK signaling mediates adult midgut epithelial homeostasis and regeneration in Drosophila. Cell Stem Cell 8:84–95

27. Shaw RL, Kohlmaier A, Polesello C, Veelken C, Edgar BA, Tapon N (2010) The Hippo pathway regulates intestinal stem cell proliferation during Drosophila adult midgut regeneration. Development 137:4147–4158

28. Lee WC, Beebe K, Sudmeier L, Micchelli CA (2009) Adenomatous polyposis coli regulates Drosophila intestinal stem cell proliferation. Development 136:2255–2264

29. Ren F, Wang B, Yue T, Yun EY, Ip YT, Jiang J (2010) Hippo signaling regulates Drosophila intestine stem cell proliferation through multiple pathways. Proc Natl Acad Sci U S A 107:21064–21069

30. Mathur D, Bost A, Driver I, Ohlstein B (2010) A transient niche regulates the specification of Drosophila intestinal stem cells. Science 327:210–213

31. Lin G, Xu N, Xi R (2008) Paracrine Wingless signaling controls self-renewal of Drosophila intestinal stem cells. Nature 455:1119–1123

32. Liu W, Singh SR, Hou SX (2010) JAK-STAT is restrained by notch to control cell proliferation of the Drosophila intestinal stem cells. J Cell Biochem 109:992–999

33. Singh SR, Zeng X, Zhao J, Liu Y, Hou G, Liu H, Hou SX (2016) The lipolysis pathway sustains normal and transformed stem cells in adult Drosophila. Nature 538(7623):109–113

34. Zeng X, Han L, Singh SR, Liu H, Neumüller RA, Yan D, Hu Y, Liu Y, Liu W, Lin X, Hou SX (2015) Genome-wide RNAi screen identifies networks involved in intestinal stem cell regulation in Drosophila. Cell Rep 10(7):1226–1238

35. Hochmuth CE, Biteau B, Bohmann D, Jasper H (2011) Redox regulation by Keap1 and Nrf2 controls intestinal stem cell proliferation in Drosophila. Cell Stem Cell 8:188–199

36. Staley BK, Irvine KD (2010) Warts and Yorkie mediate intestinal regeneration by influencing stem cell proliferation. Curr Biol 20:1580–1587

37. Korzelius J, Naumann SK, Loza-Coll MA, Chan JS, Dutta D, Oberheim J, Gläßer C, Southall TD, Brand AH, Jones DL, Edgar BA (2014) Escargot maintains stemness and suppresses differentiation in Drosophila intestinal stem cells. EMBO J 33(24):2967–2982

38. Dutta D, Dobson AJ, Houtz PL, Gläßer C, Revah J, Korzelius J, Patel PH, Edgar BA, Buchon N (2015) Regional cell-specific Transcriptome mapping reveals regulatory complexity in the adult Drosophila midgut. Cell Rep 12(2):346–358

39. Koehler CL, Perkins GA, Ellisman MH, Jones DL (2017) Pink1 and Parkin regulate Drosophila intestinal stem cell proliferation during stress and aging. J Cell Biol 216(8):2315–2327

40. Lindberg BG, Tang X, Dantoft W, Gohel P, Seyedoleslami Esfahani S, Lindvall JM, Engström Y (2018) Nubbin isoform antagonism governs Drosophila intestinal immune homeostasis. PLoS Pathog 14(3):e1006936

41. Houtz P, Bonfini A, Liu X, Revah J, Guillou A, Poidevin M, Hens K, Huang HY, Deplancke B, Tsai YC, Buchon N (2017) Hippo, TGF-β, and Src-MAPK pathways regulate transcription of the upd3 cytokine in Drosophila enterocytes upon bacterial infection. PLoS Genet 13(11):e1007091

42. Xu C, Luo J, He L, Montell C, Perrimon N (2017) Oxidative stress induces stem cell proliferation via TRPA1/RyR-mediated Ca2+ signaling in the Drosophila midgut. Elife 6. Pii: e22441. https://doi.org/10.7554/eLife.22441

43. Jin Y, Patel PH, Kohlmaier A, Pavlovic B, Zhang C, Edgar BA (2017) Intestinal stem cell pool regulation in Drosophila. Stem Cell Reports 8(6):1479–1487

44. Tian A, Wang B, Jiang J (2017) Injury-stimulated and self-restrained BMP signaling dynamically regulates stem cell pool size during Drosophila midgut regeneration. Proc Natl Acad Sci U S A 114(13):E2699–E2708

45. Lan Q, Cao M, Kollipara RK, Rosa JB, Kittler R, Jiang H (2018) FoxA transcription factor fork head maintains the intestinal stem/progenitor cell identities in Drosophila. Dev Biol 433(2):324–343

46. Resnik-Docampo M, Koehler CL, Clark RI, Schinaman JM, Sauer V, Wong DM, Lewis S, D'Alterio C, Walker DW, Jones DL (2017) Tricellular junctions regulate intestinal stem cell behaviour to maintain homeostasis. Nat Cell Biol 19(1):52–59

47. Tian A, Benchabane H, Wang Z, Ahmed Y (2016) Regulation of stem cell proliferation and cell fate specification by wingless/Wnt signaling gradients enriched at adult intestinal compartment boundaries. PLoS Genet 12(2):e1005822

48. Beehler-Evans R, Micchelli CA (2015) Generation of enteroendocrine cell diversity in midgut stem cell lineages. Development 142(4):654–664

49. Ayyaz A, Li H, Jasper H (2015) Haemocytes control stem cell activity in the Drosophila intestine. Nat Cell Biol 17(6):736–748

50. Jin Y, Xu J, Yin MX, Lu Y, Hu L, Li P, Zhang P, Yuan Z, Ho MS, Ji H, Zhao Y, Zhang L (2013) Brahma is essential for Drosophila intestinal stem cell proliferation and regulated by Hippo signaling. elife 2:e00999

51. Li Z, Zhang Y, Han L, Shi L, Lin X (2013) Trachea-derived dpp controls adult midgut

homeostasis in Drosophila. Dev Cell 24(2): 133–143

52. García Del Arco A, Edgar BA, Erhardt S (2018) In vivo analysis of centromeric proteins reveals a stem cell-specific asymmetry and an essential role in differentiated, non-proliferating cells. Cell Rep 22(8):1982–1993

53. Zhai Z, Kondo S, Ha N, Boquete JP, Brunner M, Ueda R, Lemaitre B (2015) Accumulation of differentiating intestinal stem cell progenies drives tumorigenesis. Nat Commun 6:10219

54. Lucchetta EM, Ohlstein B (2017) Amitosis of polyploid cells regenerates functional stem cells in the Drosophila intestine. Cell Stem Cell 20(5):609–620

55. Hudry B, Khadayate S, Miguel-Aliaga I (2016) The sexual identity of adult intestinal stem cells controls organ size and plasticity. Nature 530(7590):344–348

56. Liang J, Balachandra S, Ngo S, O'Brien LE (2017) Feedback regulation of steady-state epithelial turnover and organ size. Nature 548(7669):588–591

57. Singh SR, Mishra MK, Kango-Singh M, Hou SX (2012) Generation and staining of intestinal stem cell lineage in adult midgut. Methods Mol Biol 879:47–69

58. Harrison DA, Perrimon N (1993) A simple and efficient generation of marked clones in Drosophila. Curr Biol 3:424–433

59. Kirilly D, Spana EP, Perrimon N, Padgett RW, Xie T (2005) BMP signaling is required for controlling somatic stem cell self-renewal in the Drosophila ovary. Dev Cell 9:651–662

60. Wu JS, Luo L (2006) A protocol for mosaic analysis with a repressible cell marker (MARCM) in Drosophila. Nat Protoc 1:2583–2589

61. Reiff T, Jacobson J, Cognigni P, Antonello Z, Ballesta E, Tan KJ, Yew JY, Dominguez M, Miguel-Aliaga I (2015) Endocrine remodelling of the adult intestine sustains reproduction in Drosophila. elife 4:e06930

Quantitative Analysis of Intestinal Stem Cell Dynamics Using Microfabricated Cell Culture Arrays

Leigh A. Samsa, Ian A. Williamson, and Scott T. Magness

Abstract

Regeneration of intestinal epithelium is fueled by a heterogeneous population of rapidly proliferating stem cells (ISCs) found in the base of the small intestine and colonic crypts. ISCs populations can be enriched by fluorescence-activated cell sorting (FACS) based on expression of combinatorial cell surface markers, and fluorescent transgenes. Conventional ISC culture is performed by embedding single ISCs or whole crypt units in a matrix and culturing in conditions that stimulate or repress key pathways to recapitulate ISC niche signaling. Cultured ISCs form organoid, which are spherical, epithelial monolayers that are self-renewing, self-patterning, and demonstrate the full complement of intestinal epithelial cell lineages. However, this conventional "bulk" approach to studying ISC biology is often semiquantitative, low throughput, and masks clonal effects and ISC phenotypic heterogeneity. Our group has recently reported the construction, long-term biocompatibility, and use of microfabricated cell raft arrays (CRA) for high-throughput analysis of single ISCs and organoids. CRAs are composed of thousands of indexed and independently retrievable microwells, which in combination with time-lapse microscopy and/or gene-expression analyses are a powerful tool for studying clonal ISC dynamics and micro-niches. In this protocol, we describe how CRAs are used as an adaptable experimental platform to study the effect of exogenous factors on clonal stem cell behavior.

Key words Intestinal stem cell, Organoid, Enteroid, Cell raft array, Sox9EGFP, Genetically engineered mouse model

1 Introduction

1.1 Intestinal Stem Cells

The mammalian intestine is composed of three layers: (1) an outer muscular layer and nerve plexus that contracts rhythmically to move luminal contents along the proximal to distal axis, (2) a middle submucosal layer consisting of fibroblasts, blood vessels, and immune cells, and (3) an inner epithelial monolayer that is highly patterned into crypt-villus units and performs critical barrier, absorptive, and secretory functions (Fig. 1a). Intestinal epithelial cells include multiple differentiated cell types that are necessary for normal epithelial function including: nutrient-absorbing enterocytes, mucus-secreting goblet cells, antimicrobial Tuft and Paneth

Shree Ram Singh and Pranela Rameshwar (eds.), *Somatic Stem Cells: Methods and Protocols*, Methods in Molecular Biology, vol. 1842, https://doi.org/10.1007/978-1-4939-8697-2_10, © Springer Science+Business Media, LLC, part of Springer Nature 2018

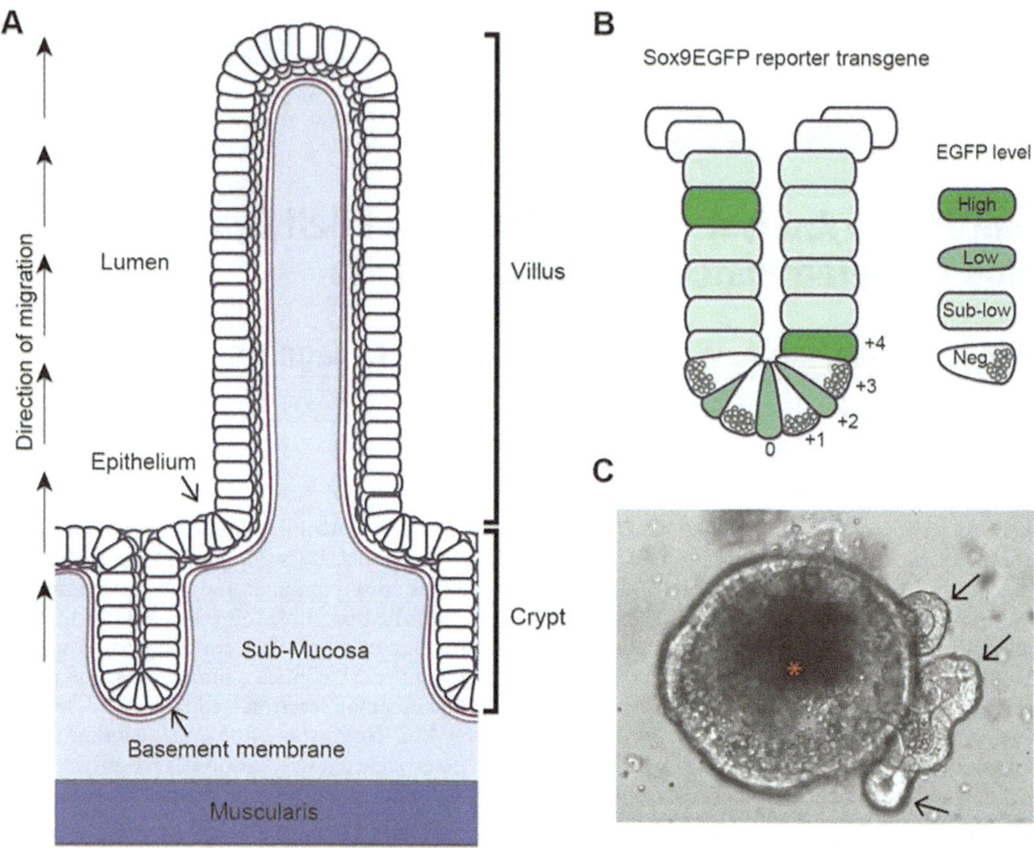

Fig. 1 (**a**) Schematic of a cross-sectional view of the small intestine. (**b**) Representation of a single crypt from a transgenic mouse expressing the Sox9EGFP reporter. Crypt cell position is counted from the base of the crypt. Sox9EGFP-neg granulated Paneth cells are intercalated among Sox9EGFP-low CBCs/aISCs, which give rise to Sox9EGFP-sublow transit amplifying cells that migrate up and out of the crypt. Sox9EGFP-high cells include enteroendocrine cells and their progenitors as wells as rISCs at the +4 positions. (**c**) Representative enteroid. Arrows point to crypt-like buds; asterisk marks the debris-filled pseudolumen

cells, and hormone-secreting enteroendocrine cells (EECs) [1]. Intestinal stem cells (ISCs) drive continual renewal of these differentiated populations, and are located at the base of the crypt where they are maintained by extrinsic signals from adjacent Paneth cells and underlying mesenchymal cells that comprise the ISC *niche* [2]. The majority of intestinal stem cells is actively dividing (aISCs), while a small minority of ISCs are "quiescent" and serve as a "reserve" ISC (rISC) population. While investigators have known for decades that actively dividing cells are located in the crypt base, the lack of ISC isolation and culture methods precluded detailed in vitro ISC studies until the emergence of advances in biomarkers and ISC niche biology [3, 4].

1.2 ISC Isolation

Genetically engineered mouse models (GEMMs) expressing fluorescent reporters have been developed which facilitate FACS-enrichment of ISCs. An Lgr5-EGFP reporter mouse is commonly used to isolate functional ISCs due to the highly restricted expression of Lgr5 to the aISC population where Lgr5-EGFP-high cells exhibit high organoid formation capacity (OFC) [5–7]. While the Lgr5-EGFP transgene demonstrates highly restricted expression to ISCs, the expression pattern is mosaic resulting in only a minority of crypts expressing EGFP [6]. Our group has developed a Sox9-EGFP transgenic mouse model (Fig. 1b), which serves as highly versatile alternative to study ISC and progenitor populations in the small intestine and colon [7–9]. In situ fluorescence and flow cytometry demonstrate Sox9-EGFP expression at three distinct levels—sublow, low, and high (Fig. 1c) [7, 9]. Sox9-EGFP 'low' cells are primarily Lgr5-expressing aISCs that proliferate and demonstrate robust OFC [10]. Other GEMMs have been developed to isolate candidate ISC populations, and many of these fluorescent reporter gene models have overlapping expression with Lgr5-EGFP and Sox9-EGFP models [2, 11]. If ISC reporter alleles are not available, a collection of cell surface markers can be used to FACS-enrich aISCs [12]. Alternatively, ISCs are also found in a side population of cells identified through their capacity to efflux Hoechst dye [13]. However, these approaches tend to be less-specific for aISCs resulting in lower ISC-enriching efficiencies.

1.3 Organoids

Historically, long-term culture of primary intestinal epithelium was not possible in large part due to the loss of stem cells and short life-span of differentiated cell in culture. GEMMs and recent conceptual advances in ISC niche biology paved the way for methodologies that enabled indefinite culture of epithelial tissues derived from the intestines of several different species [2, 5, 14–16]. When placed in the defined conditions, ISCs or crypt units develop into structures commonly called, "organoids" [5, 17]. Organoids have been further categorized into nomenclatures based on their region of derivation, and maturity of morphology [17]. Primitive spherical structures derived from the small intestine are termed "enterospheres" and are characterized by flat cells, mostly stem/progenitor lineages, that arrange in a spherical conformation with a hollow center. Enterospheres give rise to "Enteroids" over time and are characterized by mature columnar shaped cellular monolayers, comprised of stem/progenitor and all differentiated lineages, and although they exist in a spherical conformation with a hollow center, they possess self-patterning crypt buds (Fig. 1c). Likewise, "colonsphere" and "colonoid" are these equivalent structures derived from colon. Organoids can be expanded, frozen, and thawed, like transformed cells lines.

1.4 ISC Culture Media

The first long-term culture of epithelial monolayers from crypts used recombinant proteins and small molecules to regulate Wnt/Notch/TGFβ pathways with the goal of mimicking the native ISC *niche* [5]. Though multiple formulas have been reported for murine ISC culture media, most consist of a basal media (Table 1) supplemented with various factors (Table 2) depending on the starting cellular material. In vivo, Wnts and Notch-ligands are supplied by Paneth cells that intercalate between ISCs [18]. FACS-isolated ISCs placed in culture do not have neighboring Paneth cells to supply these essential factors. Jagged and Wnt3a are common factors added to the media to stimulate their cognate pathways and are essential supplements to support ISC survival and organoid forming potential. In general, conversion between enterosphere and enteroid morphologies is achieved by titrating additives that promote Wnt signaling [19].

Since the first description of organoid formation from ISCs, efforts to improve culture conditions focused on optimizing organoid formation efficiency and reproducibility as well as lowering culture costs. While recombinant proteins were used in the original description of culture conditions, these tend to be cost prohibitive for many studies. In some cases, small molecules can replace recombinant factors (Table 2). Alternatively, cell lines that produce secreted recombinant versions of Wnt-pathway agonists and TGFß-antagonists produce a "conditioned media," which is a popular choice because the media is potent, highly economical, and reduces culture costs by orders of magnitude. However, conditioned media is often poorly defined, which increases variability, and can confound experimental interpretation. In practice, the biological questions of interest and experimental design will guide selection of a particular specific media formulation. To facilitate selection, we have tabulated some of the reported media formulations compatible with ISC culture (Table 3).

1.5 ISC Extracellular Matrix

Aside from defined media requirements, organoids derived from crypts or single ISC will only develop in a three-dimensional environment composed of extracellular matrix (ECM) in the form of a hydrogel. There are several commercially available hydrogels that mimic the extracellular matrix and are capable of supporting ISC growth. Matrigel (Corning) and Cultrex (Trevigen) are derived from a mouse sarcoma line and are rich in ECM proteins, however, the exact molecular composition has not been reported. Recent work using chemically defined hydrogels has highlighted the importance of hydrogel protein composition on regulating organoid behavior in culture and in bioreactors [20, 21]. In particular, fibronectin facilitates ISC survival and proliferation while laminin-based adhesion promotes ISC differentiation and organoid formation. Likewise, high matrix stiffness promotes ISC expansion while a soft matrix promotes differentiation [20].

Table 1
ISC basal media

Basal Media [5, 12, 29]	Company	Cat #	[Working]	Function
Advanced DMEM/F12	ThermoFisher	12634-028	Varies	Basic mammalian culture media
GlutaMAX	ThermoFisher	35050-079	100 U/mL	Glutamine supplement
Penicillin–streptomycin	ThermoFisher	15140-122	100 U/ml	Antimicrobial
HEPES	ThermoFisher	15630-106	1 mM	Physiologic buffer
B27 supplement	ThermoFisher	17504-044	1× (1:100)	Chemically defined growth supplement containing BSA, transferrin, insulin and a variety of other additives designed to support neural stem cell growth
N2 supplement	ThermoFisher	17502-048	1× (1:100)	Chemically defined growth supplement containing BSA, transferrin, insulin and a variety of other additives designed to support neural stem cell growth
(NAC) N-Acetylcysteine	Sigma-Aldrich	A9165-5G	1 mM	Protects against death, antimucolytic, ROS scavenger
Primocin	Invivogen	Ant-pm-1	1× (1:100)	(Optional) Antimicrobial

1.6 Quantifying ISC/Organoid Cultures

Ever increasing use of organoids for mechanistic studies brought to the forefront several challenges pertaining to quantifying important readouts including ISC/organoid viability, OFC, growth parameters, and crypt-budding. In conventional 3D culture, OFC is extremely challenging to accurately calculate due to technical challenges in calculating an accurate numerator (organoids formed) and denominator (initial number of cells). The most straightforward route to quantifying ISC growth and organoid development is to follow individual ISCs in real time with live imaging, and then normalize organoid formation rate to the number of individual ISCs examined. However, cell counts reported by FACS instrumentation are notoriously inaccurate; and, cells can become damaged during sorting, leading to biological changes that are not accounted for prior to culture. Also, ISCs embedded in hydrogel move in three dimensions as they grow, introducing challenges to accurate quantitation due to difficulties associated with locating individual cells in different Z-planes, and detecting cells from

Table 2
ISC Media additives

Additive	Company	Cat #	[Working]	Abb.	Function
Recombinant mouse EGF	R&D Systems	2028-EG	50 ng/mL	E	Epidermal growth factor receptor agonist; necessary for epithelial cell growth
Recombinant mouse Noggin	Peprotech	250-38	100 ng/mL	N	Inhibits BMP signaling
Recombinant human R-spondin 1	R&D Systems	4645-RS/CF	500 ng/mL	R	Augments Wnt/B-catenin signaling
Recombinant mouse WNT3a	R&D Systems	GF-160	40 ng/µL	W	Wnt agonist
Recombinant mouse Jagged-1	Anaspec	61298	7.5 µM	J	Stimulates Notch signaling
Y-27632	Selleck Chemicals	S1049	10 µM	Y	Inhibits apoptosis due to anoikis
Thiazovivn[a]			2.5 µM	T	Inhibits apoptosis due to anoikis
Valproic acid[a]	Sigma-Aldrich	P4543	10 µM	V	HDAC (histone deacetylase) inhibitor and Notch activator
CHIR99021	Selleck Chemicals	S2924	5 µM	C	Stimulates canonical Wnt signaling
WNT3A conditioned media	ATCC for cells, make in house	N/A	25–50%	[W]	Conditioned media is enriched with WNT3a secreted from stable transgenic cell line
RSPO2 conditioned media	ATCC for cells, make in house	N/A	12.5%	[R2]	Conditioned media is enriched with WNT3a secreted from stable transgenic cell line
RSPO1 conditioned media	ATCC for cells, make in house	N/A	12.5%	[R1]	Conditioned media is enriched with WNT3a secreted from stable transgenic cell line
Noggin conditioned media	ATCC for cells, make in house	N/A	12.5%	[N]	Conditioned media is enriched with WNT3a secreted from stable transgenic cell line

[a]Apoptosis inhibitors are added only on the first day of culture

artifacts [22]. The requirement for manual focusing, counting, and artifact verification, substantially reduces throughput. Additionally, analysis of ISC as populations masks the inherent phenotypic heterogeneity that exists at the single ISC level. Differential biological responses of ISCs to experimental perturbations cannot be evaluated at the population level reducing the data resolution. A growing appreciation for the quantitative

Table 3
ISC efficiency in selected published media formulas

Citation	Media	Formula and Method	ISC source	Organoid efficiency (%)[a]	Quantification method
Sato et al. 2009 [5]	ENRJ+Y	• Basal Media + Growth factors: 10–50 ng/mL EGF, 500 ng/mL Rspo1, 100 ng/mL Noggin. • First plating: Also 1 µM Jagged in Matrigel, 10 µM Y-27632 • GFs added every 2 days. Entire media changed every 4 days	Lgr5GFP-high	6	Manual
Gracz et al. 2010 [8]	ENRJ+Y	• Basal Media + Growth factor: 10–50 ng/mL EGF, 500 ng/mL Rspo1, 100 ng/mL Noggin. • First plating: Also 1 µg/mL Rspo1, 1 µM Jagged in Matrigel, 10 µM Y-27632 • GFs added every other day. Entire media changed every 4 days	CD24+ Sox9EGFP-low	0.5 5	Manual
Gracz et al. 2015 [22]	ENRJ+T	• Basal Media + Growth factors: 50 ng/mL EGF, 100 ng/mL Noggin, 1 µg/mL Rspondin, 1 µM Jagged-1, • First plating: Matrigel also contained 15 µM Jagged-1, 750 ng/mL EGF, 100 ng/mL Noggin, and 50 nM LY2157299. Initial media contained 2.5 µM CHIR99021 and 2.5 µM Thiazovivin. • GFs added every other day. Entire media changed every 4 days	Lgr5GFP-high Sox9EGFP-low	0.6 0.7	Cell Raft Array
Sato et al. 2011 [18]	ENR-W+Y	• Basal Media + Growth factors: 10–50 ng/mL EGF, 500 ng/mL Rspo1, 100 ng/mL Noggin, 100 ng/mL Wnt3a. • First plating: Also 10 µM Y-27632 • GFs added every 2 days. Entire media changed every 4 days	Lgr5GFP-high	20–22	Manual

(continued)

Table 3
(continued)

Citation	Media	Formula and Method	ISC source	Organoid efficiency (%)[a]	Quantification method
Yin et al. 2014 [19]	ENR+JWY	• Basal Media + Growth factors: EGF 50 ng/mL, Noggin 100 ng/mL, Rspondin 1 500 ng/mL, Wnt3a 100 ng/mL • First plating: Matrigel had 1 µM Jagged, media had 10 µM Y-27632 • Media changed every 2 days.	Lgr5GFP-high	3–4	Manual
Yin et al. 2014 [19]	ENR-CV	• Basal Media + Growth factors: EGF 50 ng/mL, Noggin 100 ng/mL, Rspondin 1 500 ng/mL, CHIR99021 3 µM, Valproic Acid 1 mM. • First plating: Matrigel had 1 µM Jagged in it, media had 10 µM Y-27632 • Media changed every 2 days.	Lgr5GFP-high	25–40	Manual
van Es et al. 2012 [11]	ENRW+Y	• Basal Media + Growth factors: 10–50 ng/mL EGF, 500 ng/mL Rspo1, 100 ng/mL Noggin, 100 ng/mL Wnt3a. • First plating: Also 10 µM Y-27632 • GFs added every 2 days. Entire media changed every 4 days	Dll1GFP+/ CD24mid	1.2	Manual

GF growth factors

[a]Organoid forming efficiency on D6-D14 relative to # sorted cells

limitations of conventional single ISC and organoid cultures motivated the development of microfabricated culture platforms to address the challenges.

1.7 Cell Raft Arrays   Microfabricated cell culture array devices spatially separate cells and are employed to increase the specificity, resolution, and throughput of basic biomolecular cell-based assays. The Cell Raft Array (CRA) system (Qiagen, Cell Microsystems, LLC), enables cell separation by trapping cells or organoids within spatially defined microwells—facilitating identification, tracking, and retrieval of individual cells and their clonal progeny [23, 24]. Of particular interest, the four-reservoir (quad-well) Cell Raft Array (CRA) system traps cells within ~8500, 200 μm × 200 μm microwells arrayed at a depth of 100 μm within a 2.5 cm² footprint on a polydimethylsiloxane (PDMS) plate mounted on a 4-reservoir polycarbonate cassette (Fig. 2a) [22]. This four-reservoir cassette provides separate media environments containing ~2100 wells each,

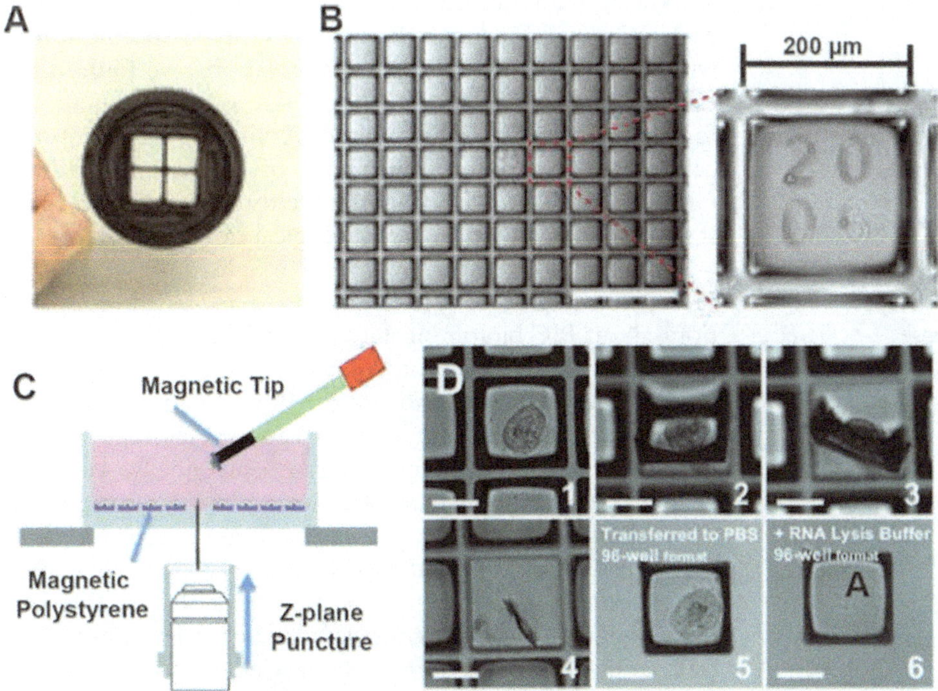

Fig. 2 (**a**) Macroscopic view of a quad-well Cell Raft Array (CRA). (**b**) Microwells are 200 μm² and arranged in a grid with the physical address stamped at 5 well intervals. (**c**) Schematic of raft retrieval. A device containing a fine needle positioned in the center of a clear window is affixed to a 10× objective lens to allow for z-plane puncture. A magnetic wand is used to collect the raft. The raft is removed from the wand when in proximity to a stronger magnet. (**d**) Time lapse image of raft retrieval where a single raft is liberated (1–4), transferred to a 96-well plate (5) then lysed to extract RNA (6). Figure modified and reproduced with permission from Nature Cell Biology. 2015 March; 17(3): 340–349. Doi: https://doi.org/10.1038/ncb3104

allowing researchers to assay different treatments or replicates. Each microwell is embedded with a ferrous (magnetic) polystyrene microraft which is stamped at regular intervals with row and column number, giving wells a physical addressed for tracking purposes (Fig. 2b). The embedded rafts enable specific cell-sampling by releasing cells from the microwell using a micro-needle device mounted to the objective of an inverted microscope. Rafts are retrieved from the media using a magnetic wand which attracts the ferrous polystyrene (Fig. 2c). As illustrated in Fig. 2d, after collection, individual microrafts can be isolated for clonal expansion or gene expression analysis (Fig. 2d).

By combining live imaging with the ability to physically retrieve discrete samples, CRAs offer a powerful approach to studying ISC biology that circumvents the major problems with conventional ISC assays. We recently used CRAs to quantify clonal ISC behaviors in high-throughput to demonstrate the importance of ISC-Paneth cell contacts in supporting Sox9EGFP-low OFC [22]. We cocultured single ISCs with differentiated niche cells at various proportions and developed the advanced computer vision analysis workflows referenced in this protocol to quantify the contents of each microwell and correlate early cell-type specific interactions with later OFC. Among other observations, we found that niche cells only supported stem cell growth and subsequent organoid formation when plated in physical contact with cocultured stem cells, highlighting the importance of juxtacrine signaling [22]. CRAs were essential for reaching the throughput necessary for the statistical power to make this biological observation.

1.8 Protocol Overview

Here, we describe how to use CRAs to ask a wide range of questions about ISC biology at the single-cell level in high throughput. In the first part of this protocol, we describe an optimized method for FACS-enriching ISCs from Sox9EGFP small intestine. Then, we conduct culture these ISCs on a multireservoir CRA and conduct time lapse, tile-scanning microscopy. Finally, we describe an analysis pipeline used to extract image-based readouts from the time-lapse imaging data which can be used to compare the discrete ISC behaviors in different treatment reservoirs. This work flow is illustrated in Fig. 3. Though not described in this protocol, individual rafts and their contents may be retrieved using commercially available CRA retrieval system (Cell Microsystems) for clonal expansion or downstream gene expression assays.

Automated image analysis requires a high contrast image to delimit cell borders. Since brightfield images do not provide sufficient contrast, we use the CAG:dsRED GEMM which produces a strong red fluorescent signal in all cells, providing contrast to the cell borders [25, 26]. An alternative approach is to stain sorted cells dyes to label the plasma membrane—an approach which may be for adapted versions of this protocol to use with other [22].

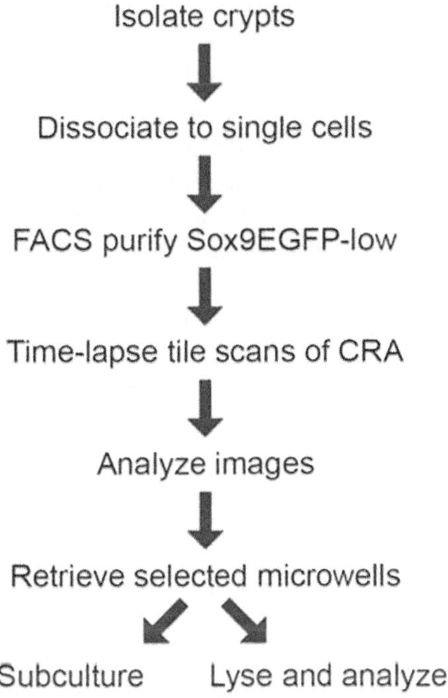

Fig. 3 Workflow

However, the effects of long term exposure to plasma membrane labeling dyes are poorly defined in ISC systems and should be optimized for each application.

Our protocol in Volume 1 of this series describes a method for using FACS to obtain an enriched population of live ISCs derived from whole epithelium [12]. The intestinal epithelium is released from the basement membrane by incubation with the calcium chelator EDTA, then is further digested into single cells by enzymatic cleavage of cell-cell bonds. Here, we describe an adapted method which produces a highly crypt-enriched population of IECs and increases this proportion of aISCs in the cell preparation. To this end, we manually scrape the luminal surface to remove most of the villus layer and digest separately the crypt a villus preparations essentially as described before [12]. The single cell suspensions are stained with an antibody recognizing the pan-epithelial marker CD326 (EPCAM, epithelial cell adhesion molecule) and the combination of a cell-impermeable dye and AnnexinV to identify and exclude dead and dying cells, respectively. Taking advantage of the fact that villi are comprised of only Sox9EGFP-high and Sox9EGFP-negative cells, the villus-enriched cells are used as "bookends" to establish gating boundaries to FACS-isolate the Sox9EGFP-low ISC population.

In the next part of this protocol, Sox9EGFP-low cells are embedded in a hydrogel, which mimics the basement membrane, and cultured on the CRA. The CRA is mounted on an inverted microscope and imaged with conventional live-imaging, tile-scanning, and time-lapse microscopy techniques. A custom Matlab script applies an algorithm to identify microwells and segment stitched images such that there is a different image file for each microwell [22]. These microwells are passed through a custom CellProfiler™ image analysis pipeline, which extracts quantitative information about well-contents (cells, debris, multimers, etc.). These analysis pipelines can be easily customized for a wide variety of applications using the CellProfiler graphical user interface [27, 28].

2 Materials

2.1 Epithelial Isolation and Dissociation

1. Equipment: IACUC approved isoflurane container for euthanasia, Chemical biosafety hood, Rocker at 4 °C, dissection tools (small scissors, large scissors, small forceps, large forceps, dissection surface, ice-cold glass plate sterilized with 70% EtOH, 10 cm ruler), pipettes, clinical centrifuge, tissue culture microscope, water bath at 37 °C.

2. Reagents and consumables: Pipette tips, Isoflurane, 10 cm petri dish, conical tubes (15 mL, 50 mL), cell strainer (40 μm, 70 μm, and 100 μm), 70% EtOH in water, Dulbecco's phosphate buffered saline DPBS, Hanks' Balanced Salt Solution without calcium and magnesium HBSS, EDTA 0.5 M, dispase, DNase I, 100 U/mL, Fetal Bovine Serum FBS, Y-27632 10 mM, Advanced DMEM/F12, N2, B27, HEPES 1 M, GlutaMAX 100×, penicillin–streptomycin 100×, N-acetylcysteine (NAC) 500 mM in ddH$_2$O.

3. Ice-cold DPBS: Per sample, two samples per mouse (25 mL in 10 cm petri dish on ice).

4. Epithelial Dissociation Solution: Per sample, two samples per mouse (15 mL ice-cold DPBS, 90 μL 0.5 M EDTA, 3 mM EDTA final (1:166), 15 μL Y-27632, 10 μM final (1:1000).

5. Crypt Dissociation Solution: Per sample (20 mL ice-cold DPBS, 20 μL Y-27632, 10 μM final (1:1000).

6. Single Cell Dissociation: Per sample (Prewarm 10 mL HBSS with 10 μM Y-27632 (1:1000 of 10 mM stock) in 15 mL conical tube to 37°. Immediately before use, add 50 μL dispase and 0.5 U/mL DNase (50 μL).

7. Mice raised in IACUC-approved facility with institutional approved husbandry protocols (Mouse 1: Sox9EGFP−; CAG:DsRED−, Mouse 2: Sox9EGFP+;CAG:DsRED−, Mouse 3: Sox9EGFP-;CAG:DsRed+, Mouse 4: Sox9EGFP+;CAG:DsRed+).

2.2 FACS

1. Additional Equipment: Hemocytometer, FACS machine with 405, 488, 561, and 633 excitation lines and emission filters compatible with SYTOX Blue, Pacific Blue, EGFP, dsRED, and APC/Cy7 fluorophores. Protocol has been tested using Sony SH800 benchtop instrument equipped with filter set #2.

1. Additional Reagents and consumables: 35 µm cell strainer topped FACS tube, Trypan Blue 0.4% (ThermoFisher 15250-061), SYTOX Blue (ThermoFisher S34857), Annexin V Pac Blue (Biolegend 640917), Anti-CD326 APC-Cy7 (Biolegend 118217) (*see* **Note 1**).

2. 2× ISC Basal Media (Store 4° for up to 7 days), recipe for 12.5 mL, 11.15 mL Advanced DMEM/F12 (volume), 250 µL P/S (2×, 1:50), 250 µL GlutaMAX (2×, 1:50), 50 µL HEPES (4 mM, 1:250), 500 µL B27 (2×, 1:25), 250 µL N2 (2×, 1:50), 50 µL 1 mM N-acetyl-cysteine (NAC) (2 µM, 1:250).

3. ISC Sorting Media: (Store 4° for up to 7 days): 5 mL 2× ISC Basal Media (see above), 5 mL/Advanced DMEM/F12, 10 µL Y27632.

2.3 Time-Lapse Imaging on Cell Raft Arrays

1. Additional Equipment and Software: Microscope with incubator, automated stage, and CCD (charge coupled device) camera. Protocol has been tested on Olympus IX81 (*see* **Note 2**), tissue culture incubator set to 37 °C with 5% CO_2, software capable of multichannel, time-lapse tile scan. The protocol described here was tested with Micro-Manager software (download at https://micro-manager.org/), and has been successfully used with Metamorph (Molecular Devices) software equipped with a "Scan Slide" application.

2. Additional reagents and consumables: Cultrex (VWR/Trevigen, 89496-092), Quad-well Cell Raft Array (Cell Microsystems)

3. ISC culture media: 500 µL per reservoir of quad-well CRA (*see* Table 3).

2.4 Data Processing and Analysis

1. Additional Software

   (a) FIJI ImageJ with Java (download at https://fiji.sc/)

   (b) Matlab (download at https://www.mathworks.com/)

   (c) Cell Profiler 2.0.1170 (download at http://cellprofiler.org/)

   (d) Microsoft Excel (download at https://www.microsoft.com/en-us/download/office.aspx)

2. Code: download from http://www.magnesslab.org

   (a) Segmenter (Matlab)

   (b) SCPipeline (Cell Profiler)

   (c) CP_InitialSheet_Blank (Excel)

3 Methods

3.1 Epithelial Isolation and Dissociation

1. Euthanize mice in chemical hood with isoflurane followed by cervical dislocation in accordance with IACUC-approved protocol. Four mice are necessary for appropriate FACS gating controls

    (a) Mouse 1: Sox9EGFP−; CAG:DsRED−

    (b) Mouse 2: Sox9EGFP+;CAG:DsRED−

    (c) Mouse 3: Sox9EGFP−;CAG:DsRed+

    (d) Mouse 4: Sox9EGFP+;CAG:DsRed+

3.2 Dissect Intestine (Fig. 4a)

1. Mount euthanized animal on dissection plate and secure in supine position with needles.

2. Spray abdomen with 70% EtOH to wet and sterilize fur. Using large scissors and large forceps, make an incision into abdominal skin ~ 2 cm above the anus.

3. Insert scissors below skin and open the blades to separate skin from muscle.

4. Cutting in a V-pattern, free the skin, lifting up and to the side to expose the abdominal muscle. Spray scissors with 70% EtOH to remove any fur that may be stuck on the blades.

5. Perform laparotomy to open abdominal cavity. Use small scissors and forceps to shift visceral organs to the left and locate the stomach underneath the liver.

6. Holding the stomach with forceps at pyloric antrum, cut through the stomach immediately proximal to the forceps.

7. Gently pull upward and to the left to remove intact intestine from duodenum to cecum (*see* **Note 3**).

8. Transfer to petri dish filled with ice-cold DPBS (Fig. 4a). Trim and remove any remaining stomach, cecum, myenteric plexus and pancreatic attachments (*see* **Note 4**).

3.3 Remove Lumenal Debris (Fig. 4b)

1. Fillet open the intestine by passing scissors through the lumen, starting at the proximal end of the tissue.

2. Rinse thoroughly in 10 cm petri dish by agitating filleted intestine twice in 25 mL ice-cold DPBS.

3.4 Loosen Epithelium and Deplete Villi (Fig. 4c)

1. Use forceps to transfer intestine from ice-cold DPBS to 50 mL conical tube containing 15 mL Epithelial Dissociation Solution.

2. Secure tube in horizontal position on rocker at 4 °C. Agitate 80 rpm for 15 min.

3. Use forceps to remove intestine and transfer to ice-cold glass plate. Carefully orient intestine to expose the luminal surface. Avoid allowing the tissue to dry out by moistening with cold DPBS.

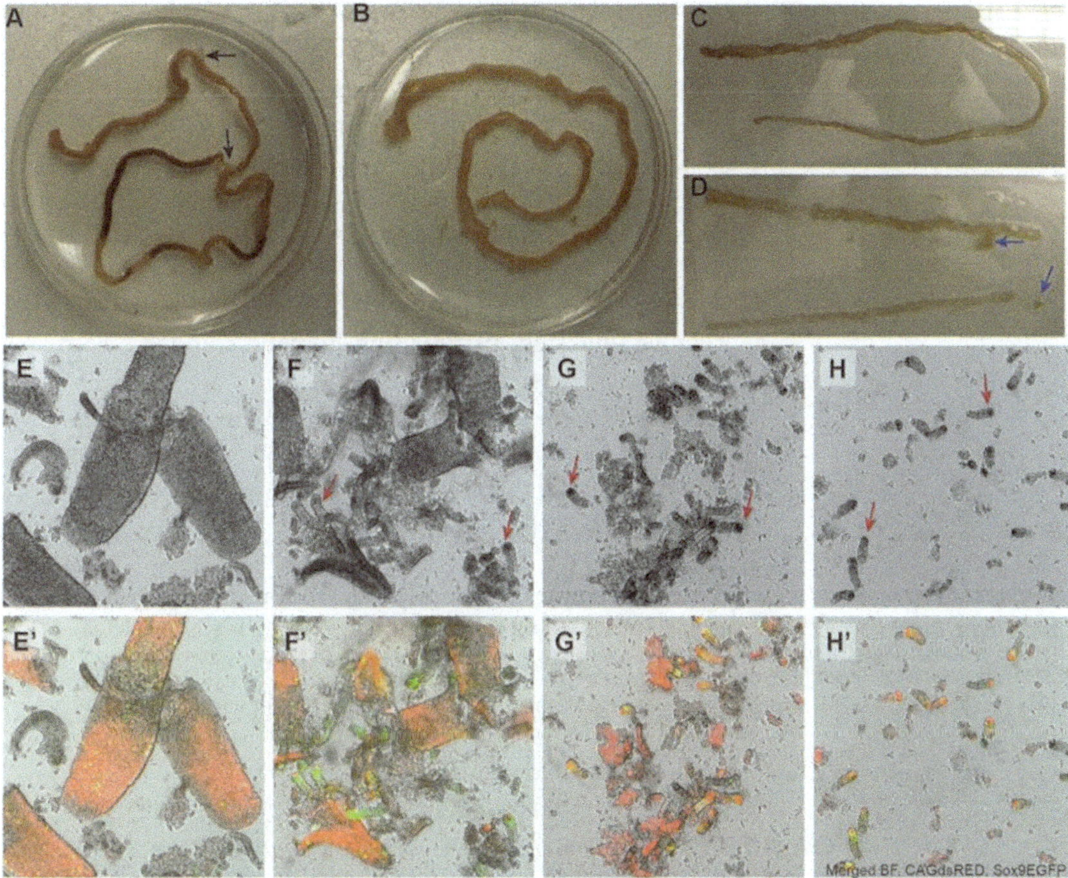

Fig. 4 (**a**) Macroscopic view of the small intestine after dissection with proximal end on left. Black arrows point to nonintestinal tissue that will be trimmed free. (**b**) Macroscopic view of small intestine after being filleted opened and rinsed. (**c, d**) Macroscopic view of filleted intestine (**c**) before and (**d**) after scraping off villi. Blue arrows point to villus remnant reserved for later FACS gating controls. (**e–h**) Representative brightfield images of dissociation steps showing (**e**) reserved villus-enriched fraction and the crypt-enriched preparation (**f**) pre-filtration and after passing through (**g**) 100 μm and (**h**) 70 μm cell strainers. Red arrows point to select crypt bases. (**e'–h'**) Bright field and merged view including CAGdsRED and Sox9EGFP signals

4. Gently scrape luminal surface with the side of a p200 tip in both directions two to four times. This should remove ~30% of the total volume of tissue. Excess scraping will reduce crypt yield. Insufficient scraping will increase FACS time.

5. Important: For the Mouse 4 Sox9EGFP;CAG:dsRED intestine, use P1000 to transfer 250–500 μL villus-enriched scrapings to 10 mL Crypt Dissociation Solution and reserve on ice (Fig. 4e).

6. Cut tissue into 2–3 cm pieces and transfer to a fresh tube with 15 mL Epithelial Dissociation Solution.

7. Return to rocker for 35–45 min until 60 min total exposure to Epithelial Dissociation Solution (including scraping time).

3.5 Isolate Crypts

1. Use sterile forceps to gently transfer tissue fragments to 15 mL Crypt Dissociation solution in 50 mL conical.

2. Manually shake the tube at 2 cycles per second for 2–3 min. Tissue remnants will float when completely depleted of crypts.

3. Confirm intact crypts and crypt removal by examining tissue remnants and 20 μL droplet of supernatant with light microscope. Intact crypts are elongated with a base that has a "bunches of grapes" morphology. Crypts remaining in the tissue appear as dense circles in the epithelium (Fig. 4f) (*see* **Note 5**).

4. Remove digested tissue using forceps sterilized with 70% ETOH.

5. Filter Crypts through 100 μm then 70 μm cell strainers fitted on 50 mL conical tubes (Fig. 4g, h).

6. Collect reserved villus prep and crypt preps by centrifugation at $600 \times g$ 4 °C for 5 min.

3.6 Dissociate into Single Cells

1. For each sample (5 samples total, Mouse 1–4 crypt fractions and Mouse 4 villus fraction), add 50 μL DNAse I and 50–100 μL dispase to 10 mL prewarmed HBSS.

2. Remove supernatant and resuspend pellet in 1 mL prewarmed Single Cell Dissociation solution and transfer to remaining 9 mL prewarmed solution.

3. Incubate in 37 °C water bath. Every 2 min, shake for 30 s at 3 cycles per second.

4. Check dissociation progress by examining 10 μL aliquots every 2 min. Move to the next step when ~75% of cells are present as single cells and most of the remaining tissue is 2–4 cells. Depending dispase efficacy, this can take as little as 5 min or up to 30 min (*see* **Note 6**).

5. Add 1 mL FBS to quench enzymes.

6. Premoisten 40 μm filter fitted on 50 mL conical tube with 1 mL DPBS. Apply cell suspension to filter.

7. Wash 2× PBS by spinning at $1800 \times g$ for 5 min. Resuspend pellet in ice cold 1 mL ISC Sorting Media Media.

3.7 Antibody Staining and FACS Sorting

1. Assess yield and viability using hemocytometer

   (a) Dilute 10 μL cells with 10 μL Trypan Blue. Load 15 μL of cell mixture onto hemocytometer and calculate concentration and yield.

   (b) Adjust cell volume to $<5 \times 10^6$ cells/mL in Sorting Media.

2. Antibody staining:

   (a) Divide samples into round bottomed tubes as described in Table 4.

Table 4
FACS conditions

	Sample	Mouse	# cells	"Live/Dead" Compromised membrane SYTOX blue	"Dying" Annexin V Pacific blue	"ISC" Sox9 EGFP	"Cell" Constitutive dsRED	"Epithelial cell" CD326 EPCAM Apc/Cy7
Control	1	#1	1×10^5	–	–	–	–	–
Single color ctrl	2	#1	1×10^5	1:2000	1 assay	–	–	–
Single color ctrl	3	#2	1×10^5	–	–	+	–	–
Single color ctrl	4	#3	1×10^5	–	–	–	+	–
Single color ctrl	5	#1	1×10^5	–	–	–	–	1 assay
High/low ctrl	6	#4 villus	1×10^5	1:2000	1 assay	+	+	1 assay
SORT	7	#4 crypt	$>1 \times 10^6$	1:2000	Adjust by cell #	+	+	Adjust by cell#

    (b) Add anti-CD326 APC/Cy7 at manufacturer's recommended concentration

    (c) Incubate on ice 45 min to stain cells

    (d) Wash 3× ($500 \times g \times 5$ min) with 1 mL 1× ISC Media.

    (e) Resuspend Samples 1–7 in 500 µL 1× ISC Media. *See* Table 4 for "Sample" designations.

    (f) Resuspend Sample 8 in 2–3 mL 1× ISC Media

3. Flow cytometry

    (a) Filter Sample 1 through a 35 µM cell strainer topped FACS tube

    (b) Load Sample 1 on FACS sorter and adjust voltage and gains to relevant ranges.

    (c) Adjust forward (FSC-A) and side or back scatter (BSC-A) to approximate the basic gating scheme shown in Fig. 5a which minimizes off-scale events while allowing for clear separation between cell and debris events (*see* **Note 7**).

    (d) Draw a positive gate to select nondebris events

    (e) Doublet Screen 1: Plot FSC-Area by FSC-Height and gate events showing a linear or near-linear relationship between FSC-Area and FSC-Height. (Fig. 5b)

    (f) Doublet Screen 2: Plot FSC-Area by BSC-Height and gate events showing a linear or near-linear relationship between BSC-Area and BSC-Height. (Fig. 5c)

    (g) Adjust individual channel voltage and gain to detect a maximum of 10^3 FU (fluorescence units) in each channel corresponding to the emission spectra for SYTOX Blue, EGFP, dsRED, and Apc/Cy7.

3.8 Set Compensation Controls

1. Filter samples 2–5 through 35 µM cell strainer topped FACS tubes

2. Add dead cell exclusion dyes (SYTOX Blue and Annexin V-PacBlue) to sample 2 and incubate at 37 °C for 1–5 min.

3. Load Samples 2–5 and adjust compensation matrix to correct for fluorescence spillover.

3.9 Set Sox9EGFP Gates

1. Filter sample 6 through a 35 µM cell strainer topped FACS tube

2. Load sample 6 and set positive gates for live and dsRED+ CD326+ cells (Fig. 5d, e).

3. Set Sox9EGFP level thresholds. Villi contain only Sox9high and Sox9negative cells, so this villus-enriched preparation facilitates establishing the "high" gate (Fig. 5f, g).

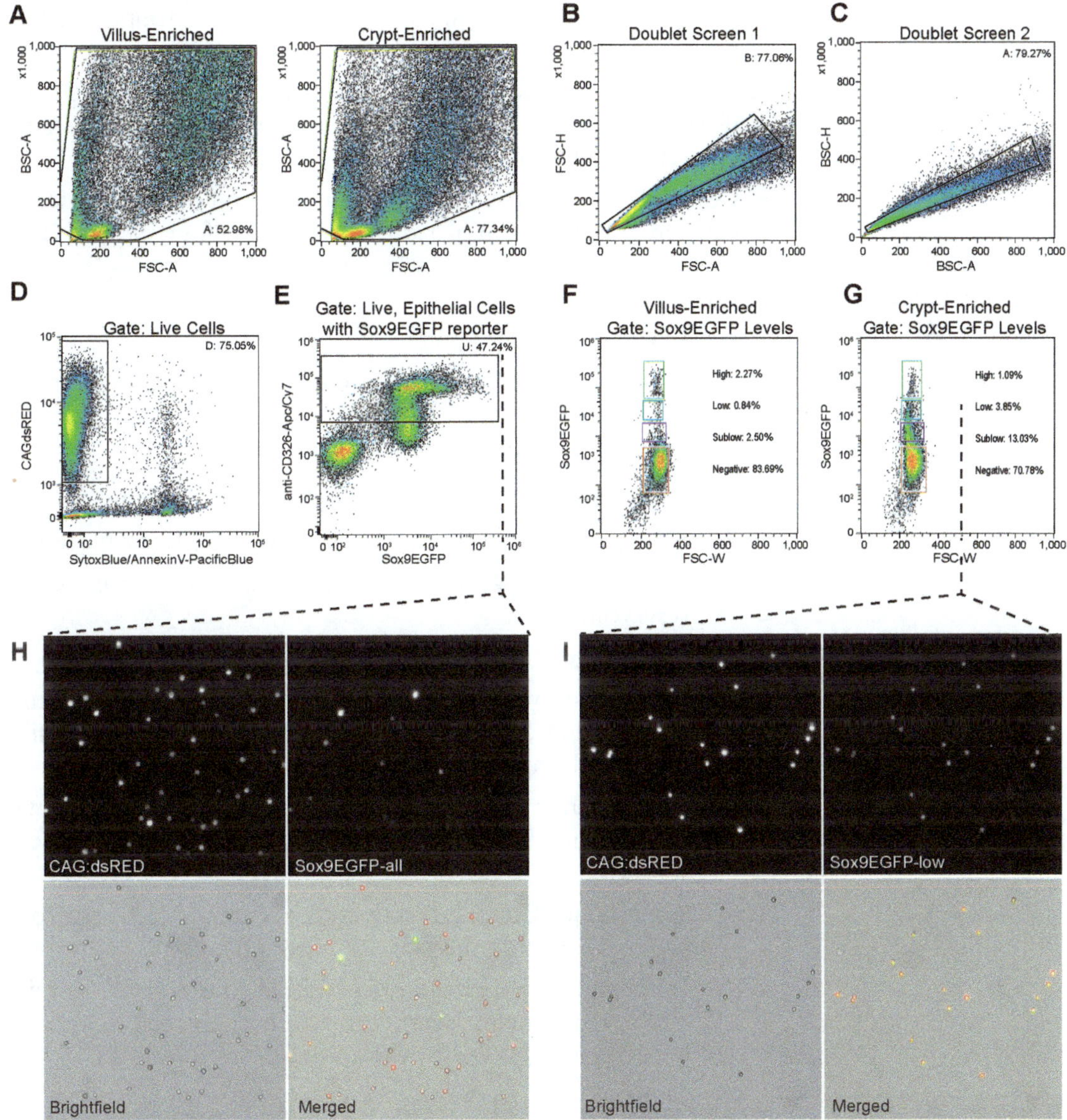

Fig. 5 Fluorescence-activated cell sorting. (**a–g**) Representative gating Schemes. (**a**) Forward scatter area under the curve (FSC-A) and back scatter area under the curve (BSC-A) profiles of villus-enriched and crypt-enriched preps. (**b**) Multimers are excluded by gating on events with a linear relationship between FSC-A and FSC-Height (FSC-H). (**c**) A second multimer exclusion is applied by gating on events with a linear relationship between BSC-A and BSC-Height (BSC-H). (**d**) Live cells are selected based negative SYTOX Blue/Annexin V-Pacific Blue and positive CAG:dsRED signals. (**e**) Epithelial cells are further separated from nonepithelial cells based on positive CD326-Apc/Cy7 (EPCAM) expression. (**f, g**) Live, epithelial cells expressing CAG-dsRED and variable levels of Sox9EGFP are plotted by FSC-Width (FSC-W) and EGFP intensities on linear and logarithmic scales, respectively. (**f**) The villus-enriched prep is used to draw Sox9EGFP negative/sublow/low/high gates—this prep is enriched for Sox9EGFP negative and high cells. (**g**) Using gates set in **f**, Sox9EGFP-low cells are sorted from the crypt-enriched prep. (**h, i**) Representative CAG:dsRED, Sox9EGFP, and bright field images from cells sorted from (**h**) gates set in **e** and (**i**) Sox9EGFP-low cells from gates set in **g**

3.10 Sort Sample for Plating on CRA

1. Filter sample 7 through a 35 μM cell strainer topped FACS tube

2. Sort 500 cells from the gate shown in Fig. 5d (single, live, dsRed+EPCAM+) and 500 cells from the Sox9EGFP-low gate onto slide and visually confirm gating (Fig. 5h, i).

3. Collect 2000 sorted cells into a 10 μL cushion of 1× ISC Media cushion in the bottom of 1.5 mL microcentrifuge tube on ice.

3.11 Time Lapse Image of CRA

1. Prepare raft.

   (a) Open sterile packaging.

   (b) Wash each well twice with 500 μL sterile DPBS to moisten (*see* **Note 8**).

2. Seed cells on raft (Fig. 6a).

   (a) Pellet cells by centrifugation at $2000 \times g$ for 5 min.

   (b) Resuspend cells in culture media at 6 cells/μL (*see* **Note 9**).

   (c) Gently pipet 200 μL cell suspension mixture into each reservoir.

   (d) Centrifuge at $50 \times g$ for 5 min at 4 °C.

   (e) Gently remove media with pipet tip and reserve in 1.5 mL tube. A small amount of media will remain in the microwells along with captured cells.

   (f) Observe raft with tissue culture microscope. If seeding density is inadequate, repeat steps X-X, above with reserved solution.

   (g) Add 100 μL Cultrex directly to the center of each reservoir without touching the rafts (*see* **Note 10**).

   (h) Immediately centrifuge $50 \times g$ for 5 min at 4 °C to mix Cultrex and media layers and recapture any cells that may have been dislodged by the addition.

   (i) Transfer CRA to the tissue culture incubator and allow to polymerize for 30 min.

   (j) Overlay 600 μL ISC culture media of choice (*see* Tables 1, 2, and 3) into each reservoir (*see* **Note 11**).

   (k) Add 1–3 mL DPBS to the space around the CRA to prevent condensation.

3. Scan CRA

   (a) Prewarm microscope incubator box to 37 °C with 5% CO_2.

   (b) Secure the CRA in an established orientation on the stage.

   (c) Set tile scan parameters (Fig. 6b).

   - Open the Multi-Dimensional Acquisition tab.

   - Click "Edit position."

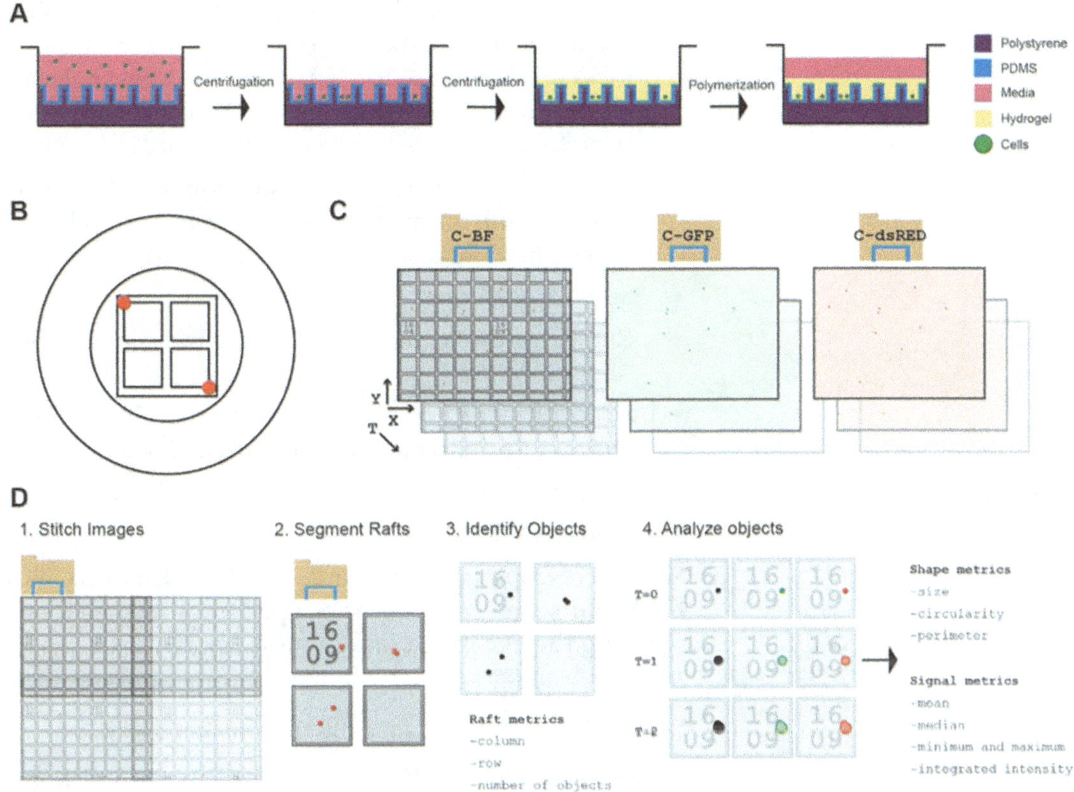

Fig. 6 Cell raft array (CRA) experimental design. (**a**) Cells are seeded into microwells by centrifugation, and then covered in a hydrogel mimicking the basement membrane. After polymerization, culture media is overlaid on top of the hydrogel. (**b**) Schematic illustrating selection points to designate XY scan parameters with points marked in red. (**c**) Schematic representation of datasets collected by time lapse, tile scanning microscopy. (**d**) Illustration of the data analysis pipeline performed off line. 1—Tiles are stitched together into a single stitched or "fused" image for each channel and time point. 2—A Matlab script crops individual rafts from the stitched image, saving each into a new file labeled with column and row annotations (download at http://www.magnesslab.org). 3—Raft files are passed through Cell Profiler pipeline (download at http://www.magnesslab.org) to identify objects based on dsRED signal and produce a binary mask (black). The pipeline referenced in this document is optimized to exclude very small and very large objects as they are unlikely to represent true cells. Pipelines need to be optimized for each experiment. 4—After restricting to objects within the defined masked area, location (microwell row and column), shape and signal metrics are extracted from the nonmasked channels and exported to a comma-separated variable format. An Excel sheet is available at http://www.magnesslab.org to facilitate delimitation of data by location rather than unique object location. Customized exclusion/inclusion criteria are applied at this stage, and statistical analysis conducted on comparisons and metrics of interest

- With the 4× objective (*see* **Note 12**), navigate to a location with bright signal in all channel. Adjust the exposure for each channel. Exposure times will vary for different microscope setups. Avoid exposures that will saturate pixels (*see* **Note 13**).

- Move the stage and focus on the upper left corner of the CRA. Select "Set to   Current."

- Move the stage to the bottom right corner. DO NOT change focus. If the bottom right corner is out of focus relative to the top left, manually adjust the stage insert fitting until top left and bottom right are in the same focal plane. Select "Set to Current."
- Return to upper left corner and confirm the z-focal plane is still acceptable. Repeat slide area selection steps if necessary.
- Select 5–20% overlap between tiles.
- Return to the main screen of the Multi-Dimensional Acquisition tab.
- Select wavelengths and exposure times 3 to acquire images of brightfield, GFP, and dsRED signals.
- Select time interval.
- Select save location.

(d) Seal microscope incubator and wait 10–30 min to allow CRA and scope component temperatures to equilibrate, reducing the likelihood of focal drift (*see* **Note 14**).

(e) Confirm top left and bottom right are still in focus

(f) Start automated time lapse tile scan imaging (Fig. 6c)

- Select a save location on a fast-writing hard disk with adequate memory.
- Collect brightfield *and* fluorescence images for each channel.
- Tile scan the CRA one channel at a time (*see* **Note 15**)

(g) End time lapse imaging.

- If imaging time exceeds media change intervals, carefully conduct media changes in the incubated microscope.
- If rafts will be retrieved for downstream analyses, return CRA to tissue culture incubator or leave in the incubator microscope.

4. Data processing

(a) Mosaic Stitching of a Composite Image displaying the entire CRA surface.

(b) In the Fiji image analysis platform, use the "Grid/collection Stitching" function found under the "Stitching" option in the "Plugins" menu to stitched together collected images into a single high-resolution image displaying the entire CRA surface for quality assurance purposes.

(c) Selecting the "Grid/collection Stitching" will open a new window prompting the user to describe the format in which the images are indexed. If images were collected as described above, select "Grid: column-by-column" for the type and "Down & Right" for the order.

(d) After confirming the image format, a new window will open prompting the user to enter the grid size in number of images, the overlap between adjacent image areas. (Set during scanning process scan), the directory locations of the images, and a text expression describing their naming (*see* **Note 16**).

(e) Images should be fused using a "Linear Blending" method to compute the image overlap at a subpixel accuracy to provide the most accurate composite image.

(f) Stitching can be computed under either memory conservative or expensive parameters depending on the computation power of the device used.

5. Microwell-based segmentation of Mosaic Images

(a) Label separate folders for each reservoir and take note of the full file path.

(b) In the MATLab environment, load the Segmentor.m script by navigating to it after selecting "open" in the "home" toolbar.

(c) When loaded, the Segmentor.m script will open in an editor window displaying the editable code to segment and name the microwell images. Avoid editing this script to ensure function.

(d) "Run" the script using the button located on the tool bar of the editor window (*see* **Note 17**)

(e) When prompted, input the full path to the folder containing all stitched images for segmentation in the command window.

(f) Input the filename of up to two images you wish to segment sequentially, pressing enter between each name. If only one mosaic image is to be segmented, enter the same name for both prompts.

6. Identify reference coordinates.

(a) Segmentor.m prompts users to open a mosaic image in Fiji to determine reference coordinates. Use the bright field or phase contrast image for this task because microwell borders are easiest to distinguish in these channels. Take note of reference coordinates as they may be used to segment from fluorescence mosaic images without the need to run parallel segmentation with the brightfield.

(b) Input coordinates for the first Reservoir.

- Determine the x-axis and y-axis coordinates for each corner well using the "Point" tool in Fiji. Coordinates will be requested in sequence by prompts for the X and Y coordinates at a center position of the top left, top right, bottom left, and bottom right microwell of the Reservoir.

- Input the number of microwells spanning the height and width of the mosaic image, when prompted for "number of rows" and "number of columns," respectively.

- Use the line tool in Fiji to measure the width of a single microwell in pixels. When prompted, input this pixel value.

- Confirm all input values by entering "y" or enter "n" to reenter dimensions.

(c) fter dimensions are confirmed, Segmentor.m prompts the user to close other memory intensity programs to reduce processing time. Input "y" to begin segmentation.

- Stitched images are loaded and segmented into multiple images over several minutes.

- Microwell images are saved in the current folder with row and column indexing in the file name.

(d) Repeat for additional channels and Reservoirs, taking care to save to unique file directories.

(e) Epithelial cell identification and quantification

- Load "SCPipeline.cp."

  - The CellProfiler Automated Analysis suite first prompts with a window displaying the current analysis settings.

  - Load "SCPipeline.cp" by "Load Pipeline..." under the "File" menu and navigating to SCPipeline.cp.

  - Successful loading will automatically add the 6 modules utilized by the pipeline into the analysis window.

- Input paths to the desired input and output folders in the CellProfiler window

  - Input the path to the folder containing the microwell images to be analyzed in the "Default Input Folder:" menu

  - Input the path to the folder where you wish to store analysis data in the "Default Output Folder:" menu

  - Add a title in "Output Filename:" menu to name the output file generated by CellProfiler

- Load images

  - Select the "LoadImages" module displaying the module settings in the CellProfiler window.

- Edit the "Text that these images have in common (case-sensitive)" menu so the text-regular expression matches the base name of the image series.

- Edit the "Regular expression that finds metadata in the file name" menu by replacing the text "BaseName" with the base name of the image series.

- Analyze images

  - Initiate image analysis by selecting "Analyze Images" in the bottom right-hand corner of the CellProfiler window (*see* **Note 18**).

- Analyze number of cells per well.

  - Data generated by SCPipeline.cp is formatted as a .csv file and is object delineated such that microwell location, shape, and fluorescence metrics of each identified object are recorded on each line. Data is retabulated to determine microwell contents using "WellContents.xls" (*see* **Note 19**).

  - In Excel, open the .csv file generated by CellProfiler

  - Copy all data points generated by SCPipeline (column A to column AA) into the WellContents.xls file in A1 of the ".csv" sheet (*see* **Note 20**).

  - Embedded formulas in the WellContents.xls file automatically tabulate the number of cells and debris objects in each microwell and displayed those values on the sheet titled "WellDelineated".

  - The addresses of microwells from each cell count are displayed in the "target wells" sheet grouped by intact cell contents with microwells containing 1 cell in the first column, microwells containing 2 cells in the second column, microwells containing 3 cells in the third column, and microwells containing 4 cell in the fourth column, etc.

  - The distribution of cells within the CRA is displayed in the sheet titled "Distribution".

  - Using this information, identify microwells and objects of interest. These will vary based on experimental design.

- Analyze Data

  - Return to the .csv file generated by the CellProfiler pipeline.

  - Restricting data analysis to cells in the identified microwells of interest, compare changes in shape and fluorescence metrics between reservoirs over time (*see* **Note 21**).

4   Notes

1. CD24 may be used in lieu of CD326. CD24 is expressed only in crypt based cells [8, 13]. CD326 is used in this protocol to ensure that villus-based Sox9EGFP high cells are not excluded when establishing Sox9EGFP levels gates.

2. It is very important that the microscope be equipped with an incubator to avoid having to move the CRA in and out of the microscope.

3. While pulling, use small scissors or a second pair of forceps to release pancreatic adhesions and mesenteric plexus. The duodenum is particularly prone to breaking near the pancreatic and bile ducts.

4. Adipose tissue from the myenteric plexus will inhibit single cell dissociation. Acinar cells from the pancreas can lyse and damage IECs.

5. If crypt yield is sub-optimal and many crypts are visible in the remaining tissue, repeat shaking step using reserved tissue. To avoid damaging crypts isolated from the first round of shaking, transfer tissue to a new tube with Crypt Dissociation Solution before reshaking. Pool crypts before proceeding to next step.

6. There may be a few large clumps remaining. These will be excluded by filtration. Dispase efficacy varies by lot and must be titrated for each lot. If the amount of tissue is overwhelming, it may be necessary to increase dispase concentration.

7. Actual flow cytometry profiles may vary between machines.

8. Add to the side of the wall. Avoid contact with the rafts and scaffolds.

9. This is sufficient for 3:2 cells–microarray for the four-resevoir array; each well of the four-reservoir array has approximately 2100 rafts.

10. Follow manufacturers' instructions for handling Cultrex and other hydrogels. Cultrex is liquid at 4 °C but solidifies at room temperature and 37 °C. Work quickly and use cold tips to avoid premature polymerization.

11. Resolution can be increased by scanning with a 10× objective instead of a 4× objective. Objective selection is a tradeoff between data collection speed, focal plane, and resolution.

12. We recommend conducting pilot experiments where ISCs are cultured in Matrigel on a multiwell plate to identify an appropriate exposure level that will produce adequate signal over the course of the experiment without overt signs of photobleaching or phototoxicity.

13. Temperature equilibration is important for minimizing focal drift caused by shrinking or swelling of physical components during temperature changes.

14. After completion of the first time point it is highly recommended to navigate to the save location and confirm files are being populated with images.

15. Avoid rotating through filters that expose cells to UV light which will degrade Matrigel and Cultrex hydrogels and damage cells.

16. Stitching is typically most successful when the overlap is set equal to the overlap used when collecting the images.

17. Users will be prompted to "Add to Path" if the Segmentor.m script is not located within the working folder.

18. Users can observe specific steps along the analysis process by opening the eyeball graphic next to corresponding module. Observation significantly increases analysis time.

19. Data formatting relies on formulas embedded in the WellContents spreadsheet. We recommend using "WellContents.xls" as a read-only file to ensure proper formatting.

20. Copying additional data or empty cells into WellContents.xls may change embedded formulas. Always transfer Columns A-AA.

21. Always consult a biostatistician to ensure appropriate statistical tests are used for the experimental design of interest. We recommend treating each array as a single biological replicate.

Acknowledgments

We would like to thank Eric Bankaitis, Ph.D. and Bailey Zwarcyz, MS for critical reading of the manuscript. Additionally, we would like to acknowledge Adam Gracz, Ph.D. for his work in developing the crypt and villus-enrichment methodology used here. We acknowledge all authors who contributed to the publication of this protocol which is based on "A high-throughput platform for stem cell-niche cocultures and downstream expression analysis" published in Nature Cell Biology 2016, 17(3): 340-349. The original work described in this chapter was funded by the National Institutes of Health, R01 DK091427 (Magness), R03 EB013803 (Y. Wang/ Magness), R01 EB012549 (Allbritton), Small Business Innovation Research R43 GM106421 (Y. Wang/Magness), U01 DK085507-01 (Li), University Cancer Research Fund of the University of North Carolina (Magness/Allbritton), and the Center for Gastrointestinal Biology and Disease P30 DK034987 (Magness, Y. Wang, Galanko).

References

1. Cheng H, Leblond CP (1974) Origin, differentiation and renewal of the four main epithelial cell types in the mouse small intestine. V. Unitarian theory of the origin of the four epithelial cell types. Am J Anat 141(4):537–561

2. Gracz AD, Magness ST (2014) Defining hierarchies of stemness in the intestine: evidence from biomarkers and regulatory pathways. Am J Physiol Gastrointest Liver Physiol 307(3): G260–G273

3. Henning SJ, von Furstenberg RJ (2016) GI stem cells - new insights into roles in physiology and pathophysiology. J Physiol 594(17):4769–4779

4. Wright, N.A., The biology of epithelial cell populations. M. Alison. 1984, Oxford: Oxford Science Publications

5. Sato T et al (2009) Single Lgr5 stem cells build crypt-villus structures in vitro without a mesenchymal niche. Nature 459(7244):262–265

6. Barker N et al (2007) Identification of stem cells in small intestine and colon by marker gene Lgr5. Nature 449(7165):1003–1007

7. Muñoz J et al (2012) The Lgr5 intestinal stem cell signature: robust expression of proposed quiescent '+4' cell markers. EMBO J 31(14): 3079–3091

8. Gracz AD, Ramalingam S, Magness ST (2010) Sox9 expression marks a subset of CD24-expressing small intestine epithelial stem cells that form organoids in vitro. Am J Physiol Gastrointest Liver Physiol 298(5):G590–G600

9. Formeister EJ et al (2009) Distinct SOX9 levels differentially mark stem/progenitor populations and enteroendocrine cells of the small intestine epithelium. Am J Physiol Gastrointest Liver Physiol 296(5):G1108–G1118

10. Roche KC et al (2015) SOX9 maintains reserve stem cells and preserves radioresistance in mouse small intestine. Gastroenterology 149(6):1553–1563.e10

11. van Es JH et al (2012) Dll1+ secretory progenitor cells revert to stem cells upon crypt damage. Nat Cell Biol 14(10):1099–1104

12. Gracz AD, Puthoff BJ, Magness ST (2012) Identification, isolation, and culture of intestinal epithelial stem cells from murine intestine. Methods Mol Biol 879:89–107

13. von Furstenberg RJ et al (2011) Sorting mouse jejunal epithelial cells with CD24 yields a population with characteristics of intestinal stem cells. Am J Physiol Gastrointest Liver Physiol 300(3):G409–G417

14. Gonzalez LM, Williamson I, Piedrahita JA, Blikslager AT, Magness ST (2013) Cell lineage identification and stem cell culture in a porcine model for the study of intestinal epithelial regeneration. PLoS One 8(6):e66465

15. Gracz AD et al (2013) Brief report: CD24 and CD44 mark human intestinal epithelial cell populations with characteristics of active and facultative stem cells. Stem Cells 31(9):2024–2030

16. Ramalingam S, Daughtridge GW, Johnston MJ, Gracz AD, Magness ST (2012) Distinct levels of Sox9 expression mark colon epithelial stem cells that form colonoids in culture. Am J Physiol Gastrointest Liver Physiol 302(1):G10–G20

17. Stelzner M et al (2012) A nomenclature for intestinal in vitro cultures. Am J Physiol Gastrointest Liver Physiol 302(12):G1359–G1363

18. Sato T et al (2011) Paneth cells constitute the niche for Lgr5 stem cells in intestinal crypts. Nature 469(7330):415–418

19. Yin X et al (2014) Niche-independent high-purity cultures of Lgr5+ intestinal stem cells and their progeny. Nat Methods 11(1):106–112

20. Gjorevski N et al (2016) Designer matrices for intestinal stem cell and organoid culture. Nature 539(7630):560–564

21. Sachs N, Tsukamoto Y, Kujala P, Peters PJ, Clevers H (2017) Intestinal epithelial organoids fuse to form self-organizing tubes in floating collagen gels. Development 144(6): 1107–1112

22. Gracz AD et al (2015) A high-throughput platform for stem cell niche co-cultures and downstream gene expression analysis. Nat Cell Biol 17(3):340–349

23. Attayek PJ et al (2015) Array-based platform to select, release, and capture Epstein-Barr virus-infected cells based on intercellular adhesion. Anal Chem 87(24):12281–12289

24. Wang Y et al (2010) Micromolded arrays for separation of adherent cells. Lab Chip 10(21): 2917–2924

25. Vintersten K et al (2004) Mouse in red: red fluorescent protein expression in mouse ES cells, embryos, and adult animals. Genesis 40(4):241–246

26. Gong S et al (2003) A gene expression atlas of the central nervous system based on bacterial artificial chromosomes. Nature 425(6961):917–925

27. Kamentsky L et al (2011) Improved structure, function and compatibility for CellProfiler: modular high-throughput image analysis software. Bioinformatics 27(8):1179–1180

28. Carpenter AE et al (2006) CellProfiler: image analysis software for identifying and quantifying cell phenotypes. Genome Biol 7(10):R100

29. Sato T, Clevers H (2013) Primary mouse small intestinal epithelial cell cultures. In: Randell SH, Fulcher ML (eds) Epithelial cell culture protocols, 2nd edn. Humana Press, Totowa, NJ, pp 319–328

Chapter 11

Detection, Labeling, and Culture of Lung Stem and Progenitor Cells

Ivan Bertoncello, Gianni Carraro, and Jonathan L. McQualter

Abstract

Identification, isolation, and clonal culture of stem cells is essential for understanding their proliferative and differentiation potential, and the cellular and molecular mechanisms that regulate their fate. Akin to development in vivo, the in vitro growth of adult lung epithelial stem cells requires support of mesenchymal-derived growth factors. In the adult mouse lung, epithelial stem/progenitor cells are defined by the phenotype CD45neg CD31neg EpCAMpos CD104pos CD24low, and mesenchymal cells are defined by the phenotype CD45neg CD31neg EpCAMneg Sca-1hi. Here we describe a method for primary cell isolation from the adult mouse lung, a flow cytometry strategy for fractionation of epithelial stem/progenitor cells and mesenchymal cells, and a three dimensional epithelial colony forming assay.

Key words Lung epithelium, Stem cells, Colony-forming assay, Flow cytometry

---

1   Introduction

The respiratory tree of the adult mouse lung comprises approximately 13 generations of branching and tapering conducting airways that transmit air to the gas-exchanging epithelium of the alveolar bed. Histomorphometric analysis has revealed significant regional differences in the cellular composition and functional specialization of respiratory cells along the proximal–distal lung axis (Fig. 1) [1]. The trachea and mainstem bronchiolar airways of mice are lined by pseudostratified epithelium comprising basal cells, secretory cells, and ciliated cells, while submucosal glands that produce the majority of serous and mucus secretions in the airway are restricted to the subepithelial layer of the upper trachea. This pseudostratified epithelium transitions into simple cuboidal epithelium in the proximal bronchiolar airways that are devoid of basal cells and predominantly comprises secretory club cells and ciliated cells. The proximal bronchiolar airways lead in turn to the distal bronchiolar airways and terminal bronchioles that give rise to alveolar ducts and alveoli lined with cuboidal surfactant secreting

Shree Ram Singh and Pranela Rameshwar (eds.), *Somatic Stem Cells: Methods and Protocols*, Methods in Molecular Biology, vol. 1842, https://doi.org/10.1007/978-1-4939-8697-2_11, © Springer Science+Business Media, LLC, part of Springer Nature 2018

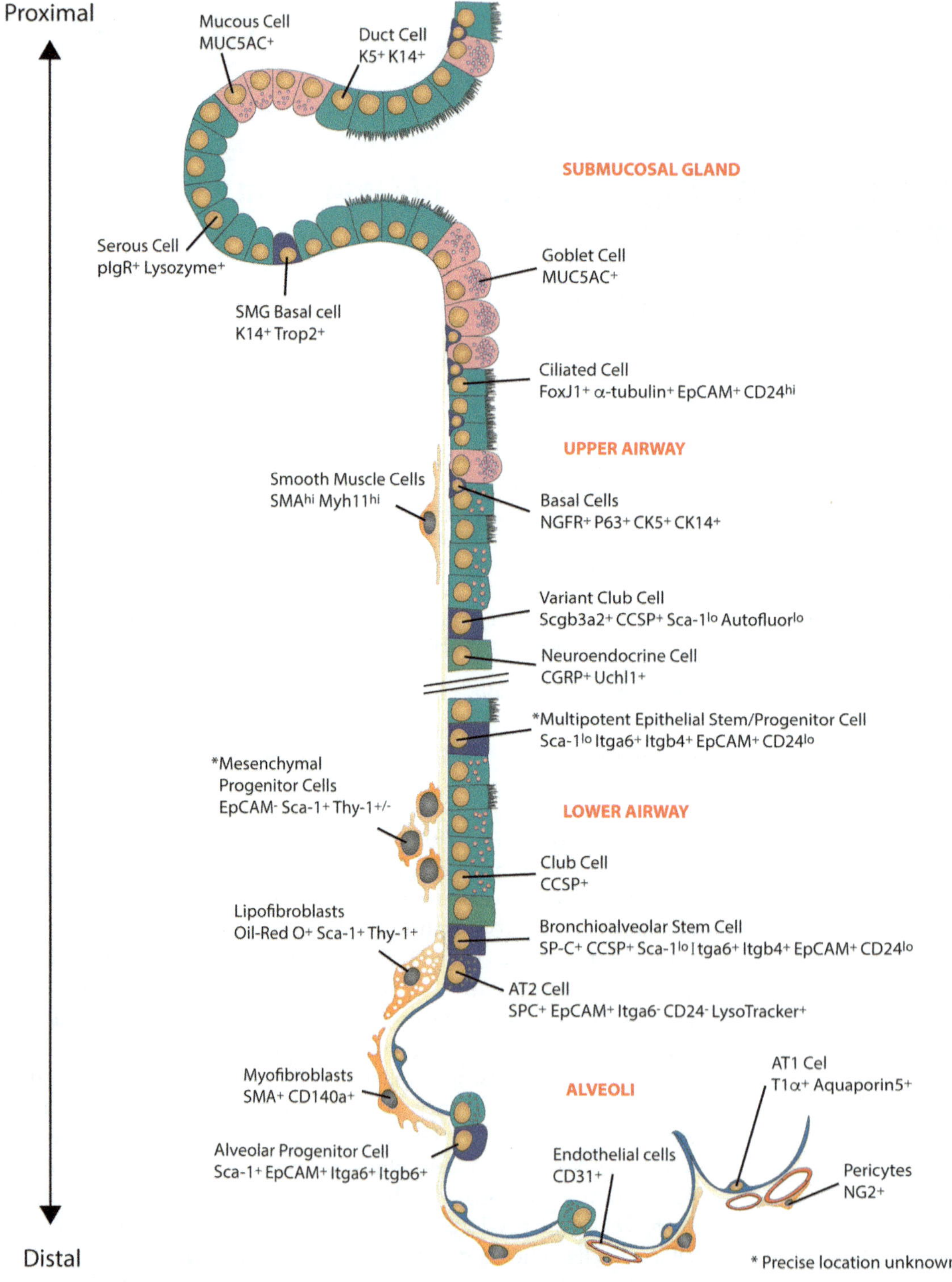

Fig. 1 Cellular composition of the adult mouse lung. Schematic representation of the cellular heterogeneity along the proximal distal axis of the adult mouse lung (not to scale)

alveolar type-2 (AT2) cells and squamous gas-exchanging alveolar type-1 (AT1) cells. Cell lineage tracing and the analysis of lung development in normal and genetically engineered mouse models; and the analysis of lung regeneration and repair following selective lung injury have shown that the epithelial cell lineages lining the adult lung are maintained by diverse pools of regional epithelial stem and progenitor cells distributed along the proximal distal lung axis and the alveolar bed [2, 3].

When grown in hydrogels, such as Matrigel, in coculture with lung mesenchymal cells, lung epithelial stem and progenitor cells proliferate to form three-dimensional multicellular spheroidal and branching organoids that recapitulate the apical-basal organization, differentiation and functionality of epithelia in the respiratory compartments from which they are derived. Analysis of the clonal dynamics of organoid formation in these culture systems has been used to more precisely classify and hierarchically order lung epithelial stem and progenitor cells on the basis of their ability to self-renew and give rise to descendent lineages [4, 5]. Cell lineage tracing, transcriptional profiling, and gene expression analysis within developing organoids has been used to analyze the role of specific gene networks and signaling pathways in stem and progenitor cell activation and recruitment. And coculture of FACS sorted lung epithelial stem cells with cytokines [4], and/or mesenchymal cells [6–8] and endothelial cells [9] has proven to be a powerful tool in determining how the regenerative capacity of lung epithelial stem and progenitor cells is modified by interaction with their niche microenvironment. Of note, clonal analyses of lung epithelial stem cells after influenza virus infection revealed that their proliferative response is significantly reduced by influenza virus-mediated impairment of β-catenin-dependent Fgf10-Fgfr2b signaling [8]. This demonstrates that the extent of lung stem cell infection by influenza virus critically impacts on lung regeneration capacity after severe influenza injury.

In the upper airways, in vitro colony-forming assays and in vivo lineage tracing studies have shown that basal cells (NGFR[pos], CD104[pos], Krt14[pos], and/or Krt5[pos]) act as lineage-restricted airway progenitor cells, giving rise to club, ciliated, and goblet cells in the trachea and proximal airways [5, 10–12]. Basal cells can be isolated from the mouse trachea and human airways based on the expression of NGFR and CD49f (integrin α6), and can be grown and propagated in vitro in Matrigel as spheroids [5, 13].

This protocol has been developed for the isolation and characterization of adult mouse lung bronchiolar and bronchoalveolar epithelial stem/progenitor cells (EpiSPC). In the bronchiolar airways of the adult mouse lung, a subset of EpiSPC (CCSP[pos] Scgb3a2[pos] CyP450[neg]) cells that colocalize with neuroepithelial bodies in the distal airways have been shown to give rise to mature club cells and ciliated cells in vivo [11, 14–17]. Similarly, at the bronchoalveolar duct junction, a subset of EpiSPC termed bronchoalveolar stem cells (CCSP[pos] SP-C[pos]) has been shown to proliferate in response to both bronchiolar and alveolar injury [18]. In the alveoli, it is accepted that type I alveolar cells are descendant from type II alveolar cells. A subset of Type II alveolar cells has also been shown to have the capacity for renewal [19–21].

We and others have demonstrated that EpiSPC can be isolated from the adult mouse distal lung by flow cytometry on the basis of EpCAM[hi], Sca-1[low], CD104[pos], CD49f[pos] and CD24[low] expression [4, 8, 18, 22–24]. By culturing these cells in a three-dimensional colony-forming assay, we have shown that the adult mouse lung contains a renewing multipotent epithelial stem cell population with the capacity to differentiate into epithelial cells of the airway (club, ciliated, and goblet cells) and alveolar (type I and type II) cell lineages. Other progenitor cells, also enriched in this cell fraction, exhibit more restricted cell fates, giving rise to only alveolar or airway cell progeny [4]. Such alveolar progenitors likely represent a subset of AT2 cells [20] while airway progenitors are likely basal cells which have recently been identified within the EpiSPC fraction [8].

Comparable dissociation protocols are also being developed to isolate primary cells from adult human lung tissue samples for the identification of analogous adult human lung stem cells (*see* **Note 1**). However, compared to the mouse lung, the adult human lung has significant disparities in organ size, extent of branching and anatomical distribution of cells. For example, mucin-secreting goblet cells are abundant in human airways, but are rare in mice. Pseudostratified basal cell containing epithelium extends to the terminal bronchioles in the human lung, but is restricted to the trachea in mice. And simple cuboidal epithelium (devoid of basal cells) lines the murine bronchiolar airways, but is only found in the respiratory bronchioles of the human lung. To date, studies suggest that basal cells, which exhibit clonal growth and multilineage differentiation in vivo and in vitro organoid assays (*see* **Note 2**), are the dominant stem cell type responsible for the maintenance and repair of the human airways. However, ongoing research is beginning to appreciate that the stem and progenitor cell population in adult human lung is far more heterogeneous than previously thought.

The dissociation and cell culture protocol described here has been developed for the isolation and characterization of adult mouse lung bronchiolar and bronchoalveolar stem/progenitor cells. Adaptations of this protocol can also be used for dissociation of primary adult human cells for cell fractionation and clonal analyses of potential human lung stem cell populations.

2   Materials

Prepare all solutions using ultrapure water (prepared by purifying deionized water to attain a sensitivity of 18 MΩ cm at 25 °C). Dispose of all waste materials as per environment, health, and safety regulations.

2.1   General Components

1. Hank's balanced salt solution (HBSS): 140 mg/L $CaCl_2$, 100 mg/L $MgCl_2$-$6H_2O$, 100 mg/L $MgSO_4$-$7H_2O$, 400 mg/L KCl, 60 mg/L KH_2PO_4, 350 mg/L $NaHCO_3$, 8000 mg/L NaCl, 48 mg/L Na_2HPO_4, 1000 mg/L D-glucose, pH 7.4. Store at 4 °C.

2. Centrifuge: 400 × g, 4 °C.

3. 50 mL sterile polypropylene tubes.

4. FACS buffer: HBSS, 0.2% bovine serum albumin BSA (*see* **Note 3**).

2.2   Tissue Dissociation Components

1. Dissecting equipment: forceps, scissors, and single-sided razor blade.

2. 30 mm petri dish.

3. 18 gauge and 21 gauge needles.

4. 20 mL syringes.

5. Liberase solution: For stock solution, reconstitute 50 mg Liberase TM Research Grade (Roche) in 10.4 mL sterile HBSS to make a stock solution at 25 Wunsch U/mL. Aliquot and store at −20 °C (*see* **Note 4**).

6. Red cell lysis buffer: 1000 mg/L $KHCO_3$, 8024 mg/L NH_4Cl, 37 mg/L EDTA, pH 7.4. Store at 4 °C.

7. Liberase wash buffer: HBSS, 5% fetal bovine serum (FBS). Store at 4 °C.

8. 40 μm cell strainer.

9. Eppendorf ThermoMixer: 50 mL tube block (*see* **Note 5**).

2.3   Cell Depletion Components

1. EasySep™ Mouse FITC Positive Selection Kit (Stem Cell Technologies).

2. EasySep™ Magnet (Stem Cell Technologies): Compatible with 5 mL tubes.

3. 5 mL (12 × 75 mm) polystyrene FACS tubes.

4. 50 mL tubes.

5. Tube rotator.

6. Centrifuge: 400 rcf, 4 °C.

7. Depletion antibody cocktail: FITC anti-mouse antibodies directed against CD31 (clone 390 or MEC13.3), CD45 (clone 30-F11), and TER119 (clone TER-119) antigens.

8. Depletion buffer: PBS (without Mg^{2+} and Ca^{2+}), 2% FBS.

2.4 Flow Cytometry Components

1. Positive selection antibody cocktail: Alexa Fluor® 647 anti-mouse CD104 (clone 346-11A), PE anti-mouse EpCAM (clone G8.8), Brilliant Violet 421 anti-mouse CD24 (clone M1/69) (*see* **Note 6**).

2. Viability dye: SYTOX™ Green Nucleic Acid Stain—5 mM Solution in DMSO (ThermoFisher Scientific) (*see* **Note 7**).

3. 5 mL (12 × 75 mm) FACS tubes.

4. 5 mL (12 × 75 mm) FACS tubes with 35 μm cell strainer cap.

5. Cell Sorter: 100 μm nozzle, 20 psi.

6. Collection Buffer: DMEM/F12 media, 10% FBS. Store at 4 °C.

2.5 Epithelial Colony-Forming Assay Components

1. Growth factor reduced Matrigel.

2. Millicell 24 well cell culture inserts (Merck Millipore): 0.4 μm pore size, hydrophilic PTFE.

3. 24 well flat bottom tissue culture plate.

4. CFU-Epi medium: DMEM/F12, penicillin, streptomycin, glutamax, insulin, transferrin, selenium, 10% FBS, 2 μg/mL heparin sodium salt.

5. Trigas incubator set at: 5% v/v O_2, 10% v/v CO_2, 85% v/v N_2 (*see* **Note 8**).

3 Methods

Carry out all procedures on ice (or at 4 °C) and in a sterile biological safety cabinet, unless otherwise specified.

3.1 Dissociation of Mouse Lung Tissue (See Note 1 for Adaptation to Human Lung Tissue)

1. Dilute Liberase stock solution 1/100 in sterile HBSS (4 mL for each mouse lung; *see* **Note 9**) in a 50-mL tube and preheat to 37 °C.

2. Exsanguinate deceased mouse (*see* **Note 10**) by severing the major arteries behind intestines. Open the thoracic cavity and excise the lungs. Remove the extralobular airways and place the lung lobes in a 50-mL tube containing 30 mL of HBSS. Shake to wash out excess blood and transfer the lungs into a fresh 50-mL tube containing 30 mL of HBSS.

3. Transfer the lungs into a sterile petri dish, and finely mince the lungs using scissors or a single sided razor blade. Transfer the minced tissue into a 50-mL tube and add 4 mL of the preheated Liberase solution per lung. Place the tube in the Thermomixer and agitate (750 rpm) at 37 °C for 30 min.

4. Triturate the sample with an 18-gauge needle attached to a 20-mL syringe until the tissue passes freely through the needle.

Place in the Thermomixer and agitate (750 rpm) at 37 °C for a further 15 min (*see* **Note 11**).

5. Triturate the sample with a 21-gauge needle attached to a 20-mL syringe until the tissue passes freely through the needle. Strain the tissue digest through a 40-μm cell strainer into a 50-mL tube to remove tissue debris and cell clumps. Top up tube to 50 mL with wash buffer and centrifuge at 400 × *g*, 4 °C for 5 min. Remove and discard supernatant.

6. Resuspend the cell pellet in Red Cell Lysis Buffer (4 mL per lung) and incubate at RT for 90 s. Top up to 50 mL with wash buffer and centrifuge at 400 × *g*, 4 °C for 5 min. Remove supernatant and resuspend the cell pellet in FACS buffer for cell counting.

7. Count the cells and calculate cell concentration.

8. Aliquot approximately 100,000 cells for each compensation and unstained control tube for FACS setup (*see* **Note 12**).

3.2 Depletion of Hematopoietic and Endothelial Cells (See Note 13)

1. Resuspend the cell pellet in 50 mL FACS buffer and centrifuge at 400 × *g*, 4 °C for 5 min and discard the supernatant. Resuspend the cell pellet in a 5-mL polystyrene tube at 1 × 10^8 cells/mL in depletion buffer containing the depletion antibody cocktail: FITC anti-mouse antibodies directed against CD31 (clone 390 or MEC13.3), CD45 (clone 30-F11), and TER119 (clone TER-119) antigens. Incubate on ice for 20 min.

2. Top up with depletion buffer and centrifuge at 400 × *g*, 4 °C. Carefully aspirate and discard supernatant. Resuspend in the same volume as **step 1**.

3. Add selection cocktail (100 μL/mL of sample) and incubate at RT for 15 min.

4. Premix magnetic particles (pipette up and down more than five times), add to sample (50 μL/mL of sample) and incubate at RT for 10 min.

5. Top up to 2.5 mL to with depletion buffer, mix gently (pipette up and down 2–3 times), place the tube onto the magnet and incubate at RT for 5 min.

6. Pick up the magnet, and in one continuous motion invert the magnet and tube, pouring off the supernatant into a 50-mL collection tube (the supernatant contains the enriched cells devoid of hematopoietic and endothelial cells).

7. Optional wash: Repeat **steps 5** and **6** to ensure all unbound cells collected.

8. Top up collection tube to 50 mL with FACS buffer and centrifuge at 400 × *g*, 4 °C for 5 min. Resuspend cell pellet in FACS buffer for cell counting.

9. Count the cells and calculate cell concentration.

10. Aliquot some cells into 5-mL tubes for fluorescence minus one (FMO) controls (50,000–200,000 cells is ideal). Store on ice.

3.3 Flow Cytometry for Isolation of Adult Mouse Lung Epithelial Stem/Progenitor Cells and Resident Mesenchymal Cells (See Note 2 for Adaptation to Human Lung Tissue)

1. Resuspend cells in a 50-mL tube at 2.5×10^7 cells/mL in FACS buffer containing the positive selection antibody cocktail: Alexa Fluor® 647 anti-mouse CD104 (clone 346-11A), PE anti-mouse EpCAM (clone G8.8), Brilliant Violet 421 anti-mouse CD24 (clone M1/69), and PECy7 anti-mouse Sca-1 (clone D7). Incubate on ice in the dark for 20 min (*see* **Note 14**).

2. Top up tube to 50 mL with FACS buffer and centrifuge at $400 \times g$, 4 °C for 5 min. Repeat this wash step two more times.

3. Resuspend cells at 1×10^7 cells/mL in FACS buffer containing SYTOX™ Green Nucleic Acid Stain (diluted 1/5000). Transfer the cells to a 5-mL FACS tubes with 35 μm cell strainer cap.

4. Set up individual FMO control tubes for each antibody by staining cells with all antibodies replacing each antibody in turn in each control tube with the relevant fluorochrome-matched isotype control antibody (*see* **Note 15**).

5. Set up the flow cytometer with 100 μm nozzle at 20–30 psi. Set voltages and compensation using unstained and single colour controls.

6. Isolate epithelial stem/progenitor cells (EpiSPC) by setting up sequential gates for selection of single (SSC-W vs. SSC-H, FSC-W vs. FSC-H), viable, nonhematopoietic, nonendothelial (FITC vs. SSC-H), EpCAM[pos] CD104[pos] (PE vs. Alexa Fluor® 647), CD24[low] (PE vs. Brilliant Violet 421) cells as shown in Fig. 2. Isolate resident mesenchymal cells (rMC) by setting up sequential gates for selection of single (SSC-W vs. SSC-H, FSC-W vs. FSC-H), viable, nonhematopoietic, nonendothelial (FITC vs. SSC-H), EpCAM[neg] Sca-1[pos] (PE vs. PECy7) cells as shown in Fig. 2.

7. Collect EpiSPC and rMC in separate 5-mL collection tubes containing 1 mL collection buffer.

3.4 Epithelial Colony-Forming Assay (See Note 2 for Adaptation to Human Lung Tissue)

1. Centrifuge sorted cells at $400 \times g$, 4 °C for 5 min. Resuspend cell pellets in 1 mL chilled CFU-Epi medium and store on ice.

2. Take a small aliquot for cell counting and calculate cell concentration.

3. Mix sorted EpiSPC with sorted rMC cells so that the final cell concentrations are at 2×10^4 cells/mL and 2×10^6 cells/mL, respectively (*see* **Note 16**). Centrifuge at $400 \times g$, 4 °C for 5 min.

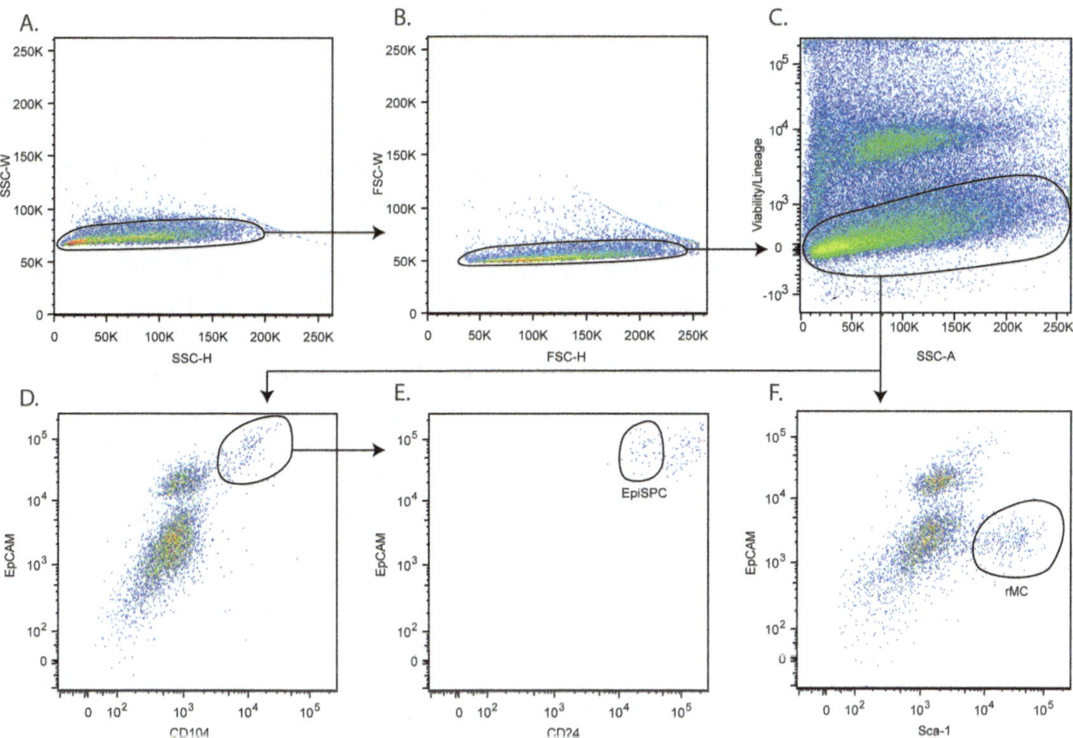

Fig. 2 Flow cytometry gating strategy for adult mouse lung EpiSPC. Set up sequential gates (indicated by black arrows) for (**a**) SSC-W doublet exclusion, (**b**) FSC-W doublet exclusion, (**c**) viability, CD45[neg] CD31[neg] TER119[neg], (**d**) EpCAM[pos] CD104[pos], (**e**) CD24[low] EpiSPC, and (**f**) EpCAM[neg] Sca-1[pos] rMC

4. Resuspend the cell pellet in chilled Matrigel diluted at 1:1 ratio with CFU-Epi medium so that the final concentration of the EpiSPC suspension is 2×10^4 cells/mL, and rMC is 2×10^6 cells/mL (*see* **Note 17**). Mix the Matrigel cell suspension (*see* **Note 18**).

5. Place Millicell inserts in a 24 well culture plate.

6. Add 100 µL drops of Matrigel cell suspension atop of the filter membrane of a Millicell insert. Incubate cultures at 37 °C for 5 min to allow Matrigel to set.

7. Add 400 µL of CFU-Epi medium around the insert in each well (*see* **Note 19**).

8. Incubate cultures at 37 °C, 5% O_2, 10% CO_2, 85% N_2 and change media three times weekly.

9. Score colonies using a stereomicroscope under phase contrast as shown in Fig. 3 (*see* **Note 20**).

10. Analyze cell fate by gene expression (RNA isolated from live cultures) or immunohistochemistry (cultures fixed with 4% paraformaldehyde).

4   Notes

1. Human lung tissue samples can be dissociated using the same mechanical and enzymatic digestion protocol described for mouse lung cells except for volume of digestion buffer and incubation times. Samples should be finely mince the lungs using scissors and transferred into a 50-mL tube containing 10 mL of the preheated Liberase TM solution (0.25 Wunsch U/mL) per cm^3 of minced lung tissue. Place the tube in the Thermomixer and agitate (750 rpm) at 37 °C for 60 min before proceeding with trituration and red blood cell lysis.

2. There is still a lack of consensus about the protocols for isolation, identification, and culture of human epithelial cells. However, it is well accepted that basal cells represent a major epithelial stem cell population in the adult human lung. To isolate primary human airway epithelial basal cells, dissociate fresh lung tissue using an adaptation of the protocol described here for mouse EpiSPC (*see* **Note 1**). Deplete CD45[+] hematopoietic and CD31[+] endothelial cell populations as described for Subheading 3.2 using a depletion antibody cocktail containing FITC anti-human antibodies directed against CD31 (clone WM59) and CD45 (clone HI30) antigens. Isolate human basal cells by phenotype: viable, nonhematopoietic, nonendothelial, EpCAM[pos] NGFR[pos]. Expand and Passage sorted basal cells in culture flasks using PneumaCult-Ex media containing specific supplements as described in the manufacturer protocol (Stemcell Technologies). Differentiation can be achieved by transferring basal cells to a transwell and growing them in PneumaCult-Ex media, till confluent, then transitioning to an Air Liquid Interfase (ALI) environment, with Pneumacult-ALI media (Stemcell Technologies). Basal cell-derived 3D organoids can be achieved by plating a single cell suspension of basal cells in a media containing Matrigel:PneumaCult-Ex (mixed in a 1:1 ratio). Organoids can be dissociated using Liberase and maintained in 3D (Fig. 4). Note that unlike mouse lung EpiSPC, the supplements provided in Pneumacult Media are sufficient for basal cells to grow without the need of mesenchymal support cells.

3. HBSS, 2% FBS can be used as an alternative FACS buffer.

4. Liberase TM Research Grade (Roche) contains highly purified Collagenase I and II blended in a precise ratio with a medium concentration of Thermolysin. We use this enzyme mix because the high purity provides higher lot-to-lot consistency. However, other sources of Collagenase are equally effective for tissue dissociation using this protocol.

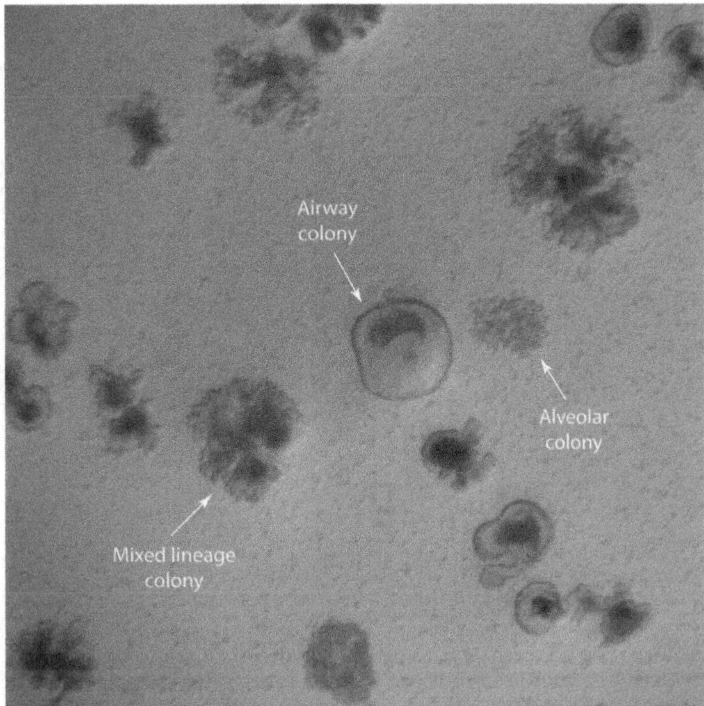

Fig. 3 Representative phase contrast image of mouse CFU-Epi. Image of mouse lung EpiSPC grown in 3D Matrigel coculture with rMC support cells

5. Any apparatus that maintains a constant 37 °C with agitation can be used. Agitation is important for the penetration of Liberase into the tissue fragments.

6. Different fluorochrome conjugates can be used. The optimal combination of fluorochromes will depend on the laser and filter configuration of the flow cytometer.

7. Alternative viability dyes can be used (e.g., DAPI, propidium iodide, 7AAD) if compatible with the laser and filter configuration of the flow cytometer and the flourochromes selected for antibodies used.

8. We use a trigas incubator set at low oxygen tension (5% v/v O_2, 10% v/v CO_2, 85% v/v N_2) which has been shown to be optimal for the growth of stem/progenitor cells at clonal density in vitro [25]. However, CFU-Epi can be grown under standard oxygen tension (10% v/v CO_2 in air) but the cloning efficiency may be lower.

9. Up to five lungs can be pooled per 50 mL tube for tissue dissociation. We have found that tissue dissociation efficiency is compromised if more than five lungs per tube are processed.

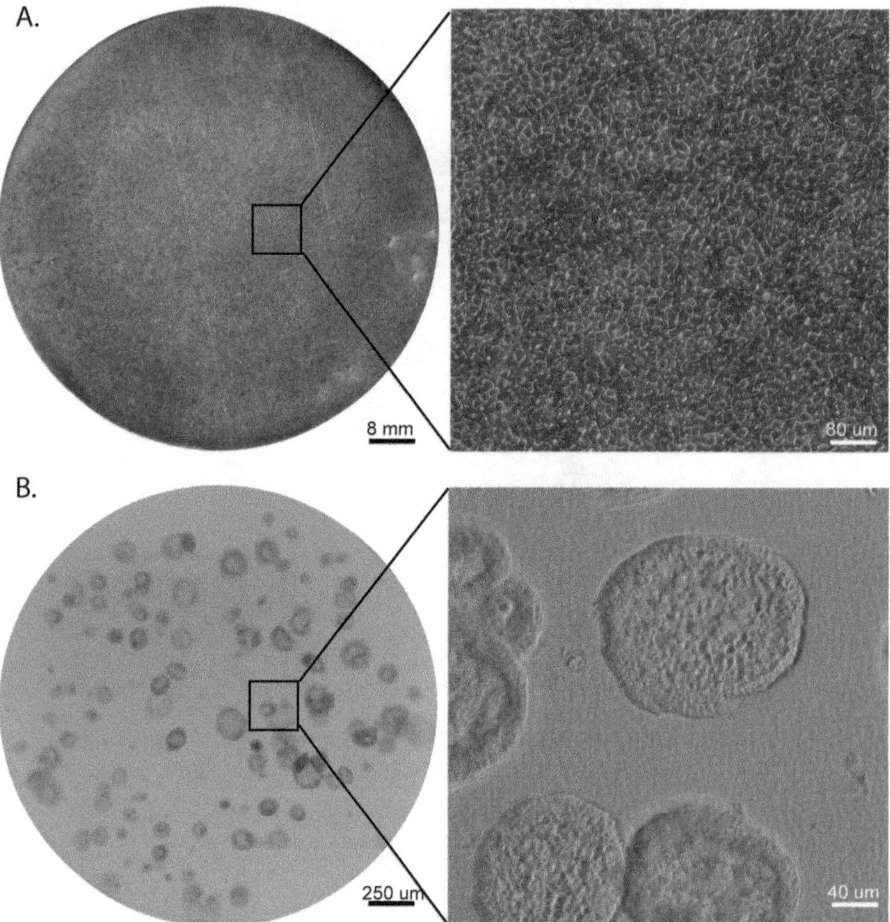

Fig. 4 Representative phase contrast images of human basal cell air–liquid interface and three-dimensional organoid cultures. Human basal cells grown in (**a**) ALI interface culture and (**b**) 3D Matrigel organoid cultures

10. Mice should be killed in accordance with ethical standards at individual institutes and only as approved by institutional animal ethics committees.

11. When tissue digestion is complete you should be unable to see chunks of pink lung tissue, however clumps of extracellular matrix will be visible as white strands in the suspension.

12. To avoid unnecessary use of valuable samples, this step can be excluded and CompBeads (BD) can be used for fluorescence compensation.

13. We use magnetic bead-based depletion step to remove contaminating hematopoietic and endothelial cells by immunomagnetic selection prior to cell sorting. Alternatively, cells can be labeled with fluorochrome-conjugated primary anti-CD45 and anti-CD31 antibodies for excluding hematopoietic and endothelial cells flow cytometrically in the sort gat-

ing strategy. However, because CD45[pos] and CD31[pos] cells are major contaminants, this will significantly extend the cell sorter time required to isolate EpiSPC and rMC.

14. Antibodies should be titrated to determine optimal concentrations.

15. FMO controls should be used to set gates for positive events. The addition of isotype control antibodies accounts for any nonspecific staining.

16. As an alternative to using primary rMC, the Mlg 2908 fetal mouse lung fibroblast cell line (ATCC CCL-206) or CFU-Epi medium supplemented with FGF10 (100 ng/mL) and HGF (30 ng/mL) can be used to support epithelial stem cells [4, 8, 22]. Mlg 2908 cells should be used at the same concentration as that used for rMC (2×10^6 cells/mL). Mlg 2908 cells can be maintained as a monolayer culture in DMEM/F12, penicillin, streptomycin, GlutaMAX (Gibco), 10% FBS. Cells exhibit optimal EpiSPC support when harvested in log-phase growth before adding to coculture. Typically we seed 5×10^4 cells in a T75 tissue culture flask and harvest with TrypLE Express 3 days later. However, the growth kinetics of Mlg 2908 cells will vary with cell batch, passage number and incubation conditions. Investigators should therefore carefully monitor the growth properties of the Mlg 2908 cell line to ensure that cells are routinely harvested in log-phase growth to provide optimal support for EpiSPC colony growth.

17. The concentration of sorted cells seeded for detection of EpiSPC depends on their level of enrichment in the sorted cell fraction. Ideally, cells should be seeded at a concentration which will generate about 20-40 colonies per well. The colony-forming efficiency of CD45[neg]CD31[neg]EpCAM[pos]CD24[low] EpiSPC from adult (8–12 week old) C57Bl/6 mice is typically 1 in 23. The concentration of rMC is optimal for support of EpiSPC under these conditions.

18. Matrigel must be kept on ice as it will begin to gel at slightly elevated temperatures. It is best to use precooled pipettes, tips, and tubes when preparing Matrigel for use. It is also important to avoid creating bubbles when mixing Matrigel. This can be achieved by gently pipetting to mix the sample.

19. This volume is sufficient to allow the medium to wet the bottom of the insert enabling diffusion of medium into the Matrigel without submerging the Matrigel culture. Colonies will not grow when submerged.

20. Temporal analysis of colony formation in this Matrigel culture system reveals the emergence of colonies after 5 days, and their continued expansion and differentiation over a 2-week period. Optimal colony formation for serial propagation and

reseeding is achieved when colonies are harvested after 1 week. Cultures can be maintained for up to 3 weeks after which colonies deteriorate and cells begin to die.

Acknowledgments

This work was supported by grants from RMIT University and the Australian National Health and Medical Research Council (NHMRC).

References

1. Kotton DN, Morrisey EE (2014) Lung regeneration: mechanisms, applications and emerging stem cell populations. Nat Med 20(8):822–832. https://doi.org/10.1038/nm.3642

2. Bertoncello I (2016) Properties of adult lung stem and progenitor cells. J Cell Physiol 231(12):2582–2589. https://doi.org/10.1002/jcp.25404

3. McQualter JL, Laurent GJ (2015) Delineating the hierarchy of lung progenitor cells and their response to influenza. Eur Respir J 46(2):315–317. https://doi.org/10.1183/09031936.00011915

4. McQualter JL, Yuen K, Williams B, Bertoncello I (2010) Evidence of an epithelial stem/progenitor cell hierarchy in the adult mouse lung. Proc Natl Acad Sci U S A 107(4):1414–1419. https://doi.org/10.1073/pnas.0909207107

5. Rock JR, Onaitis MW, Rawlins EL, Lu Y, Clark CP, Xue Y, Randell SH, Hogan BLM (2009) Basal cells as stem cells of the mouse trachea and human airway epithelium. Proc Natl Acad Sci U S A 106(31):12771–12775. https://doi.org/10.1073/pnas.0906850106

6. McQualter JL, McCarty RC, Van der Velden J, O'Donoghue RJ, Asselin-Labat ML, Bozinovski S, Bertoncello I (2013) TGF-beta signaling in stromal cells acts upstream of FGF-10 to regulate epithelial stem cell growth in the adult lung. Stem Cell Res 11(3):1222–1233. https://doi.org/10.1016/j.scr.2013.08.007

7. Tropea KA, Leder E, Aslam M, Lau AN, Raiser DM, Lee JH, Balasubramaniam V, Fredenburgh LE, Alex Mitsialis S, Kourembanas S, Kim CF (2012) Bronchioalveolar stem cells increase after mesenchymal stromal cell treatment in a mouse model of bronchopulmonary dysplasia. Am J Physiol Lung Cell Mol Physiol 302(9):L829–L837. https://doi.org/10.1152/ajplung.00347.2011

8. Quantius J, Schmoldt C, Vazquez-Armendariz AI, Becker C, El Agha E, Wilhelm J, Morty RE, Vadasz I, Mayer K, Gattenloehner S, Fink L, Matrosovich M, Li X, Seeger W, Lohmeyer J, Bellusci S, Herold S (2016) Influenza virus infects epithelial stem/progenitor cells of the distal lung: impact on Fgfr2b-driven epithelial repair. PLoS Pathog 12(6):e1005544. https://doi.org/10.1371/journal.ppat.1005544

9. Lee JH, Bhang DH, Beede A, Huang TL, Stripp BR, Bloch KD, Wagers AJ, Tseng YH, Ryeom S, Kim CF (2014) Lung stem cell differentiation in mice directed by endothelial cells via a BMP4-NFATc1-thrombospondin-1 axis. Cell 156(3):440–455. https://doi.org/10.1016/j.cell.2013.12.039

10. Rawlins EL, Okubo T, Xue Y, Brass DM, Auten RL, Hasegawa H, Wang F, Hogan BL (2009) The role of Scgb1a1+ Clara cells in the long-term maintenance and repair of lung airway, but not alveolar, epithelium. Cell Stem Cell 4(6):525–534

11. Hong KU, Reynolds SD, Watkins S, Fuchs E, Stripp BR (2004) In vivo differentiation potential of tracheal basal cells: evidence for multipotent and unipotent subpopulations. Am J Physiol Lung Cell Mol Physiol 286(4):L643–L649. https://doi.org/10.1152/ajplung.00155.2003

12. Hegab AE, Ha VL, Gilbert JL, Zhang KX, Malkoski SP, Chon AT, Darmawan DO, Bisht B, Ooi AT, Pellegrini M, Nickerson DW, Gomperts BN (2011) Novel stem/progenitor cell population from murine tracheal submucosal gland ducts with multipotent regenerative potential. Stem Cells 29(8):1283–1293. https://doi.org/10.1002/stem.680

13. Tesei A, Zoli W, Arienti C, Storci G, Granato AM, Pasquinelli G, Valente S, Orrico C, Rosetti M, Vannini I, Dubini A, Dell'Amore D, Amadori D, Bonafe M (2009) Isolation of

stem/progenitor cells from normal lung tissue of adult humans. Cell Prolif 42(3): 298–308

14. Hong KU, Reynolds SD, Watkins S, Fuchs E, Stripp BR (2004) Basal cells are a multipotent progenitor capable of renewing the bronchial epithelium. Am J Pathol 164(2):577–588. https://doi.org/10.1016/S0002-9440(10)63147-1

15. Giangreco A, Reynolds SD, Stripp BR (2002) Terminal bronchioles harbor a unique airway stem cell population that localizes to the bronchoalveolar duct junction. Am J Pathol 161(1):173–182. https://doi.org/10.1016/S0002-9440(10)64169-7

16. Reynolds SD, Giangreco A, Power JH, Stripp BR (2000) Neuroepithelial bodies of pulmonary airways serve as a reservoir of progenitor cells capable of epithelial regeneration. Am J Pathol 156(1):269–278. https://doi.org/10.1016/S0002-9440(10)64727-X

17. Guha A, Vasconcelos M, Cai Y, Yoneda M, Hinds A, Qian J, Li G, Dickel L, Johnson JE, Kimura S, Guo J, McMahon J, McMahon AP, Cardoso WV (2012) Neuroepithelial body microenvironment is a niche for a distinct subset of Clara-like precursors in the developing airways. Proc Natl Acad Sci U S A 109(31):12592–12597. https://doi.org/10.1073/pnas.1204710109

18. Kim CF, Jackson EL, Woolfenden AE, Lawrence S, Babar I, Vogel S, Crowley D, Bronson RT, Jacks T (2005) Identification of bronchioalveolar stem cells in normal lung and lung cancer. Cell 121(6):823–835

19. Chapman HA, Li X, Alexander JP, Brumwell A, Lorizio W, Tan K, Sonnenberg A, Wei Y, Vu TH (2011) Integrin alpha6beta4 identifies an adult distal lung epithelial population with regenerative potential in mice. J Clin Invest 121(7):2855–2862. https://doi.org/10.1172/JCI57673

20. Barkauskas CE, Cronce MJ, Rackley CR, Bowie EJ, Keene DR, Stripp BR, Randell SH, Noble PW, Hogan BL (2013) Type 2 alveolar cells are stem cells in adult lung. J Clin Invest 123(7):3025–3036. https://doi.org/10.1172/JCI68782

21. Chen Q, Suresh Kumar V, Finn J, Jiang D, Liang J, Zhao YY, Liu Y (2017) CD44high alveolar type II cells show stem cell properties during steady-state alveolar homeostasis. Am J Physiol Lung Cell Mol Physiol 313(1): L41–L51. https://doi.org/10.1152/ajplung.00564.2016

22. Teisanu RM, Chen H, Matsumoto K, McQualter JL, Potts E, Foster WM, Bertoncello I, Stripp BR (2011) Functional analysis of two distinct bronchiolar progenitors during lung injury and repair. Am J Respir Cell Mol Biol 44(6):794–803. https://doi.org/10.1165/rcmb.2010-0098OC

23. Chen H, Matsumoto K, Brockway BL, Rackley CR, Liang J, Lee JH, Jiang D, Noble PW, Randell SH, Kim CF, Stripp BR (2012) Airway epithelial progenitors are region specific and show differential responses to bleomycin-induced lung injury. Stem Cells 30(9):1948–1960. https://doi.org/10.1002/stem.1150

24. Lee JH, Kim J, Gludish D, Roach RR, Saunders AH, Barrios J, Woo AJ, Chen H, Conner DA, Fujiwara Y, Stripp B, Kim CF (2012) SPC H2B-GFP mice reveal heterogeneity of surfactant protein C-expressing lung cells. Am J Respir Cell Mol Biol 48(3):288–298. https://doi.org/10.1165/rcmb.2011-0403OC

25. Wion D, Christen T, Barbier EL, Coles JA (2009) PO(2) matters in stem cell culture. Cell Stem Cell 5(3):242–243

Chapter 12

Isolation, Characterization and Differentiation of Mouse Cardiac Progenitor Cells

Santosh Kumar Yadav and Paras Kumar Mishra

Abstract

Despite several strategies developed for replenishing the dead myocardium after myocardial infarction (MI), stem cell therapy remains the leading method to regenerate new myocardium. Although induced pluripotent stem cells (iPS) and transdifferentiation of the differentiated cells have been used as novel approaches for myocardial regeneration, these approaches did not yield very successful results for myocardial regeneration in in vivo studies. Asynchronous contractility of newly formed cardiomyocytes with the existing cardiomyocytes is the most important issue with iPS approach, while very low yield of transdifferentiated cardiomyocytes and their less chances to beat in the same rhythm as existing cardiomyocytes in the MI heart are important caveats with transdifferentiation approach. CSCs are present in the heart and they have the potential to differentiate into myocardial cells. However, the number of resident CSCs is very low. Therefore, it is important to get maximum yield of CSCs during isolation process from the heart. Increasing the number of CSCs and initiating their differentiation ex vivo are crucial for CSC-based stem cell therapy. Here, we present a better method for isolation, characterization and differentiation of CSCs from the mouse heart. We also demonstrated morphological changes in the CSCs after 2 days, 3 days, and 7 days in maintenance medium and a separate group of CSCs cultured for 12 days in differentiation medium using Phase-Contrast microscopy. We have used different markers for identification of CSCs isolated from the mouse heart such as marker for mouse CSC, Sca-1, cardiac-specific markers NKX2–5, MEF2C, GATA4, and stemness markers OCT4 and SOX2. To characterize the differentiated CSCs, we used CSCs maintained in differentiation medium for 12 days. To evaluate differentiation of CSCs, we determined the expression of cardiomyocyte-specific markers actinin and troponin I. Overall; we described an elegant method for isolation, identification, differentiation and characterization of CSCs from the mouse heart.

Key words Cardiac stem cells, CSC differentiation, CSC characterization, Sca-1, Actinin, Troponin I

1   Introduction

Loss of myocardial tissue due to myocardial infarction, diabetes mellitus or other pathological conditions leads to heart failure, which is the leading cause of morbidity and mortality in the world [1, 2]. Although annual turnover of cardiomyocyte decreases with age, the number of cardiomyocytes remains constant during the

Shree Ram Singh and Pranela Rameshwar (eds.), *Somatic Stem Cells: Methods and Protocols*, Methods in Molecular Biology, vol. 1842, https://doi.org/10.1007/978-1-4939-8697-2_12, © Springer Science+Business Media, LLC, part of Springer Nature 2018

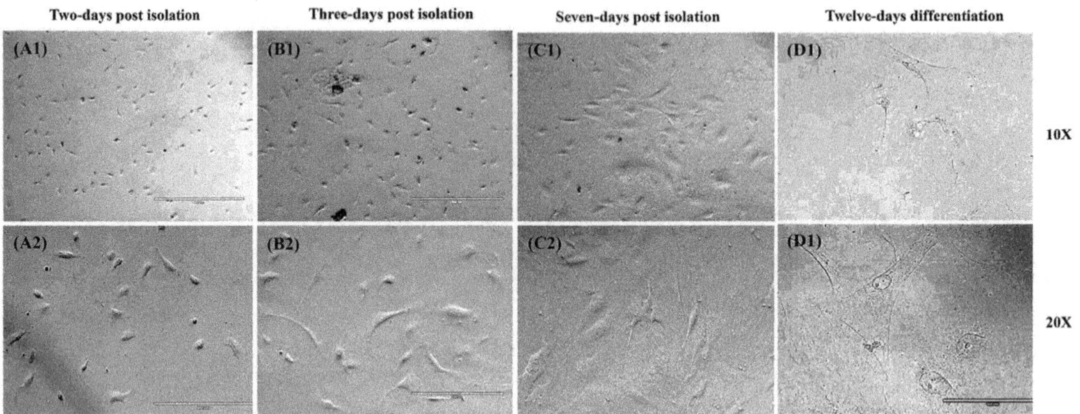

Fig. 1 Phase-contrast images (10X or 100 times and 20X or 200 magnifications) of isolated mice cardiac stem cells (CSCs) at different days of culture and after 12 days of differentiation. (**a1** and **a2**). CSCs after 2 days of culture in CSC maintenance medium after isolation. (**b1** and **b2**) CSCs after 3 days of culture in CSC maintenance medium after isolation. (**c1** and **c2**) CSCs after 7 days of culture in CSC maintenance medium after isolation. (**d1** and **d2**). Cardiomyocyte like cell morphology after 12-day culture of CSCs in CSC differentiation medium. Scale bar is 400 μm for 10X and 200 μm for 20X

human lifespan [3]. It is because of the unique features of the mammalian heart to maintain turnover of cardiomyocytes (by renewal of cardiomyocytes) and microvasculature (smooth muscle and endothelial cells) throughout life [4, 5]. The mammalian heart contains distinct types of endogenous stem and progenitor cells, which have potential for self-renewal, clonogenecity, and multilineage differentiation [5–14]. However, the number of CSCs or cardiac progenitor cells decreases with ageing [15]. The regenerative approach to increase the number of CSCs for cardiac repair provides a promising strategy for ischemic and/or ageing hearts [16, 17]. For regenerative therapy, it is crucial to characterize CSCs and trace CSC-derived cardiomyocytes [18–20]. The morphology of cultured CSCs changes with time (Fig. 1). After 2 days in culture, CSCs increase in number (Fig. 1a). On day three, they started increasing in size (Fig. 1b) and continued increasing in size until day seven (Fig. 1c). CSCs cultured in a differentiation medium for 12-days showed cardiomyocyte like phenotype (Fig. 1d). At day 7 of post-isolation, we stained CSCs with different markers. For stem cells, we used well-validated mouse stem cell marker Sca-1 (Fig. 2a). For cardiac origin, we used Nkx2-5 (Fig. 2B), Gata4 (Fig. 2c), and Mef2c (Fig. 2d) markers. We also validated the stemness of these cells using pluripotency markers Oct4 (Fig. 2e) and Sox2 (Fig. 2f). To determine whether these CSCs were differentiated into cardiomyocytes, we used

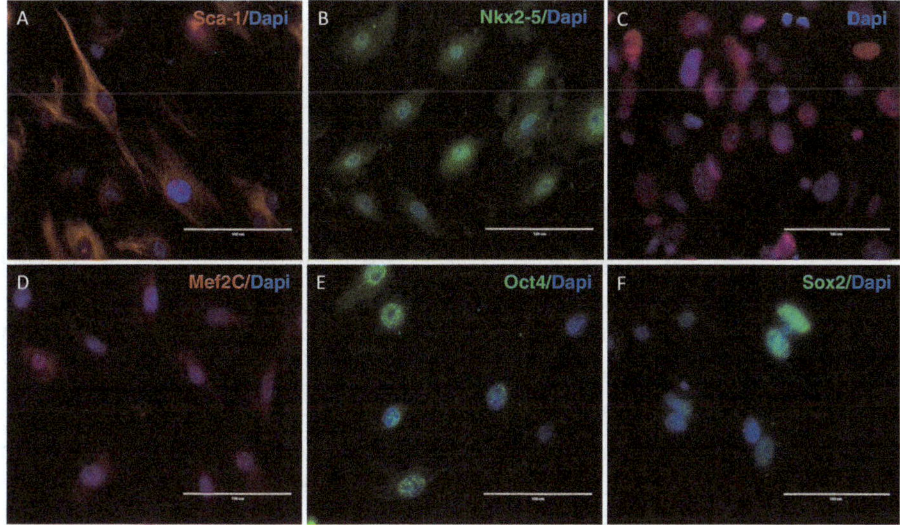

Fig. 2 Immunofluorescence images of mice CSCs stained with different cardiac stem cell and pluripotent stem cell specific markers. Progenitor cells were stained with cardiac stem cell- specific markers such as Sca-1, Nkx2-5, Mef2c, and Gata4. These cells were also positive for pluripotent stem cell markers Oct4 and Sox2. Blue color Dapl Is used to stain the nucleus. All the representative images are merged image of specific markers and Dapi. Scale bar is 100 μm

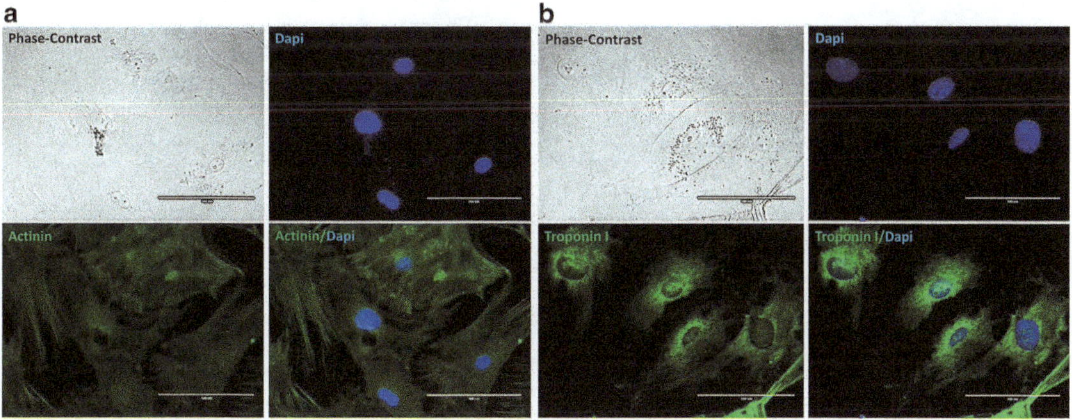

Fig. 3 Immunofluorescence images of cardiomyocyte like cells after 12 days of CSC differentiation. Cells were immunostained with cardiomyocyte-specific markers Actinin (**a**) and Troponin I (**b**). Blue color DAPI is used to stain the nucleus. All the representative images are merged images of cardiomyocyte markers with DAPI. Scar bar is 100 μm

CSCs cultured for 12 days in differentiation medium and determined the expression of cardiomyocyte markers actinin (Fig. 3a) and troponin I (Fig. 3b). The characterization of CSCs and CSC-derived cardiomyocytes validates our successful isolation of CSCs from the mouse heart.

2   Materials

Mice are ordered from an authorized vendor such as The Jackson Laboratories.

1. Mice: C57BL/6J is procured from The Jackson Laboratories, USA.

2. Mouse CSC culture medium components (maintenance and differentiation media, Millipore, USA).

3. Mouse cardiac progenitor cell isolation buffer components.

4. Histopaque solution.

5. Filter-sterilized 1% collagenase I stock solution in Hank's Balanced Salt Solution (HBSS). (0.2% working solution diluted in HBSS).

6. 1× HBSS.

7. Surgical equipment:

    (a) Surgical scissors.

    (b) Fine surgical scissors.

    (c) Curve shank forceps.

    (d) Surgical blade.

8. Instruments: Cell counter; Microscope (EVOS, USA); Centrifuge machine (Thermo Scientific, USA).

9. Tissue culture:

    (a) 10 mm petri dish.

    (b) 6-well plate.

    (c) 24-well plate.

    (d) T-25 and T-75 culture flask.

    (e) 50 and 15 mL conical tubes.

    (f) 10, 200, and 1000 µL pipette tips.

    (g) 10, 200, and 1000 µL pipette man.

    (h) Pasture pipette.

    (i) 5, 10, and 25 mL serological disposable pipettes.

    (j) 0.22 µm filter.

    (k) 20 and 40 µm sterile cell strainers.

    (l) 10 mL syringe.

    (m) Disposable hemocytometer slide and Trypan Blue dye.

10. Others:

    (a) Powder- free nitrile gloves.

    (b) 10× PBS (pH 7.4, cell culture grade).

    (c) 70% ethanol.

    (d) 10% bleach (disinfectant).

3   Method

***3.1   Isolation Method
of Cardiac Progenitor
Cells***

1. Ten to fourteen-week old, four or five adult mice are euthanized using CO_2 chamber.

2. Mice are sterilized with 70% ethanol under the hood.

3. Thoracic cavity of each mouse is opened inside the hood, washed with ice cold PBS, and blood removed.

4. The mouse heart is surgically removed and placed into a sterile 10 cm petri dish containing 10 mL of ice cold 1× PBS.

5. The heart is gently palpitated using curved shanked forceps to remove the blood left inside the heart. Atria of the heart are dissected out and ventricles are used for CSC isolation.

6. The extracted ventricles are transferred onto a 50-mL conical tube containing 10 mL chilled HBSS on ice and kept until further processing (*see* **Note 1**).

7. The ventricles containing tube is placed in a biosafety cabinet for further processing (*see* **Note 2**).

8. Under the biosafety cabinet, the ventricles are transferred into a sterile 10 cm petri dish containing 5–10 mL cold HBSS.

9. The ventricles are minced into very small pieces using fine surgical scissors (*see* **Note 3**).

10. Ventricular minced tissue is centrifuged at 500 *g* at 4 °C for 5 min and supernatant removed.

11. The pellet from the previous step is transferred into a 50-mL conical tube.

12. Five to six milliliter of 0.2% collagenase solution, which is prepared in HBSS, is added to the pellet (*see* **Note 4**).

13. The pellet is mixed thoroughly with the collagenase solution by agitating the tube or rocking the tubes on a incubator shaker for 60–90 min at 37 °C at 150 rpm(*see* **Note 5**).

14. Using either wide-orifice 1000 μL pipette tip or cut 1000 μL narrow-orifice pipette tip, the ventricle tissue pellet is triturated for proper dissociation of the pellet (*see* **Note 6**).

15. The lysis process of collagenase is then stopped by adding cardiac stem cell maintenance medium (*see* **Note 7**).

16. Filter the cell suspension through a 100 μm cell strainer to remove the larger undigested tissue pieces.

17. Further, passed filtrate through 40 μm cell strainer to remove any endothelial cells.

18. Gently overlaid the equal volume of filtrate over Histopaque solution. For example, in a 50-mL tube, add 20 mL Histopaque

solution first and then add 20 mL filtrate solution over the Histopaque solution (*see* **Note 8**).

19. Centrifuged the tube containing filtrate and Histopaque solution in a swing bucket centrifuge at 500 g for 20 min at RT (*see* **Note 9**).

20. After centrifugation, the tube is carefully taken out. The tube has three layers: upper medium layer, middle buffy coat-containing CSCs, and lower Histopaque layer containing mixed cell population.

21. Using 1 mL pipette, the upper and the middle layers are transferred into a fresh 15 mL conical tube (*see* **Note 10**).

22. An equal volume of pre-warm CSC maintenance medium is added to the cell suspension and mixed properly.

23. The tube containing the CSCs with the CSC maintenance medium is centrifuged at 500 g/5 min/4 °C. The supernatant is discarded and the pellet containing CSCs is washed with incomplete DMEM medium to remove residual Histopaque. Finally, the purified CSC pellet is collected.

24. The CSC containing pellet is resuspended in 1 mL cardiac stem cell maintenance medium.

25. These CSCs are then seeded into a culture plate (*see* **Note 11**). They are cultured into cell culture incubator maintained at 37 °C with 5% CO_2.

26. Change the medium of cells every day for at least 3 days with fresh pre-warm cardiac stem cell maintenance medium. After that, change the medium of the cells on alternate days.

27. Observe the cell morphology and growth condition under the microscope and capture images of the CSCs.

3.2 Maintenance of Cardiac Stem Cells

1. After reaching confluent growth of the cells in flask, transfer the cells from one T-25 flask into a new T-25 flask by trypsinization. At this stage, the cells are considered as passage 1 (P1).

2. The confluent P1 cells is passaged into two T-25 flasks (P2). At this stage, the cells either can be propagated for experimentation by transferring them into a T-75 flask or can be stored into liquid nitrogen for future experiments.

3. To maintain the cells in the culture medium, one million CSCs is seeded in a T-25 flask or two million CSCs is seeded in a T-75 flask (*see* **Note 12**).

3.3 Characterizations of Isolated Cardiac Progenitor Cells

To characterize isolated CSCs

1. Observe the isolated CSCs under a Phase-Contrast microscope. Initial cell morphology will look like spindle shape or mesenchymal stem cell like morphology (Fig. 1a, b).

2. Stain the cells with immunofluorescence antibodies for detection of CSC markers Sca-1, cardiac markers NKx2.5, GATA4, MEF2C, and pluripotency markers OCT4 and SOX2 (Fig. 2a–f).

3. For fluorescent staining of CSCs, seed 10,000 cells in a 24-well culture plate.

4. After one day, fix the cells with 4% formalin solution for 15 min.

5. After fixation, wash the cells with ice cold PBS for 5 min, three times.

6. Permeabilize the cells with 0.2% Tritox-X-100 in PBS for 10 min (*see* **Note 13**).

7. Wash the cells with ice cold PBS for 5 min, three times.

8. Block the cells with 1% BSA in PBST (PBS+ 0.1% Tween 20) for 1 h.

9. After the blocking, incubate the cells with primary antibody diluted in 0.5% blocking solution for overnight (*see* **Note 14**).

10. After overnight primary antibody incubation, wash the cells with ice cold PBS with 5 min, three times.

11. Incubate the cells with secondary antibody diluted in 0.5% blocking solution for 1 h (*see* **Note 15**).

12. After competition of incubation, remove the secondary antibody solution and wash the cells with cold PBS for 5 min, three times.

13. Counterstain the cells with DAPI, 1 μg/mL for 5 min.

14. Capture the images using fluorescent microscope.

15. Store the plate in dark at 4 °C.

3.4 Differentiation of CSCs into Cardiomyocyte

For differentiation of CSCs

1. Seed 50,000 CSCs in a 6-well plate in CSC maintenance medium.

2. At 85–90% confluency, replace the medium with CSC differentiation medium and culture them in a cell- culture incubator for 2–3 weeks (*see* **Note 16**).

3. Replace the differentiation medium every 2–3 days.

4. After 2–3 weeks of differentiation, stain the differentiated cells with cardiomyocyte markers such as Troponin I and Actinin (Fig. 3a, b).

4   Notes

1. Replace the HBSS after each dissection to ensure almost complete clearance of blood.

2. Make biosafety cabinet completely sterile using 70% ethanol before placing the ventricles. Wipe the tubes with 70% ethanol before placing it in the biosafety cabinet.

3. Ventricles are held tight by forceps and cut into small pieces (1–2 mm in diameter) using scissors or surgical blade.

4. The volume of the 0.2% collagenase solution should be double the volume of the pellet.

5. Shake the tube vigorously by hand every 30 min to enhance the lysis.

6. To get the maximum recovery of cells, triturate the large pieces for at least 3–5 min.

7. The volume of the maintenance medium should be 2–3 times the volume of the collagenase.

8. It is a very important step of the protocol where care should be taken to avoid mixing of the two solutions. For that, the filtrate solution should be added very slowly onto the Histopaque solution in a tube.

9. Keep the setting of centrifuge with lowest acceleration/deceleration speed to keep proper separation and to avoid mixing of the two solutions.

10. Take precaution during removal of buffy layer. Avoid sucking Histopaque layer because Histopaque solution is toxic to CSCs.

11. Depending on CSC number/density, they will be cultured in a six-well plate or a T25 flask. For low density CSCs, a six-well plate culture is better. Cell number can be determined using a hemocytometer.

12. It is necessary to passage CSCs every 3–4 days or when they attain 85% confluency. Cells can be maintained in CSC maintenance medium for up to 3–4 weeks.

13. No need of permeabilization if surface antigen or receptor is investigated.

14. Antibody should be diluted as per data sheet specification.

15. Dilution of secondary antibodies should be at least two-times the dilution of primary antibody .

16. Always pre-warm the medium before adding to the CSCs.

Acknowledgments

This work is supported, in part, by the National Institutes of Health grants HL-113281 and HL116205 to Paras Kumar Mishra.

References

1. Lloyd-Jones D et al (2010) Executive summary: heart disease and stroke statistics—2010 update: a report from the American Heart Association. Circulation 121(7):948–954

2. Schnell O et al (2013) Type 1 diabetes and cardiovascular disease. Cardiovasc Diabetol 12:156

3. Bergmann O et al (2015) Dynamics of cell generation and turnover in the human heart. Cell 161(7):1566–1575

4. Bergmann O et al (2009) Evidence for cardiomyocyte renewal in humans. Science 324(5923):98–102

5. Monsanto MM et al (2017) Concurrent isolation of 3 distinct cardiac stem cell populations from a single human heart biopsy. Circ Res 121(2):113–124

6. Beltrami AP et al (2003) Adult cardiac stem cells are multipotent and support myocardial regeneration. Cell 114(6):763–776

7. Ellison GM et al (2013) Adult c-kit(pos) cardiac stem cells are necessary and sufficient for functional cardiac regeneration and repair. Cell 154(4):827–842

8. Ellison GM et al (2011) Endogenous cardiac stem cell activation by insulin-like growth factor-1/hepatocyte growth factor intracoronary injection fosters survival and regeneration of the infarcted pig heart. J Am Coll Cardiol 58(9):977–986

9. Hou X et al (2012) Isolation, characterization, and spatial distribution of cardiac progenitor cells in the sheep heart. J Clin Exp Cardiolog S6. 1–16

10. Chong JJ et al (2011) Adult cardiac-resident MSC-like stem cells with a proepicardial origin. Cell Stem Cell 9(6):527–540

11. Bu L et al (2009) Human ISL1 heart progenitors generate diverse multipotent cardiovascular cell lineages. Nature 460(7251):113–117

12. Smits AM et al (2009) Human cardiomyocyte progenitor cells differentiate into functional mature cardiomyocytes: an in vitro model for studying human cardiac physiology and pathophysiology. Nat Protoc 4(2):232–243

13. Ellison GM, Nadal-Ginard B, Torella D (2012) Optimizing cardiac repair and regeneration through activation of the endogenous cardiac stem cell compartment. J Cardiovasc Transl Res 5(5):667–677

14. Oh H et al (2003) Cardiac progenitor cells from adult myocardium: homing, differentiation, and fusion after infarction. Proc Natl Acad Sci U S A 100(21):12313–12318

15. Laugwitz KL et al (2005) Postnatal isl1+ cardioblasts enter fully differentiated cardiomyocyte lineages. Nature 433(7026):647–653

16. Soonpaa MH et al (1994) Formation of nascent intercalated disks between grafted fetal cardiomyocytes and host myocardium. Science 264(5155):98–101

17. Sekine H et al (2008) Endothelial cell coculture within tissue-engineered cardiomyocyte sheets enhances neovascularization and improves cardiac function of ischemic hearts. Circulation 118(14 Suppl):S145–S152

18. Wu SM, Chien KR, Mummery C (2008) Origins and fates of cardiovascular progenitor cells. Cell 132(4):537–543

19. Martin-Puig S, Wang Z, Chien KR (2008) Lives of a heart cell: tracing the origins of cardiac progenitors. Cell Stem Cell 2(4):320–331

20. Matsuura K et al (2013) Cell sheet transplantation for heart tissue repair. J Control Release 169(3):336–340

Chapter 13

Isolating and Characterizing Adipose-Derived Stem Cells

Guangpeng Liu and Xi Chen

Abstract

The research and application of regeneration medicine will require a reliable source of stem cells. Adipose tissue has proven to be an easily accessible and rich source of adult stem cells, termed adipose-derived stem cells (ASCs). ASCs have the most important advantage over stem cells from other available sources. There is no other human tissue as abundant as adipose tissue, making it possible to isolate adequate numbers of ASCs for potential clinical applications. Here, we describe detailed methods for isolating and characterizing ASCs. These procedures can be applied to adipose tissue not only for humans but also for other species.

Key words Adipose tissue, Adult stem cells, Multilineage differentiation, Isolation, Characterization

1　Introduction

The ability of nonhematopoietic stem cells to undergo self-renewal and multilineage differentiation has shown great promise for the repair and regeneration of various tissues and organs, and may provide potential therapeutic solutions for many inherited and acquired diseases. Candidates for the stem-cell-based strategies include embryonic stem cells (ESCs), induced pluripotent stem cells (iPSCs) and adult (postnatal) stem cells. Although ESCs and iPSCs are totipotent, there are many limitations to their practical use, such as the ethical, legal, political, and safety concerns. In contrast, adult stem cells offer an alternative approach that can circumvent these concerns.

Adult stem cells can be isolated from many tissue sources, including bone marrow, adipose tissue, skeletal muscle, skin, periosteum, tooth pulp, synovial membrane, hair follicles, umbilical cord, and even amniotic fluid [1]. But the emerging field of regenerative medicine will require a reliable stem cell source. Gimble et al. have suggested that an ideal stem cell source for clinical applications should meet the following requirements: (1) can be found in abundant quantities (millions to billions of cells); (2) can be

Shree Ram Singh and Pranela Rameshwar (eds.), *Somatic Stem Cells: Methods and Protocols*, Methods in Molecular Biology, vol. 1842, https://doi.org/10.1007/978-1-4939-8697-2_13, © Springer Science+Business Media, LLC, part of Springer Nature 2018

harvested by a minimally invasive procedure; (3) can be differentiated along multilineage pathways in a reproducible manner; (4) can be safely and effectively transplanted to either autologous or allogeneic host; (5) can be manufactured in accordance with the current Good Manufacturing Practices (cGMP) [2]. Adipose tissue can fulfill all these criteria, since it is abundant and can be easily obtained in large quantities to isolate adequate numbers of adipose-derived stem cells (ASCs) with little donor site morbidity and patient discomfort. In vitro and in vivo, ASCs can differentiate along multiple pathways, including osteogenic, adipogenic, chondrogenic, endothelial, epithelial, hepatic, myogenic, and neuronal-like lineages. They can also act through the paracrine mechanism of releasing growth factors to accelerate tissue repair and regeneration [3]. Furthermore, large-scale production of ASCs required for cellular therapies can be achieved under cGMP condition, and ASCs' application has been proven safe and efficacious in preclinical and clinical studies [4, 5].

The defining characteristics of stem cells are inconsistent among different investigators and laboratories. To address this issue, the International Society for Cellular Therapy (ISCT) recommended three identifying criteria to define human multipotent mesenchymal stromal cells [6], which is adopted to characterize ASCs in this study (Table 1). With their mesodermal origin, ASCs possess the ability to differentiate into osteoblasts, adipocytes as well as chondrocytes, which are all derived from the embryonic mesoderm. Although recent studies have shown that ASCs are able to differentiate along the ectodermal and endodermal lineages [1], the plasticity of ASCs is not the concern of this chapter.

The subcutaneous fat tissue is ubiquitous and readily accessible due to the increased incidence of obesity in modern populations. For example, over 400,000 liposuction surgeries are performed each year in the USA [3]. The lipoaspirate containing large quantities of ASCs is routinely discarded as medical waste. In this chapter, we describe methods for isolating and characterizing ASCs obtained from human liposuction fat aspirates. These protocols are also applicable (with minor modifications) to adipose samples excised by surgical procedures as well as fat tissues harvested from other species.

---

2 Materials

2.1 ASC Isolation and Expansion

1. Dulbecco's modified Eagle's medium with L-glutamine (DMEM), low glucose (LG-DMEM) and high glucose (HG-DMEM).

2. Fetal bovine serum (FBS).

Table 1
Three criteria to identify ASCs

1. Adherence to plastic substrate in standard culture conditions	
2. Specific cell surface marker expression	
Positive ($\geq$95%)	Negative ($\leq$2%)
CD105	CD45
CD73	CD34
CD90	CD14 or CD11b
HLA-ABC	CD79a or CD19
	HLA-DR
3. In vitro multilineage differentiation into osteoblasts, adipocytes, and chondrocytes	

3. Trypsin–EDTA solution, 0.25% trypsin and 0.02% EDTA in Hank's balanced salt solution.

4. Phosphate buffered solution (PBS), pH 7.4.

5. Collagenase type I solution, 0.075%.

6. Penicillin and streptomycin solution (P/S), 1%.

7. NH_4Cl solution, 160 mM.

8. Trypan Blue solution.

9. Centrifuge tubes, 15 and 50 mL.

10. Cell culture dishes, Φ 100 mm.

11. Nylon cell strainers, 100 μm.

2.2 ASC Characterization

1. For flow cytometry analysis using fluorescein isothiocyanate (FITC)- *and/or phycoerythrin (PE)-conjugated antibodies: CD45, CD34, CD14 or CD11b, CD79a or CD19, CD105, CD73, CD90, HLA-ABC, and HLA-DR. Nonspecific IgG-FITC or IgG-PE are used as control.*

2. Flow cytometry buffer (FCB): PBS containing 0.5% bovine serum albumin (BSA).

3. Human monoclonal antibody of collagen type II for immunocytochemistry staining.

4. Adipogenic differentiation medium (AM): LG-DMEM with 10% FBS, 0.5 mM isobutyl-methylxanthine (IBMX), 1 μM dexamethasone, 10 μM insulin, 200 μM indomethacin, and 1% P/S.

5. Osteogenic differentiation medium (OM): LG-DMEM with 10% FBS, 0.1 μM dexamethasone, 50 μM ascorbate-2-phosphate, 10 mM β-glycerophosphate, and 1% P/S.

6. Chondrogenic differentiation medium (CM): HG-DMEM with 1% FBS, 6.25 μg/mL insulin, 10 ng/mL TGF-β1, 50 nMascorbate-2-phosphate, and 1% P/S.

7. Microfuge tubes, 1.5 mL.

8. Nylon cell strainers, 70 μm.

9. Paraformaldehyde solution (PFA), 1 and 4%.

10. Oil red O reagent.

11. 70% ethanol solution.

12. Distilled water.

13. Silver nitrate solution, 1%.

3   Methods

All the experimental procedures are performed under sterile condition in a biological safety cabinet.

3.1   ASC Isolation and Expansion (Fig. 1)

1. Human adipose tissue is obtained by liposuction aspiration or surgical procedures (*see* **Note 1**). Fat samples can be preserved on ice under aseptic condition for no more than 6 hours prior to processing.

2. The tissue sample is placed into sterile 50-mL centrifuge tubes and washed extensively with equal volumes of prewarmed PBS containing 1% P/S for three times to remove the fat debris, oil, blood cells and the local anesthetics.

3. Transfer 10 mL of the fat sample into a new 50-mL tube with equal volume of 0.075% collagenase type I prepared in PBS and digested in water bath (37 °C) for 30–60 min (*see* **Note 2**).

4. Enzyme is neutralized with 5 mL of DMEM containing 10% FBS and 1% P/S. Pipette the sample up and down several times to disintegrate the adipose aggregates.

5. The digested sample is centrifuged at 1500 rpm (1300 × g) for 5 min to yield the high-density stromal vascular fraction (SVF) layer containing ASCs.

6. Shake the tube vigorously to disrupt the SVF and mix the cells. Repeat the centrifugation step.

7. Aspirate the collagenase and DMEM solution above SVF carefully.

8. Add 1 mL of NH_4Cl solution to resuspend the SVF pellet and incubate at room temperature for 10 min to lyse red blood cells.

9. Pipette with 10 mL of PBS containing 1% P/S and centrifuge at 1500 rpm for 5 min.

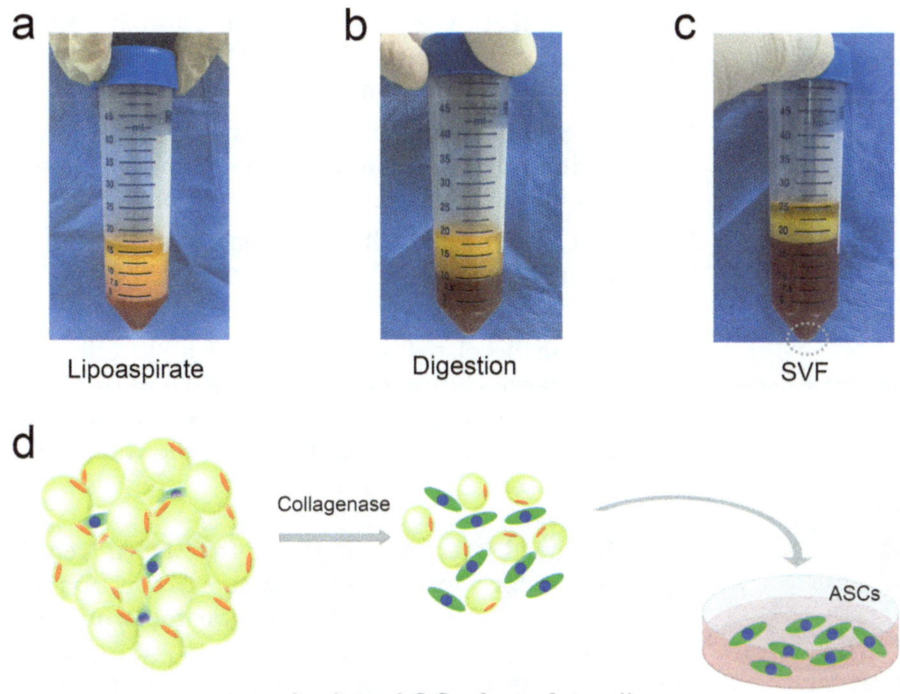

Fig. 1 Isolation of ASCs from adipose tissue. (**a**) Raw liposuction aspirates contain oil (upper layer), fat tissue (middle layer) and mixture of blood and the local anesthetics (lower layer). (**b**) 10 mL of fat tissue is digested with equal volume of collagenase type I solution. (**c**) After digestion and centrifugation, SVF (dotted circle) is obtained for cell culture. (**d**) Scheme for processing and isolating ASCs from adipocytes

10. Aspirate all the supernatant without disturbing the cells, add 5 mL of DMEM containing 10% FBS and 1% P/S to resuspend SVF and filter through a 100-μm cell strainer to remove cellular debris.

11. Plate the filtered cells into a Φ 100 mm cell culture dish and add an additional 5 mL of DMEM containing 10% FBS and 1% P/S. Incubate the cells in a humidified CO_2 incubator at 37 °C and 5% CO_2.

12. 48 hours after plating, remove the culture medium and wash the cells with prewarmed PBS containing 1% P/S thoroughly to clean tissue fragments and nonadherent cells. Replace 10 mL of fresh DMEM containing 10% FBS and 1% P/S, and maintain the cells at 37 °C and 5% CO_2.

13. Change the medium every three days until the remaining cells reach 80–90% confluence. Then aspirate the medium and wash the cells with 2 mL of PBS for two times.

14. Replace the PBS with 1 mL of 0.25% Trypsin-EDTA solution and place the culture dish in the incubator for 5 min.

15. Verifying that most of the adherent cells have detached from the plastic substrate, add 2 mL of DMEM containing 10% FBS to neutralize the trypsin reaction.

16. Pipette the medium gently and transfer it into a 15 mL tube. Centrifuge at 1500 rpm for 5 min. Aspirate the supernatant and suspend the cell pellet using 1 mL of DMEM.

17. Take 12.5 μL of the cell suspension and dilute with 12.5 μL of Trypan Blue solution. Count the cell number using a hemocytometer.

18. Cells can be plated at a density of 5000 cells/cm² into Φ 100 mm cell culture dishes, expanded and passaged similarly according to the above procedures (*see* **Note 3**). ASCs of passage 2 are usually used for the following characterizing procedures.

3.2 ASC Characterization

3.2.1 Flow Cytometry Analysis of Cell Surface Markers

1. ASCs are harvested using trypsin/EDTA, centrifuged at 1500 rpm for 5 min. Aspirate the supernatant and suspend the cell pellet with 5 mL of PBS. Repeat the centrifugation step (*see* **Note 4**).

2. Remove PBS and prepare single cell suspension with FCB. Adjust the cell density to 10^7/mL.

3. Place 50–100 μL of cell suspension into a 1.5 mL microtube. Add 1 μg of primary antibody directly and incubate on ice for 30 min (*see* **Note 5**).

4. Add 1 mL of PBS to rinse nonbound antibody, and centrifuge at 1500 rpm for 5 min. Repeat the rinse and centrifugation step again (*see* **Note 6**).

5. Decant the supernatant and resuspend the cells in 50–100 μL of PBS. Add 1 μg of labeled appropriate secondary antibody directly and incubate on ice for 30 min.

6. Repeat **step 4** (*see* **Note 7**).

7. Remove the supernatant from the cell pellet. Resuspend the cells in 500 μL of PBS.

8. Filter the cell suspension through a 70 μm mesh filter prior to flow cytometry to prevent clogging of the instrument. Then the sample is ready for flow analysis.

3.2.2 Multilineage Differentiation of ASCs

Adipogenesis

Adipogenic differentiation is induced by culturing ASCs in AM for two weeks and assessed using Oil Red O staining to indicate the intracellular lipid accumulation (*see* **Note 8**).

1. Seed the cells onto sterile cover glasses placed in culture dishes. Cells are maintained in LG-DMEM with 10% FBS and 1% P/S until 80% confluent. Then the medium is replaced by AM, which is changed every 3 days thereafter.

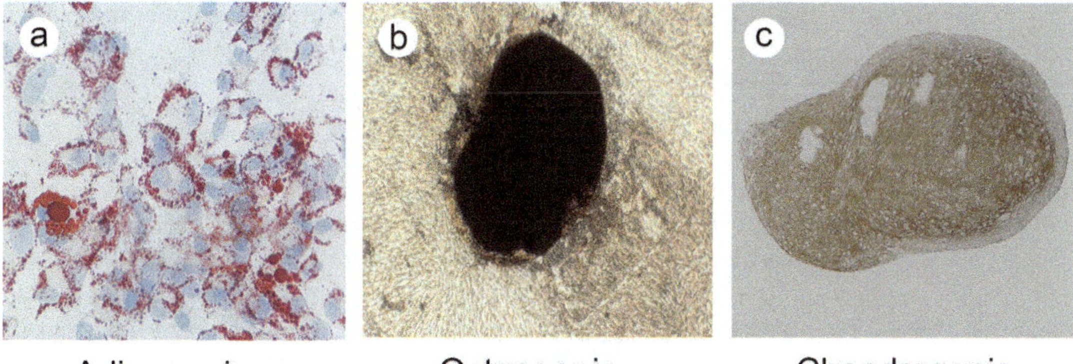

Fig. 2 In vitro multilineage differentiation of ASCs. (**a**) Adipogenic differentiation is assessed by Oil Red O staining, which demonstrates the lipid accumulation within the cells in small vacuoles (red). (**b**) Osteogenic differentiation is determined by von Kossa staining, which shows the dense nodules of extracellular matrix mineralization (black). (**c**) Chondrogenic differentiation is evaluated by immunohistochemical staining of Collagen type II (brown). Under pellet culture condition, cells are condensed into nodules and differentiated into chondrocytes embedded in lacunae, and they synthesize the proteoglycans and collagen type II necessary for cartilage formation

2. After 2 weeks of induction, the cover glasses are washed by PBS for two times and fixed in 4% PFA at room temperature for 30 min.

3. Remove the fixation solution and wash the cells with 70% ethanol.

4. Incubate the cells in 2% Oil red O staining solution for 5 min at room temperature.

5. Remove the excessive stain by washing with 70% ethanol, followed by rinsing extensively with distilled water.

6. The cells are counterstained for 2 min with hematoxylin and visualized under microscopy (Fig. 2a).

Osteogenesis

Osteogenic differentiation is induced by culturing ASCs in OM for 2 weeks and evaluated for von Kossa staining to indicate the extracellular matrix calcification (*see* **Note 9**).

1. Seed ASCs onto the sterile cover glasses in culture dishes. Cells are maintained in LG-DMEM with 10% FBS and 1% P/S until 80% confluent. Then the medium is replaced by OM, which is changed every 3 days thereafter.

2. After 2 weeks of induction, the cover glasses are washed with PBS for two times and fixed in 4% PFA at room temperature for 30 min.

3. The glasses are rinsed with distilled water and overlaid with a 1% silver nitrate solution in the dark for 30 min.

4. The cells are washed with distilled water thoroughly and placed under ultraviolet (UV) light for 60 min.

5. Cells are counterstained with 0.1% eosin in ethanol for 5 min and observed under microscopy.

Chondrogenesis

Chondrogenic differentiation is induced using the cell pellet culture method and assessed for immunohistochemical staining of the cartilage-specific collagen type II expression [8] (*see* **Note 10**).

1. Cells are harvested using trypsin/EDTA, counted and centrifuged at 1500 rpm for 5 min. Cell pellets (5×10^6 cells/pellet) are allowed to settle to the bottom of 15 mL tubes.

2. Pellets are cultured in LG-DMEM with 10% FBS and 1% P/S overnight at 37 °C in the CO_2 incubator, then chondrogenesis is induced in CM. The tube lids are loosely fastened and half media are changed twice a week.

3. After three weeks of induction, the cell pellets are washed with PBS and then fixed in 4% PFA for 30 min at room temperature.

4. Pellets are dehydrated through an ethanol series, embedded in paraffin and sliced into 10 μm thick sections.

5. The immunohistochemical staining of collagen type II is performed according to the manufacturer's instructions for the commercial antibody kit of collagen type II.

4 Notes

1. Liposuction aspiration can provide the investigators with a finely minced fat specimen without significantly altering ASC viability [7]. For the surgically excised fat sample, it should be minced thoroughly using sterile scissors before use.

2. The appropriate digestion time depends on the particle size of fat aspirates and the activity of collagenase. Enzymatic digestion is finished when obvious fat particles are no longer visible and the enzyme solution becomes turbid.

3. Cells freshly isolated from SVF are heterogenous populations containing mast cells, endothelial cells, pericytes, fibroblasts, preadipocytes and ASCs. Continuous expansion and passaging can delete these contaminations to purify ASC population.

4. Cells should be rinsed with PBS to remove serum proteins prior to antibody staining. If staining with more than one antibody, prepare a pool of antibodies together.

5. Typically, $0.5–1.0 \times 10^6$ cells in 50–100 μL of FCB are used for one sample (a test) of the flow cytometry analysis. All incu-

bations should be performed on ice with minimal light exposure.

6. If the primary antibody is labeled with a flurorchrome, skip to **step 8**.

7. If the cells cannot be analyzed by flow cytometry within 4–6 h of antibody staining, they can be fixed in 250 µL of 1% PFA at **step 4** or **6**, and stored in the dark in the 4 °C refrigerator for no more than 1 week.

8. Besides Oil Red O staining, adipogenesis can be confirmed using Nile Red staining and detection of adipose-specific genes and proteins.

9. In addition to von Kossa staining, there are several other methods to detect osteogenesis, such as mineralization staining of Alizarin Red S, staining and quantitative evaluation of alkaline phosphatase (AKP), measurement of calcium content, and detection of bone-specific genes and proteins.

10. Chondrogenesis can also be assessed by specific staining of sulfated proteoglycans (glycosaminoglycan, GAG) present in the cartilaginous matrices, such as Alcian Blue and Toluidine Blue, measurement of GAG content, and evaluation of cartilage-specific genes and proteins.

Acknowledgments

This work was financially supported by the Natural Science Foundation of China (Research number 31271027 and 81171475).

References

1. Zuk PA (2010) The adipose-derived stem cell: looking back and looking ahead. Mol Biol Cell 21(11):1783–1787

2. Gimble JM, Nuttall ME (2011) Adipose-derived stromal/stem cells (ASC) in regenerative medicine: pharmaceutical applications. Curr Pharm Des 17(4):332–339

3. Gimble JM, Bunnell BA, Guilak F (2012) Human adipose-derived cells: an update on the transition to clinical translation. Regen Med 7(2):225–235

4. Santos FD, Andrade PZ, Abecasis MM, Gimble JM, Chase LG, Campbell AM, Boucher S, Vemuri MC, Silva CL, Cabral JM (2011) Toward a clinical-grade expansion of mesenchymal stem cells from human sources: a microcarrier-based culture system under xeno-free conditions. Tissue Eng Part C Methods 17(12):1201–1210

5. Mizuno H, Tobita M, Uysal AC (2012) Adipose-derived stem cells as a novel tool for future regenerative medicine. Stem Cells 30(5):804–810

6. Dominici M, Le Blanc K, Mueller I, Slaper-Cortenbach I, Marini F, Krause D, Deans R, Keating A, Dj P, Horwitz E (2006) Cytotherapy 8(4):315–317

7. Bunnell BA, Flaat M, Gagliardi C, Patel B, Ripoll C (2008) Adipose-derived stem cells: isolation, expansion and differentiation. Methods 45(2):115–120

8. Kretlow JD, Jin YQ, Liu W, Zhang WJ, Hong TH, Zhou G, Baggett LS, Mikos AG, Cao Y (2008) Donor age and cell passage affects differentiation potential of murine bone marrow-derived stem cells. BMC Cell Biol 9:60

Chapter 14

Enzyme-Free Isolation of Adipose-Derived Mesenchymal Stem Cells

Lauren S. Sherman, Alexandra Condé-Green, Vasanth S. Kotamarti, Edward S. Lee, and Pranela Rameshwar

Abstract

Mesenchymal stem cells (MSCs) are a population of multipotent cells that can be isolated from various adult and fetal tissues, including adipose tissue. These cells contain enormous clinical and basic research appeal due to their plasticity to differentiate into cells of all germ layers in vitro, cross allogeneic barriers in vivo, and suppress inflammation. Methods to isolate adipose-derived MSCs (ADSCs) primarily rely on enzymatic digestion of the adipose tissue using harsh enzymes such as collagenase. However, these harsh enzymes are expensive and can have detrimental effects on the ADSCs, including risks of using xenograft components in clinical application. This chapter focuses on methods of isolating ADSCs from adipose tissue without enzymatic digestion.

Key words Mesenchymal stem cell, Adipose-derived stem cells

1   Introduction

The stromal vascular fraction (SVF) of adipose tissue is a heterogeneous population composed of endothelial cells, erythrocytes, fibroblasts, lymphocytes, monocytes, macrophages, pericytes, hematopoietic stem and progenitor cells, and adipose-derived stem cells (ADSCs) [1]. SVF is an attractive source of ADSCs due to the ease of collection by lipoaspiration and the large number of ADSCs recoverable. Like other populations of mesenchymal stem cells (MSCs), ADSCs are reported to be positive for CD90, CD73, CD105, and CD44; and negative for CD45 and CD31 [2–4]. ADSCs can be distinguished from their bone marrow counterparts by their expressing CD36 and lacking CD106 [2]. While the differences in antigen presentation likely indicate differences in function, the various MSC sources are considered largely similar for clinical applications [5]. This is a topic for future research.

The most common method for isolating ADSCs is a series of enzymatic digestions with collagenase, followed by centrifugation

Shree Ram Singh and Pranela Rameshwar (eds.), *Somatic Stem Cells: Methods and Protocols*, Methods in Molecular Biology, vol. 1842, https://doi.org/10.1007/978-1-4939-8697-2_14, © Springer Science+Business Media, LLC, part of Springer Nature 2018

and red blood cell lysis [6]. While efficient, digestion with animal-derived enzymes, such as collagenase, is considered more than "minimally manipulated" by the US Food and Drug Administration, altering the characteristics of the cells. Further, these xenogeneic components may result in immune reactions when the prepared cells are administered to patients [7–9]. To abate the risks of xenogeneic components, many groups are now using non-animal derived, manufactured enzymes. However, these enzymes are harsh as well, and can alter cell phenotype [10].

A non-enzymatic approach to ADSC isolation relies on the cells migrating out of the solid pieces of adipose—likely as they are drawn to the nutrient rich media. Like other populations of MSCs, the ADSCs adhere to plastic, permitting their isolation from the adipose tissue and larger SVF population.

2 Materials

2.1 Adipose Tissue Samples

1. Lipoaspirates and abdominoplasty sections are collected by plastic surgeons during routine clinical procedures. Samples are stored at room temperature and processed within 16 h.

2. Sterile razor blades, scalpels, and/or needles.

3. Falcon 3003 Vacuum-Gas Plasma Treated Tissue Culture Petri Dishes (Falcon 3003 plates).

4. Tissue Culture Media (MSC media): Dulbeccos minimal essential medium (DMEM) with high glucose with penicillin–streptomycin 10,000 U/mL (1 mL/100 mL of DMEM) and 10% defined fetal calf serum (FCS).

3 Method

3.1 Isolation of ADSCs from Adipose Tissue

1. Ensure that tissue sample is homogeneous.

2. For lipoaspirates, gently invert or shake the sample 5–6 times to ensure nonseparation of layers (*see* **Note 1**).

3. For abdominoplasty samples, carefully mince the tissue into ≤1 mm pieces. Mincing can be accomplished using a sterile needle to hold a piece of tissue in place, and a sterile scalpel to mince pieces off. Return tissue and exudate to a fresh tube.

4. Transfer 2.5–5 mL of tissue *and* exudate to a Falcon 3003 plate (*see* **Note 2**).

5. Add an equivalent volume of MSC media. Gently swirl the plate to mix the contents around the plate.

6. Incubate at 37 °C in 5% CO_2 until enough cells are seen on the base of the plate: over time, the cells will migrate from within the tissue onto the plate surface.

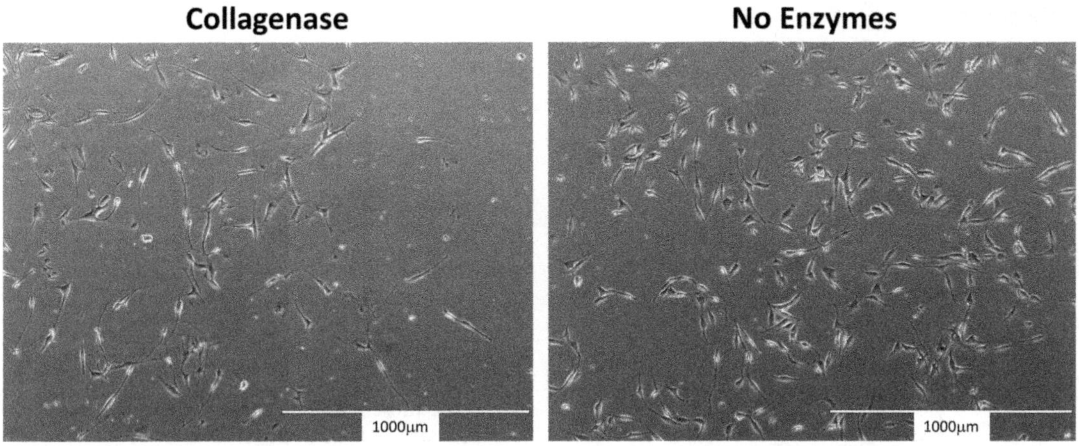

Collagenase   **No Enzymes**

1000µm   1000µm

Fig. 1 The ADSCs isolated by collagenase treatment (left panel) yield a longer, narrower morphology as compared to the traditional spindle shape seen in the nonenzymatic isolation treatment (right panel)

7. Every 24 h, remove as much fluid as possible, leaving the pieces of tissue in the plate. Replace with a similar amount of media.

8. Once enough cells are noted on the plate or after 7 days, whichever is sooner, remove all remaining pieces of tissue.

9. Culture the ADSCs as any other MSC population (*see* **Note 3**).

3.2 Summary

1. The method to isolate ADSCs in the absence of enzymatic digestion provides an opportune way to culture ADSCs without harsh enzymes or centrifugation steps. While this method may produce an initially lower cell recovery than enzymatic digestion, taking up to one week, the cells migrate out of the adipose tissue of their own accord and are thus not manipulated by the enzymatic digestion.

2. When comparing the ADSCs isolated from lipoaspirates using collagenase (0.075% Collagenase Type II, 30 min at 37 °C) and centrifugation, or as described in this chapter, both groups express similar phenotypes (CD90$^+$, CD73$^+$, CD105$^+$, CD45$^-$; data not shown). Both methods yield ADSCs capable of multiple lineage differentiation (data not shown). However, the ADSCs isolated in the absence of enzymatic digestion yield a healthier morphology than collagenase treatment (Fig. 1).

4 Notes

1. If the lipoaspirate has fully settled, the oil layer may be removed from the sample prior to mixing the sample.

2. The tissue can be measured by transferring to a centrifuge tube by pouring. If necessary, a sterile spatula may be used to assist in transferring the tissue.

3. A protocol for culturing MSCs can be found in Chapter 6.

References

1. Mitchell JB et al (2006) Immunophenotype of human adipose-derived cells: temporal changes in stromal-associated and stem cell-associated markers. Stem Cells (Dayton, Ohio) 24:376–385

2. Bourin P et al (2013) Stromal cells from the adipose tissue-derived stromal vascular fraction and culture expanded adipose tissue-derived stromal/stem cells: a joint statement of the International Federation for Adipose Therapeutics and Science (IFATS) and the International Society for Cellular Therapy (ISCT). Cytotherapy 15:641–648

3. Baptista LS et al (2009) An alternative method for the isolation of mesenchymal stromal cells derived from lipoaspirate samples. Cytotherapy 11:706–715

4. Condé-Green A et al (2014) Comparison between stromal vascular cells' isolation with enzymatic digestion and mechanical processing of aspirated adipose tissue. Plast Reconstr Surg Glob Open 134:4

5. Mattar P, Bieback K (2015) Comparing the immunomodulatory properties of bone marrow, adipose tissue, and birth-associated tissue mesenchymal stromal cells. Front Immunol 6:560

6. Conde-Green A et al (2016) Shift toward mechanical isolation of adipose-derived stromal vascular fraction: review of upcoming techniques. Plast Reconstr Surg Glob Open 4:e1017

7. Chang H et al (2013) Safety of adipose-derived stem cells and collagenase in fat tissue preparation. Aesthet Plast Surg 37:802–808

8. Spees JL et al (2004) Internalized antigens must be removed to prepare hypoimmunogenic mesenchymal stem cells for cell and gene therapy. Mol Ther 9:747–756

9. Horwitz EM et al (2002) Isolated allogeneic bone marrow-derived mesenchymal cells engraft and stimulate growth in children with osteogenesis imperfecta: Implications for cell therapy of bone. Proc Natl Acad Sci U S A 99:8932–8937

10. Tsuji K et al (2017) Effects of different cell-detaching methods on the viability and cell surface antigen expression of synovial mesenchymal stem cells. Cell Transplant 26:1089–1102

Chapter 15

Identification and Characterizations of Annulus Fibrosus-Derived Stem Cells

Qianping Guo, Pinghui Zhou, and Bin Li

Abstract

Annulus fibrosus (AF) injuries are common in degenerative disc disease (DDD) and can lead to substantial deterioration of the intervertebral disc. However, repair or regeneration of AF remains challenging. Recently, we have found that there exists a subpopulation of cells, which form colonies in vitro and could self-renew, in AF tissue. These cells express typical surface antigen molecules of mesenchymal stem cells, including CD29, CD44, and CD166. They also express common stem cell markers such as Oct-4, nucleostemin, and SSEA-4. In addition, they can be induced to differentiate into adipocytes, osteocytes, and chondrocytes. Being AF tissue-specific, such AF-derived stem cells may potentially be an ideal candidate for DDD treatments using stem cell-based cell therapies or tissue engineering approaches.

Key words Degenerative disc disease, Annulus fibrosus, Annulus fibrosus-derived stem cells, Colony-forming, Stemness, Differentiation

1   Introduction

The intervertebral disc (IVD) lies between adjacent vertebrae, jointing vertebrae and allowing their slight movements. Degenerative disc degeneration (DDD) constitutes a leading cause of low back pain which affects the vast majority of population worldwide [1]. As a major component of IVD, the annulus fibrosus (AF) plays a critical role by confining the nucleus pulposus (NP) and maintaining biomechanical properties and physiological intradiscal pressure of the disc [2]. However, AF hardly heals upon injury. Therefore, AF regeneration appears imperative for DDD treatment.

Cell therapies and tissue engineering are promising ways to AF regeneration. Due to the ageing of differentiated cells, low cellularity, limited proliferation capacity, and the intrinsic phenotype heterogeneity of AF cells, application of AF cells for AF regeneration is largely limited. Stem cells, on the other hand, are believed to be the ideal cell source for tissue engineering [3–5].

Shree Ram Singh and Pranela Rameshwar (eds.), *Somatic Stem Cells: Methods and Protocols*, Methods in Molecular Biology, vol. 1842, https://doi.org/10.1007/978-1-4939-8697-2_15, © Springer Science+Business Media, LLC, part of Springer Nature 2018

Recently, it has been suggested that stem/progenitor cells exist in AF [6–9]. Considering their AF tissue specificity, such stem/progenitor cells may be valuable for AF regeneration.

In this protocol, we isolated a population of colony-forming AF-derived cells from rabbits using cell colony forming technique. Depending on the initial cell seeding density, a fraction of cells formed colonies. These colony-forming cells have strong self-renewing capacity and expressed common markers of mesenchymal stem cells (MSCs), including CD29, CD44, CD166, Oct-4, nucleostemin (NS), and SSEA-4. They were also able to be induced to undergo adipogenic, chondrogenic, and osteogenic differentiation [9]. In brief, these cells possess clonogenicity, self-renewing capability, and multidifferentiation potential, the common characteristics of MSCs, and are therefore termed AF-derived stem cells (AFSCs). Recently, we have also found that AFSCs were sensitive to the mechanics of culture substrate and showed diversified differentiation in response to the elasticity of scaffolding materials [10, 11]. Being capable of in vitro expansion and multidifferentiation, AFSCs may be an ideal cell source for DDD treatments using cell therapy or tissue engineering approaches.

2 Materials

2.1 AFSC Isolation and Culture

1. Female New Zealand white rabbits (6–8 weeks old).
2. DMEM-LG.
3. Penicillin–streptomycin.
4. 150 U/mL collagenase I.
5. 100 U/mL collagenase II.
6. Fetal bovine serum (FBS).
7. 0.25% trypsin–EDTA.
8. 0.36% crystal violet.
9. Cell Counting Kit 8.

2.2 AFSC Characterizations

2.2.1 Immunostaining

1. 4% paraformaldehyde.
2. 4% bovine serum albumin (BSA).
3. Mouse anti-human Oct-4 antibody (1:500, Millipore, Cat. No. MAB4401).
4. Cy3-conjugated goat anti-mouse secondary antibody (1:1000, Invitrogen Molecular Probes, Cat. No. A10521).
5. 5 µg/mL DAPI.
6. Goat anti-human nucleostemin antibody (1:250, Neuromics, Cat. No. GT15050).

7. Cy3-conjugated donkey anti-goat IgG secondary antibody (1:1000, Millipore, Cat. No. AP180C).

8. Mouse anti-human SSEA-4 antibody (1:200, Invitrogen Immunodetection, Cat. No. 41-4000).

9. Cy3-conjugated goat anti-mouse secondary antibody (1:1000, Invitrogen Molecular Probes, Cat. No. A10521).

2.2.2 Differentiation

1. Adipogenic medium consisting of basic culture medium supplemented with insulin–transferrin–sodium selenite mix (ITS) containing 12 mg/mL human insulin, 11.4 mg/mL transferrin, and 15 ng/mL sodium selenite.

2. Osteogenic medium consisting of basic culture medium supplemented with 0.1 µM dexamethasone, 0.2 mM ascorbic-2-phosphate, and 10 mM glycerol 2-phosphate.

3. Chondrogenic medium consisting of basic culture medium supplemented with 40 µg/mL proline, 39 ng/mL dexamethasone, 10 ng/mL TGF-β 3, 50 µg/mL ascorbate 2-phosphate, 100 µg/mL sodium pyruvate, and 4 mg/mL human insulin, 3.8 mg/mL transferrin, and 5 ng/mL sodium selenite (ITS mix) when cells reached 80% confluence.

4. Oil Red O.

5. Alizarin Red S.

6. Safranin O.

2.2.3 Polymerase Chain Reaction (PCR)

1. TRIzol.

2. cDNA reverse transcription kit.

3. DNase I.

---

3 Methods

3.1 Isolation and Culture of AFSCs

3.1.1 Preparation

1. Prepare PBS containing 5% penicillin and streptomycin using sterile technique in a biological safety cabinet (BSC).

2. Prepare DMEM-LG containing 15% FBS and 1% penicillin–streptomycin.

3. Prepare collagenase I and collagenase II solution in DMEM-LG, filter the solution using sterile syringe filter (0.45 µm pore hydrophilic PVDF membrane).

3.1.2 Isolation and Culture of AFSCs

1. Sacrifice the rabbit using air injection. After sacrifice, dissect IVDs and put them in DMEM-LG supplemented with 5% penicillin–streptomycin.

2. In an aseptic operating board, wash the tissue using PBS containing 5% penicillin–streptomycin in 6-well plate. Harvest AF

tissue after removing the nucleus pulposus, end plates and soft tissues surrounding AF.

3. Mince the AF tissue into small pieces and digest them using 150 U/mL collagenase I and 100 U/mL collagenase II in DMEM-LG medium.

4. Incubate plate in incubator (37 °C, 5% CO_2) for about 3–4 h to digest the tissue, mix the solution one time per hour.

5. After digestion, filter the cell–tissue mixture with 200 mesh filter, collect suspensions and centrifuge cells at 1000 rpm ($100 \times g$) for 10 min.

6. Resuspend the cell pellet in culture medium (DMEM-LG supplemented with 15% FBS, 1% penicillin–streptomycin) (*see* **Notes 1** and **2**) and seed in 100 mm dish or 6-well plate at a density of 200 cells/cm² (*see* **Note 3**). Cells are maintained in humiliated incubator at 37 °C with 5% CO_2 (*see* **Note 4**).

7. Change medium to remove unattached cells after 24 h. Change medium every 2 days until the cells reach subconfluent.

3.1.3 Colony Formation

To observe the colony formation of AFSCs, P0 cells are stained with crystal violet after being cultured for 5–7 days in a 6-well plate.

1. Remove medium and wash P0 cells twice with PBS.

2. Fix cells with 4% paraformaldehyde for 15 min at room temperature.

3. Rinse cells with PBS for three times and stain cells with 0.07% crystal violet in the dark at 4 °C.

4. View cells using an inverted microscope and count the colony number (Fig. 1).

3.1.4 Cell Proliferation

1. To test the proliferation capacity of AFSCs, P2 cells are used.

2. Seed the cells in 96-well plate at a density of 5000 cells/well.

3. Measure the OD values of cells at 1st, 3rd, 5th, and 7th day using Cell Counting Kit 8.

3.2 Characterizations of AFSCs

3.2.1 Immuno fluores- cence

To characterize the stemness of AFSCs, the expression of typical MSC markers, including Oct-4, nucleostemin (NS), and SSEA-4, is examined using immunofluorescence. In this assay, P0–to P3 AFSCs are used.

1. Remove medium and wash cells with PBS for three times.

2. Fix cells in 4% polyformaldehyde for 15 min at room temperature.

3. Wash cells with PBS three times and treat cells with methanol at −20 °C for 5 min.

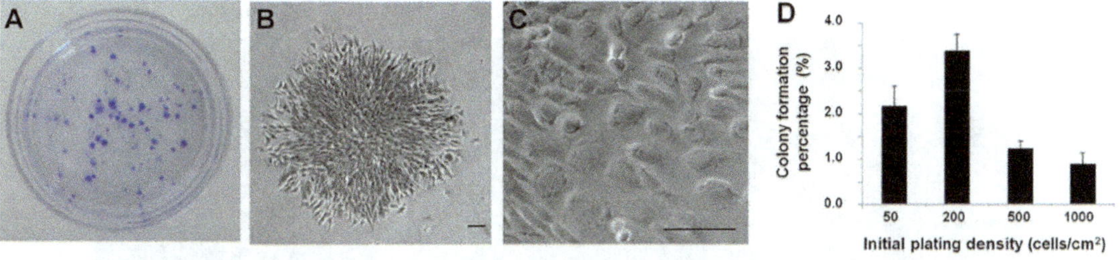

Fig. 1 The colony formation capacity of rabbit AFSCs. (**A**) Total colonies formed at 10 days and stained using crystal violet. (**B**) A typical cell colony. (**C**) Morphology of cells in a colony. (**D**) Colony forming unit assay for AFSCs. Scale bars, 100 μm

4. For Oct-4 immunofluorescence, wash cells with PBS three times and block cells with 4% BSA for 30 min.

5. Incubate cells with mouse anti-human Oct-4 antibody (1:500) overnight at 4 °C.

6. After washing cells with PBS for three times, incubate cells with Cy3-conjugated goat anti-mouse secondary antibody (1:1000) for 1 h at room temperature.

7. Wash cells with PBS for three times, incubate cells with DAPI solution for 15 min.

8. Wash cells with PBS for three times and observe cells using fluorescence inverted microscope.

9. For nucleostemin and SSEA-4 staining, goat anti-human nucleostemin antibody (1:250) along with Cy3-conjugated donkey anti-goat IgG secondary antibody (1:1000, Millipore, Cat. No. AP180C), mouse anti-human SSEA-4 antibody (1:200, Invitrogen Immunodetection, Cat. No. 41–4000) along with Cy3-conjugated goat anti-mouse secondary antibody (1:1000, Invitrogen Molecular Probes, Cat. No. A10521) are used, respectively (Fig. 2).

3.2.2 qPCR

To further confirm that the cells derived from AF tissue are stem cells, typical MSC-associated surface antigens CD29, CD44, and CD166 are checked. The expression of antigens is examined using qPCR (*see* **Note 5**).

1. Harvest cells at P0-P3.

2. Extract total RNA using TRIzol.

3. Treat mRNA with DNase I to remove residual DNA sequences.

4. Reverse-transcribe the total RNA to cDNA.

5. In a 20 μL PCR reaction mixture, use cDNA product of 200 ng mRNA as template. Evaluate the expression levels of CD29, CD44, CD166, CD4, CD8, CD14, and housekeeping gene glyceraldehydes-3-phosphate dehydrogenase (GAPDH). All the primer sequences were listed in Table 1.

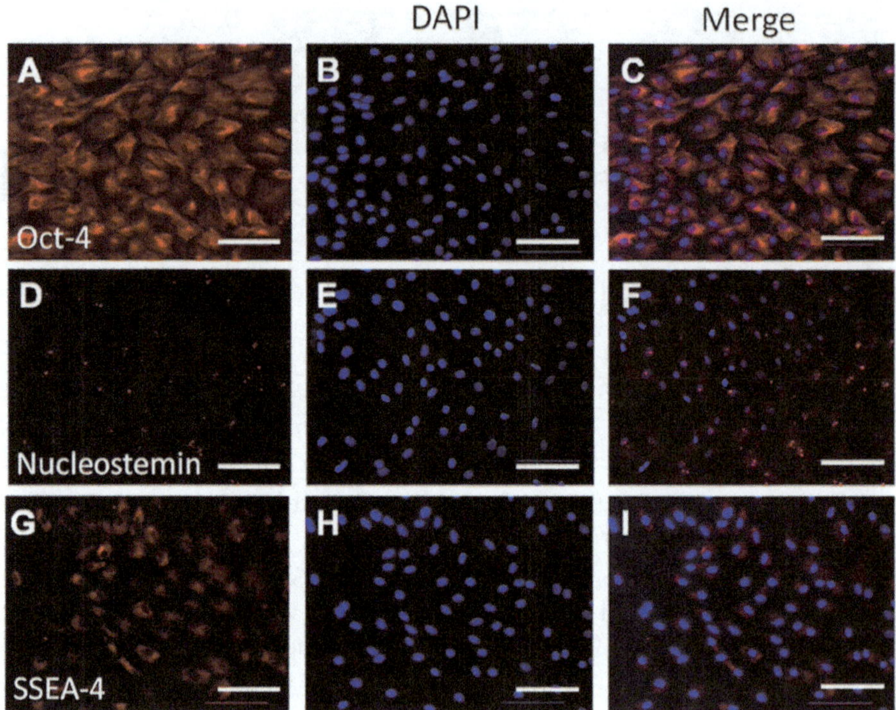

Fig. 2 Immunofluorescence for stem cell marker expression in AFSCs. As seen, AFSCs are positive for Oct-4 (**A–C**), nucleostemin (**D–F**) and SSEA-4 (**G–I**). Scale bars, 200 μm

3.2.3 Induced Differentiation

1. The differentiation potential of AFSCs is evaluated in vitro for adipogenesis, osteogenesis, and chondrogenesis.

2. P2-P4 AFSCs are seeded at a density of 4×10^4 cells/well in a 24-well plate in basic culture medium (DMEM-LG supplemented with 10%FBS, 1% penicillin–streptomycin).

3. In experiment groups cells are induced with respective induction media, while in control groups all cells are cultured in basic culture medium. Medium are changed every 3 days.

3.2.4 Osteogenesis and Alizarin Red S Assay

1. Induce cells with osteogenic medium when cells reach 80% confluence. Change medium every 3 days.

2. After induction for 3 weeks, remove medium wash cells with PBS for three times.

3. Fix cells in 4% paraformaldehyde for 40 min at room temperature.

4. Wash cells with ddH_2O for three times and subsequently stain cells with Alizarin Red S solution for 1 h.

5. Wash cells with ddH_2O carefully till the background was clean.

6. Observe the stained samples using an inverted microscope (Fig. 3a).

Table 1
Primers used for RT-PCR

Gene	Size (bp)	Primers	Type	Tm (°C)	Gene Bank#
CD29	242	5′-GTCACCAACCGTAGCAA-3′	Forward	58	AY195896.1
		5′-CTCCTCATCTCATTCATCAG-3′	Reverse		
CD44	191	5′-CGATTTGAATATAACCTGCCGC-3′	Forward	63	FJ360436.1
		5′-CGTGCCCTTCTATGAACCCA-3′	Reverse		
CD166	200	5′-GGACAGCCCGAAGGAATACGAA-3′	Forward	63	Y13243.1
		5′-GACACAGGCAGGGAATCACCAA-3′	Reverse		
GAPDH	107	5′-ACTTTGTGAAGCTCATTTCCTGGTA-3′	Forward	58	L23961
		5′-GTGGTTTGAGGGCTCTTACTCCTT-3′	Reverse		
PPARγ	130	5′-CATTTTCTCAAGCAACAGTC-3′	Forward	54	NM_001082148.1
		5′-CAAAGGAGTGGGAGTGGT-3′	Reverse		
LPL	487	5′-GGCGAGACGCACGAACA-3′	Forward	54	FJ429312.1
		5′-CACCCGCAGTACAAACCCA-3′	Reverse		
Collagen I	81	5′-CTGACTGGAAGAGCGGAGAGTAC-3′	Forward	58	AY633663
		5′-CCATGTCGCAGAAGACCTTGA-3′	Reverse		
Runx-2	154	5′-CAGGCAGTTCCCAAGCATTTCA-3′	Forward	67	AY598934
		5′-TGGTGGCAGGTAGGTATGGTAGT-3′	Reverse		
Collagen II	84	5′-TGGGTGTTCTATTTATTTATTGTCTTCCT-3′	Forward	62	S83370
		5′-GCGTTGGACTCACACCAGTTAGT-3′	Reverse		
Sox-9	261	5′-TACGACTGGACGCTGGTGC-3′	Forward	67	AY598935
		5′-CGGGTGGTCTTTCTTGTGCT-3′	Reverse		

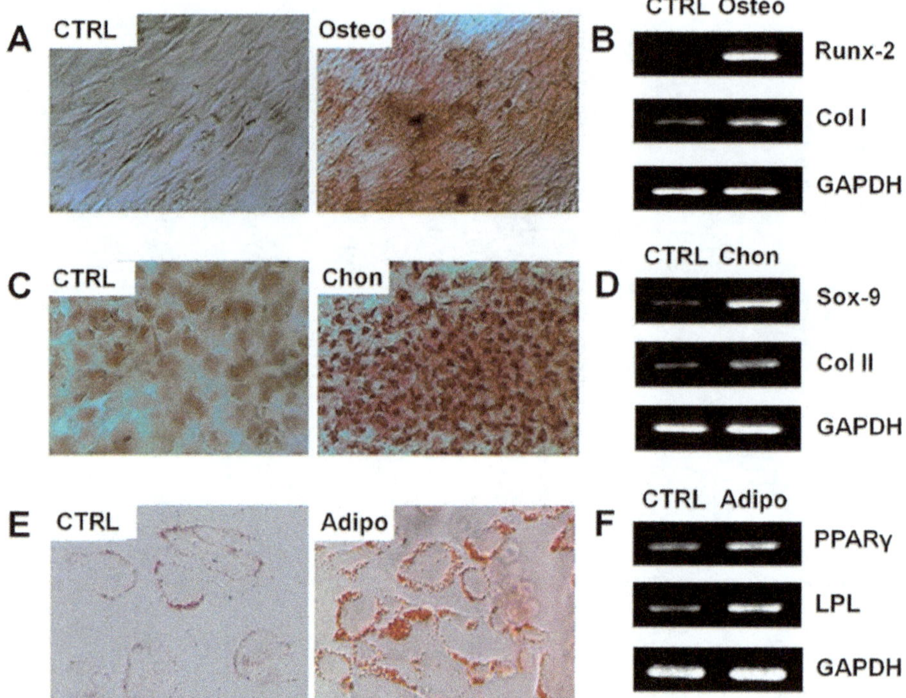

Fig. 3 Induced differentiation of AFSCs. (**A**, **B**) Osteogenic differentiation: mineralization is stained with Alizarin red S (**A**) and expression of Runx-2 and Col I genes is analyzed using RT-PCR (**B**); (**C**, **D**) Chondrogenic differentiation: cells are stained with Safranin O (**C**) and expression of chondrocyte-specific genes Sox-9 and Col II levels is analyzed using RT-PCR (D); (**E**, **F**) Adipogenic differentiation: secretion of oil droplets is visualized using Oil Red O staining (**E**) and expression of adipocyte-specific genes PPAR-c and LPL is analyzed using RT-PCR (**F**)

3.2.5 Chondrogenesis and Safranin O Assay

1. Induce cells with chondrogenic medium when cells reach 100% confluence. Change medium every 3 days.

2. After induction for 3 weeks, remove medium wash cells with PBS for three times.

3. Fix cells in 4% paraformaldehyde for 40 min at room temperature.

4. Wash cells with PBS for three times and subsequently stain cells with Safranin O solution for 1 h.

5. Wash cells with ddH_2O carefully till the background was clean.

6. Observe the stained samples using an inverted microscope (Fig. 3c).

3.2.6 Adipogenesis and Oil Red O Assay

1. Culture cells with adipogenic medium when cells are full confluent. Change medium every 3 days.

2. After induction for 2 weeks, remove medium wash cells with PBS for three times.

3. Fix cells in 4% paraformaldehyde for 40 min at room temperature.

4. Wash cells with PBS for three times and subsequently stain cells with 0.36% Oil Red O solution for 1 h.

5. Wash cells with ddH$_2$O carefully till the background is clean.

6. Observe the stained samples using an inverted microscope (Fig. 3e).

3.2.7 Gene Analysis

1. Extract total RNA using TRIzol.

2. Treat mRNA with DNase I to remove residual DNA sequences.

3. Reverse-transcribe the total RNA to cDNA.

4. In a 20 μL PCR reaction mixture, use cDNA product of 200 ng mRNA as template. Evaluate the expression levels of bone-specific genes (collagen type I, Runx 2), chondrocyte specific genes (collagen type II, Sox-9), and housekeeping gene glyceraldehydes-3-phosphate dehydrogenase (GAPDH) (Fig. 3b, d, f). All the primer sequences are listed in Table 1.

4 Notes

1. AFSC growth is dependent on the formulation of culture media. For example, when cultured in DMEM-HG, oil droplet formation is easily seen in cells. However, this phenomenon can be avoided by using DMEM-LG or a-MEM.

2. AFSCs are highly capable of proliferation. Fifteen percent, instead of 10%, of FBS, should be used for culturing them.

3. Colony formation capacity is affected by initial cell seeding density. An initial seeding density of 200 cells/cm^2 results in the best colony forming efficiency, in which about 3.4% of plated cells form colonies.

4. Oxygen concentration also affects cell behavior. Normal oxygen concentration (21%) leads to cell ageing and reduces their proliferation capacity, while low oxygen concentration (5%) can help maintain the spindle-like morphology of cells and promote their growth.

5. Due to limited availability of anti-rabbit antibodies, the expression of stem cell markers is examined using qPCR instead of flow cytometry.

Acknowledgments

This work was supported by the National Natural Science Foundation of China (81171479, 31530024, and 81672213), National Key R&D Program of China (2016YFC1100203),

References

1. Luo X, Pietrobon R, Sun SX, Liu GG, Hey LA (2004) Estimates and patterns of direct health care expenditures among individuals with back pain in the United States. Spine 29:79–86

2. Hudson KD, Alimi M, Grunert P, Hartl R, Bonassar LJ (2013) Recent advances in biological therapies for disc degeneration: tissue engineering of the annulus fibrosus, nucleus pulposus and whole intervertebral discs. Curr Opin Biotechnol 24:872–879

3. Paesold G, Nerlich AG, Boos N (2007) Biological treatment strategies for disc degeneration: potentials and shortcomings. Eur Spine J 16:447–468

4. Horner HA, Roberts S, Bielby RC, Menage J, Evans H, Urban JPG (2002) Cells from different regions of the intervertebral disc: effect of culture system on matrix expression and cell phenotype. Spine 27:1018–1028

5. Sabatino M, Ren J, David-Ocampo V, England L, McGann M, Tran M et al (2012) The establishment of a bank of stored clinical bone marrow stromal cell products. J Transl Med 10:23

6. Henriksson H, Thornemo M, Karlsson C, Hagg O, Junevik K, Lindahl A et al (2009) Identification of cell proliferation zones, progenitor cells and a potential stem cell niche in the intervertebral disc region: a study in four species. Spine 34:2278–2287

7. Henriksson HB, Svala E, Skioldebrand E, Lindahl A, Brisby H (2012) Support of concept that migrating progenitor cells from stem cell niches contribute to normal regeneration of the adult mammal intervertebral disc: a descriptive study in the New Zealand white rabbit. Spine 37:722–732

8. Henriksson HB, Lindahl A, Skioldebrand E, Junevik K, Tangemo C, Mattsson J et al (2013) Similar cellular migration patterns from niches in intervertebral disc and in knee-joint regions detected by in situ labeling: an experimental study in the New Zealand white rabbit. Stem Cell Res Ther 4:104

9. Liu C, Guo Q, Li J, Wang S, Wang Y, Li B et al (2014) Identification of rabbit annulus fibrosus-derived stem cells. PLoS One 9:e108239

10. Guo Q, Liu C, Li J, Zhu C, Yang H, Li B (2015) Gene expression modulation in TGF-beta3-mediated rabbit bone marrow stem cells using electrospun scaffolds of various stiffness. J Cell Mol Med 19:1582–1592

11. Zhu C, Li J, Liu C, Zhou P, Yang H, Li B (2016) Modulation of the gene expression of annulus fibrosus-derived stem cells using poly(ether carbonate urethane)urea scaffolds of tunable elasticity. Acta Biomater 29:228–238

Chapter 16

Maintenance of Tendon Stem/Progenitor Cells in Culture

Jianying Zhang and James H.-C. Wang

Abstract

Tendon stem/progenitor cells (TSCs) are tendon-specific adult stem cells, which play crucial roles in tendon homeostasis, repair or regeneration once tendons are injured. Additionally, their pathological role in the development of tendinopathy in response to excessive mechanical loading placed on the tendon is also implicated. Similar to other adult stem cells, TSCs also exhibit universal characteristics of stem cells including colony formation in culture, self-renewal and multidifferentiation potential. Nevertheless, once TSCs are isolated from tendinous tissues and cultured in vitro, they may quickly lose stemness by undergoing differentiation. To maintain and prolong the stemness of TSCs in culture, we have developed two effective methods, namely culturing TSCs in hypoxic condition, or in growth media supplemented with low levels of PGE_2. Here we present these methods in detail, along with the detailed description of the procedures to isolate TSCs from tendon samples and to culture them in vitro.

Key words Tendon stem cells, Tenocytes, Self-renewal, Differentiation, Hypoxia, Prostaglandin E_2

1 Introduction

Tendons are a band of connective tissues responsible for transmitting muscular forces to bone to enable joint movements. Tendon injuries are very common in both occupational and athletic settings affecting millions of people. The restoration of tendon structure and function after injury still remains as one of the greatest challenges in orthopedic surgery and sports medicine.

In pursuit of new strategies to promote tendon regeneration after injury, there has been extensive progress and interest in tendon research. One of the significant progresses in tendon research field is the identification of a new population of tendon cells termed tendon stem/progenitor cells (TSCs) from humans, mice, rats, and rabbits [1–4]. Although TSCs represent less than 5% of the tendon cell population, they can self-renew and differentiate into tenocytes, which are the dominant cell type in tendons that maintain tendon homeostasis and repair of tendons once injured. It is now known that while TSCs are capable of differentiating into tenocytes, under certain pathophysiological conditions they can

Shree Ram Singh and Pranela Rameshwar (eds.), *Somatic Stem Cells: Methods and Protocols*, Methods in Molecular Biology, vol. 1842, https://doi.org/10.1007/978-1-4939-8697-2_16, © Springer Science+Business Media, LLC, part of Springer Nature 2018

differentiate into nontenocytes, including adipocytes, chondrocytes, and osteocytes [1, 2].

There are many inherent differences between TSCs and tenocytes. Compared to tenocytes, TSCs proliferate more quickly in culture and possess multidifferentiation potential. Both types of cells also differ in morphology: under confluent conditions in vitro, TSCs appear as small, cobblestone-shaped cells with large nuclei, as compared to tenocytes that grow as highly elongated, fibroblast-like cells with smaller nuclei [2]. TSCs also express several stem cell markers including Oct-4, SSEA-1/4, and nucleostemin as opposed to tenocytes.

Like any other adult stem cells, TSCs tend to lose their stemness in culture. This particular limitation presents a challenge in using TSCs in cell therapy approach to repair or regenerate injured tendons. Additionally, it raises challenges to study TSC mechanobiology in vitro, because such an in vitro research approach requires "authentic" TSCs in order to understand how tendons maintain their homeostasis and how they develop tendinopathy, a debilitating tendon disorder that affects millions of Americans every year. In this chapter, we present the two methods we have developed to maintain the stemness of TSCs in culture for a prolonged time [5, 6]. The description of techniques to isolate and culture TSCs in vitro is also provided in detail.

2   Materials

2.1 Isolation of TSCs and Cell Culture

1. Collagenase type I.
2. Dispase.
3. Penicillin and streptomycin (×100).
4. Fetal bovine serum (FBS).
5. Dulbecco's modified Eagle's medium (DMEM), high glucose, GlutaMAX™, Pyruvate.
6. Centrifuge tubes (15 mL).
7. Culture plates: 96-well plate, 6-well plate, 12-well plate, 24-well plate. Syringe filters (0.22 μm).
8. Culture flasks: Nunc™ cell culture treated Eas Y Flasks™ (25 cm²), Nunc™ cell culture treated Eas Y Flasks™ (75 cm²).
9. Mesenchymal stem cell adipogenesis kit.
10. Mesenchymal stem cell osteogenesis kit.
11. StemPro Chondrogenesis Differentiation kit.
12. Trigas incubator.

2.2 Immunostaining of Stem Cell Markers

1. Paraformaldehyde solution (4% in PBS).
2. Triton X-100 (0.1% working solution).

3. Goat anti-octamer binding protein-4 (Oct-4) antibody (Santa Cruz Biotechnology, Inc., Cat. No sc-8629).

4. Goat anti-nucleostemin antibody (Neuromics, Cat. No GT15050).

5. Rabbit anti-Nanog antibody (Abcam, Cat. No ab106465).

6. Mouse anti-stage-specific embryonic antigen-4 (SSEA-4) antibody (Invitrogen, Cat. No 414000).

7. Mouse anti-stro-1 antibody (Invitrogen, Cat. No 398401).

8. Mouse anti-CD90 antibody (Novus Biological, Cat. No NB100–65543).

9. Rabbit anti-CD44 antibody (Abcam, Cat. No ab157107).

10. Rabbit anti-CD34 antibody (Abcam, Cat. No ab81289).

11. Rabbit anti-CD45 antibody (Abcam, Cat. No ab10558).

12. Mouse anti-collagen type I antibody (Abcam, Cat. No ab 90,395).

13. Mouse anti-collagen II antibody (Abcam, Cat. No 3092).

14. Mouse anti-collagen III antibody (Abcam, Cat. No 7778).

15. Mouse anti-PPARγ antibody (Santa Cruz Biotechnology, Inc., Cat. sc-271,392).

16. Mouse anti-osteocalcin antibody (Santa Cruz Biotechnology, Inc., Cat. sc-74,495).

17. Cyanine-(Cy3-) conjugated goat anti-mouse immunoglobulin G (IgG) secondary antibody (Invitrogen, Cat. No A10521).

18. Cyanine-(Cy3-) conjugated donkey anti-goat immunoglobulin G (IgG) secondary antibody (Millipore, Cat. No AP180C).

19. Cyanine-(Cy3-) conjugated goat anti-rabbit immunoglobulin G (IgG) secondary antibody (EMD Millipore, Cat. No AP132C).

20. FITC-conjugated goat anti-rabbit IgG antibody (Abcam, Cat. No 6717).

21. Prostaglandin E_2 (PGE_2) (Santa Cruz Biotechnology, Inc., Cat. No 363-24-6).

22. Hoechst 33,342 (Sigma, Cat. No 33270).

23. Nikon Eclipse TE2000-U Inverted BF PhaseCont Fluorescence microscope.

24. Automated cell counter.

25. BD LSR II flow cytometer.

2.3 Quantitative Real-Time PCR (qRT-PCR)

1. RNeasy Mini Kit.

2. SuperScript™ First-Strand Synthesis System for RT-PCR kit.

3. QuantiTect SYBR Green PCR Kit.

4. Applied Biosystems StepOnePlus™ Real-Time PCR system.

5. Primers for stem cell marker genes including Oct-4 and Nanog; tenocyte-related genes including collagen I, collagen III, tenascin-C, and tenomodulin; non-tenocyte-related genes, including PPARγ for adipocyte, Sox-9 for chondrocyte, and Runx-2 for osteocyte, and Glyceraldehyde-3-phosphate dehydrogenase (GAPDH) are synthesized by Invitrogen.

2.4 Preparation

1. 1% Triton X-100 solution: dissolve 1 mL of Triton X-100 in 99 mL of PBS solution.

2. Tissue washing buffer with 1% of penicillin and streptomycin each in PBS (PBS-P/S buffer).

3. Tissue culture medium: add 20% of FBS, 1% of penicillin and streptomycin each in DMEM (20% FBS-1% P/S-DMEM).

4. Digestion solution: dissolve 3 mg of collagenase and 4 mg of dispase in 1 mL of PBS and sterilized collagenase–dispase solution using a sterile syringe filter (0.22 μm) (*see* **Note 1**).

3   Methods

All procedures are carried out in a biological safety cabinet under sterile conditions.

3.1 Isolation of TSCs from Patellar and Achilles Tendons of Animals

1. Sacrifice the animals, spray the skin of the tendon areas with 70% ethanol, then carefully dissect the patellar tendon (Fig. 1a) and Achilles tendon (Fig. 1b) immediately (*see* **Note 2**).

2. Place the tendon tissues in a tissue culture dish and wash the tendon tissues once with PBS-P/S solution.

3. Remove the paratenons from the tendon tissues, cut the core portions of the patellar tendon or Achilles tendon into small pieces and weigh the tendon pieces (*see* **Note 3**).

4. Digest 100 mg of tissue pieces with 1 mL of collagenase–dispase solution in a 15 mL of sterile centrifuge tube at 37 °C for 1 h. The appropriate time depends on the size of the tendon pieces and the collagenase activity.

5. Centrifuge tendon digestion solution at 1500 g for 15 min and discard the enzyme-containing supernatant.

6. Resuspend the pellet to make a single-cell suspension and culture the cells with tissue culture medium.

3.2 Clonal Culture

The colony formation should be the first examined as one of the stem cell characteristics.

1. Seed the single-cell suspension (Fig. 2a) in either a 96-well plate (1 cell/well, Fig. 2b) or T25 flask (5×10^5/flask, Fig. 2c)

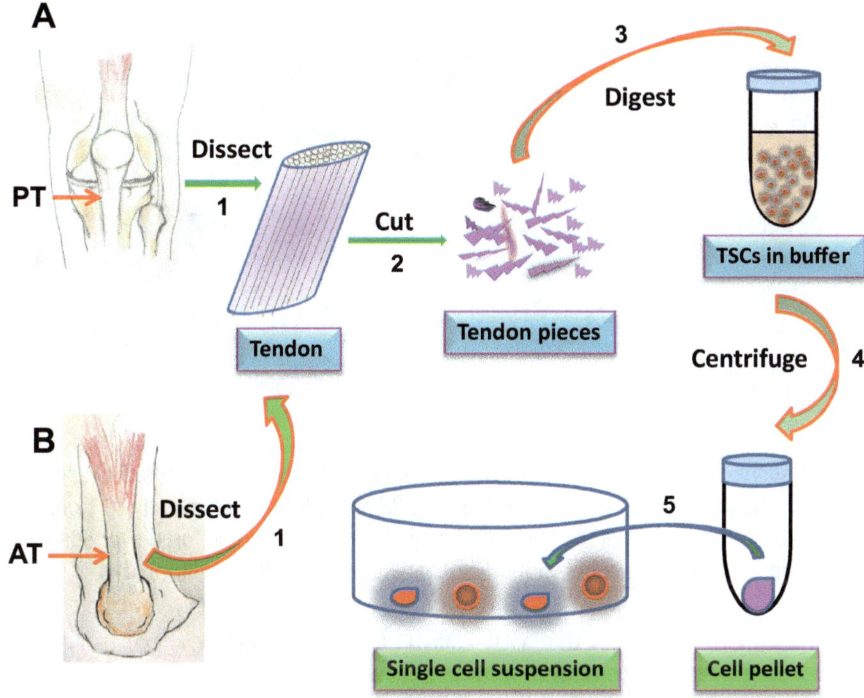

Fig. 1 Isolation of TSCs from the patellar and Achilles tendons of animals. (**a**) A drawn image of patellar tendon (PT). (**b**) A drawn image of Achilles tendon (AT). TSCs are isolated from tendon tissues by five steps. (**1**) Tendon is dissected from the animal legs. (**2**) The dissected tendon is cut into small pieces. (**3**) The tendon pieces are digested with collagenase (3 mg/mL) and dispase (4 mg/mL) in a centrifuge tube at 37 °C for 1 h. (**4**) The enzyme solution is removed by centrifugation at 1500 g for 15 min. (**5**) The cell pellet is resuspended in 20% FBS-1% P/S-DMEM to make a single-cell suspension

and culture in tissue culture medium (150 μL/well for 96-well plate and 4 mL for T25 flask)) at 37 °C with 5% CO_2.

2. Add fresh medium (100 μL/well for 96-well plate and 2 mL/ flask for T25 flask) at day-5 and change the medium every three days after 10 days of culture.

3. Monitor the colony formation every day on a microscope (Fig. 2d) and collect the individual cell colonies from each well of 96-well plate at day-21. Stain the cell colonies with methyl violet and count colony numbers and total cell number of all colonies using an automated cell counter for continued culture (*see* **Note 4**).

3.3 Immunostaining Stem Cell Markers Expressed in TSCs

The expression of stem cell markers by TSCs is examined by immunocytochemistry using the following staining protocol (Fig. 3a).

1. Culture the cells isolated from individual cell colonies at passage 1 in a 12-well plated at a density of 3×10^4/well with tissue culture medium for 5 days.

2. Remove the medium from each well and wash the cells with PBS once.

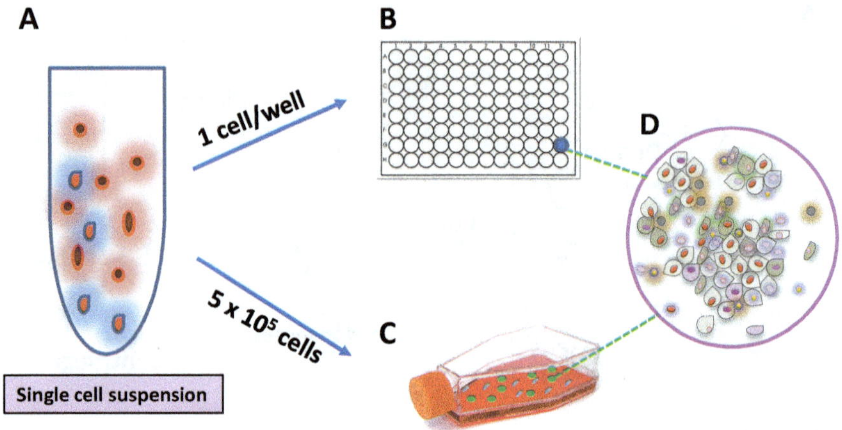

Fig. 2 Colony formation of TSCs isolated from the patellar and Achilles tendons of rats. A single-cell suspension isolated from rat patellar tendon (**a**) is seeded either in a 96-well plate at the density of 1 cell/well (**b**) or in a T25 flask at a density of 5×10^5/flask (**c**). The morphology of the TSCs is monitored every day under a microscope and TSCs form colonies with cobblestone-like shape after 2 weeks in culture (**d**)

3. Fix the cells with PBS-buffered 4% paraformaldehyde for 20 min.

4. Treat the fixed cells with 0.1% Triton X-100 for 15 min for testing the expression of nucleostemin (NS), Oct-4, Nanog, SSEA-4, Stro-1, CD90, CD44, CD34, and CD45 in TSCs.

5. Incubate the cells with the first antibodies including goat anti-Oct-4 antibody (1:300 dilution), goat anti-nucleostemin (1:350 dilution), rabbit anti-Nanog antibody (1:350 dilution), mouse anti-SSEA-4 antibody (1:350 dilution), mouse anti-stro-1 antibody (1:350 dilution), mouse anti-CD90 antibody (1:350 dilution), rabbit anti-CD44 antibody (1:200 dilution), rabbit anti-CD34 antibody (1:250 dilution), and rabbit anti-CD45 antibody (1:200 dilution) overnight at 4 °C (*see* **Note 5**).

6. Wash the cells three times with PBS, incubate the cells with appropriate species-specific secondary antibodies (1:500 dilution) at room temperature for 2 h, such as cy3-conjugated donkey anti-goat IgG antibody, to detect nucleostemin and Oct-4; or cy3-conjugated goat anti-rabbit antibody, to detect Nanog; or cy3-conjugated goat anti-mouse IgG antibody, to detect SSEA-4, CD90, and Stro-1; or FITC-conjugated goat anti-rabbit IgG, to detect CD44, CD34, and CD45.

7. Wash the cells with PBS for another three times and count total cell number after counterstaining with Hoechst 33,342.

8. Take the fluorescent images of the stained cells using a CCD camera on an inverted fluorescent microscope. The stemness is presented by the percentage of positively stained cell number in total cell number (Fig. 3b–j).

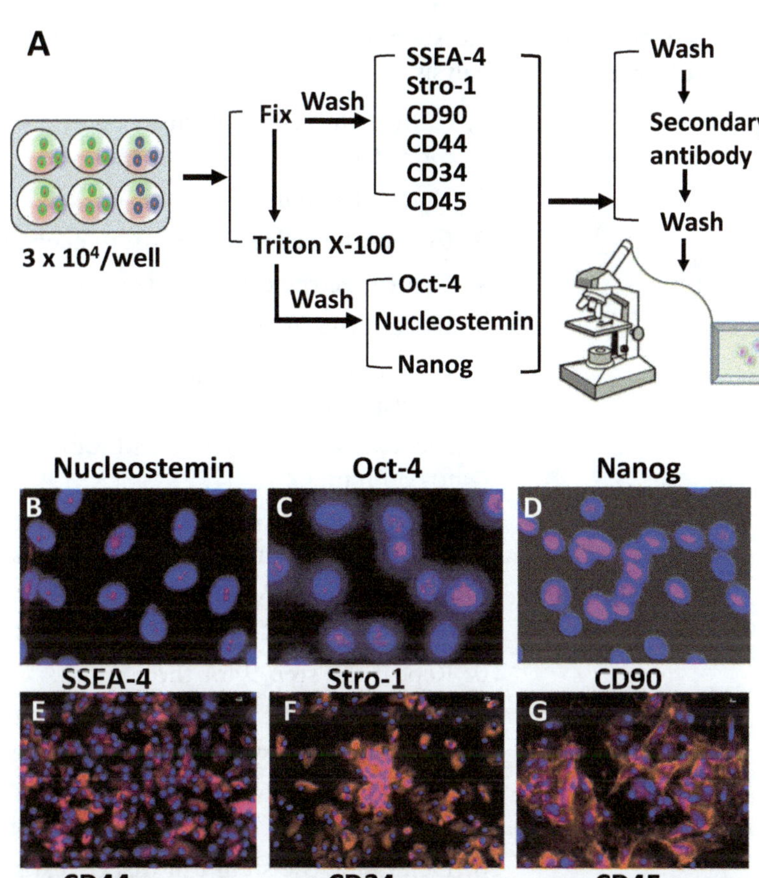

Fig. 3 Immunostaining of stem cell markers. (**a**) A brief protocol used for stem cell marker testing. The TSCs at passage 1 are seeded in a 12-well plate at a density of 3×10^4/well and cultured in 20% FBS-1% P/S-DMEM for 5 days. After washing the cells with PBS, the TSCs are fixed with 4% paraformaldehyde for 20 min at room temperature. For testing stem cell markers including Oct-4, Nanog, and nucleostemin, the cells are treated with 0.1% Triton X-100 for 15 min at room temperature. The cells are washed again and incubated with primary antibodies at 4 °C overnight. Then the cells are washed three times with PBS and incubated either with Cy3-conjugated or FITC-conjugated secondary antibodies at room temperature for 2 h. The cells are washed with PBS for 3–5 times and counted the total cell numbers by a counterstaining with Hoechst 33,342. The fluorescent images of the stained cells are taken by a CCD camera on an inverted fluorescent microscope. The stemness of the TSCs are expressed by positive staining on nucleostemin (**b**), Oct-4 (**c**), Nanog (**d**), SSEA-4 (**e**), Stro-1 (**f**), CD90 (**g**), CD44 (**h**) and negative staining on CD34 (**i**) and CD45 (**j**)

3.4 Flow Cytometry Analysis of TSCs

The stemness of TSCs is also investigated using flow cytometry analysis.

1. Culture the cells (10^6) in a T75 flask with tissue culture medium for 5 days.

2. Detach the cells with 5 mL of 0.25% trypsin for 5 min.

3. Collect the cells in a 15 mL of centrifuge tube.

4. Centrifuge the cells at 1000 *g* for 5 min.

5. Discard the supernatant and wash the cells once with PBS.

6. Centrifuge the cells again at 1000 *g* for 5 min and discard the supernatant.

7. Incubate the cells with 50 µL of the appropriate serum at 4 °C for 30 min.

8. Add 2 µL of primary antibody into the cells and incubate the cells at room temperature for 1 h (*see* **Note 6**).

9. Wash the cells with 0.5 mL of 2% FBS-PBS and remove the supernatant by centrifuge at 1000 *g* for 5 min (*see* **Note 7**).

10. Repeat the **step 9** for three times.

11. Add 1 µL of the appropriate secondary antibody into the cells and incubate the cells at room temperature for 1 h.

12. Wash the cells twice with 0.5 mL of PBS each time and centrifuge the cells at 1000 *g* for 5 min to remove the supernatant.

13. Determine the stem cell marker expression in TSCs using BD LSR II Flow Cytometer.

3.5 Maintenance of TSCs

3.5.1 Culturing TSCs in Growth Medium Supplemented with PGE$_2$

A brief protocol of PGE$_2$ for enhancing TSC stemness is shown in Fig. 4 and described below:

1. Seed the cells either in a 12-well plate at a density of 3×10^4 cells/well or a T75 flask at a density of 10^6/flask and culture the cells with low concentrations of PGE$_2$ (0.01 and 0.1 ng/mL) for 6 days.

2. Change the PGE$_2$-containing medium every day.

3. Determine the effect of PGE$_2$ on the stemness of TSCs cultured in a 12-well plate using immunostaining and in a T75 flask using flow cytometry analysis.

4. Collect the cells treated with 0.01 ng/mL of PGE$_2$.

3.5.2 Culturing TSCs Under Hypoxic Conditions

Tendons are collagen-rich tissues with only a few blood vessels on and near the tendon surface. The oxygen levels of the tendons in vivo are much lower than vascular organs and tissues. In other words, TSCs in vivo are in hypoxic conditions. Therefore, using hypoxic culture conditions can better maintain the stemness of TSCs.

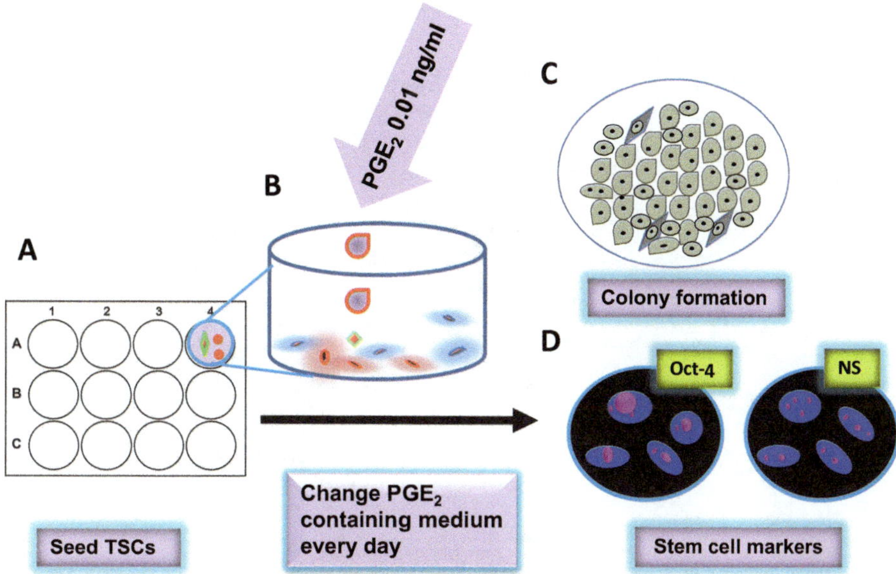

Fig. 4 Maintenance of TSCs in medium with low concentrations of PGE$_2$. (**a**) TSCs at passage 1 are seeded in a 12-well plate at a density of 3 × 10^4/well and cultured in 20% FBS-1% P/S-DMEM plus 0.01 ng/mL of PGE$_2$ for 6 days. (**b**) The fresh PGE$_2$-containing medium is changed every day. (**c**) The stemness of TSCs is tested by colony formation. (**d**) The stemness of TSCs is also tested by immunostaining of stem cell markers such as nucleostemin (NS) and Oct-4

1. Use a dedicated trigas incubator (Fig. 5a) to achieve hypoxic conditions (5% O$_2$) and maintain the oxygen concentration at a constant level with the following control devices.

2. Control the oxygen concentrations in the incubator by two gas controllers (Fig. 5b, c) and one oxygen concentration controller (Fig. 5d).

3. Connect a nitrogen gas controller to two nitrogen tanks (Fig. 5b), and connect a carbon dioxide gas controller to two carbon dioxide tanks (Fig. 5c). Connect both controllers to the tri-gas incubator (Fig. 5a). The supply of gas can be automatically switched from the first to the second tank when the first tank is empty.

4. To avoid air flow into the incubator during brief openings of the door, the incubator is separated into three isolated chambers with each chamber closed by double doors (Fig. 5e).

5. Seed the TSCs either in 6-well plates or T25 flask in 20% FBS-1% P/S-DMEM and culture the cells under hypoxia condition (5% O$_2$). The same cells cultured under normoxic condition (20% O$_2$) are used as control.

6. Put the replacement medium for the cell culture in the tri-gas incubator for at least 30 min before use (Fig. 5f) (*see* **Note 8**).

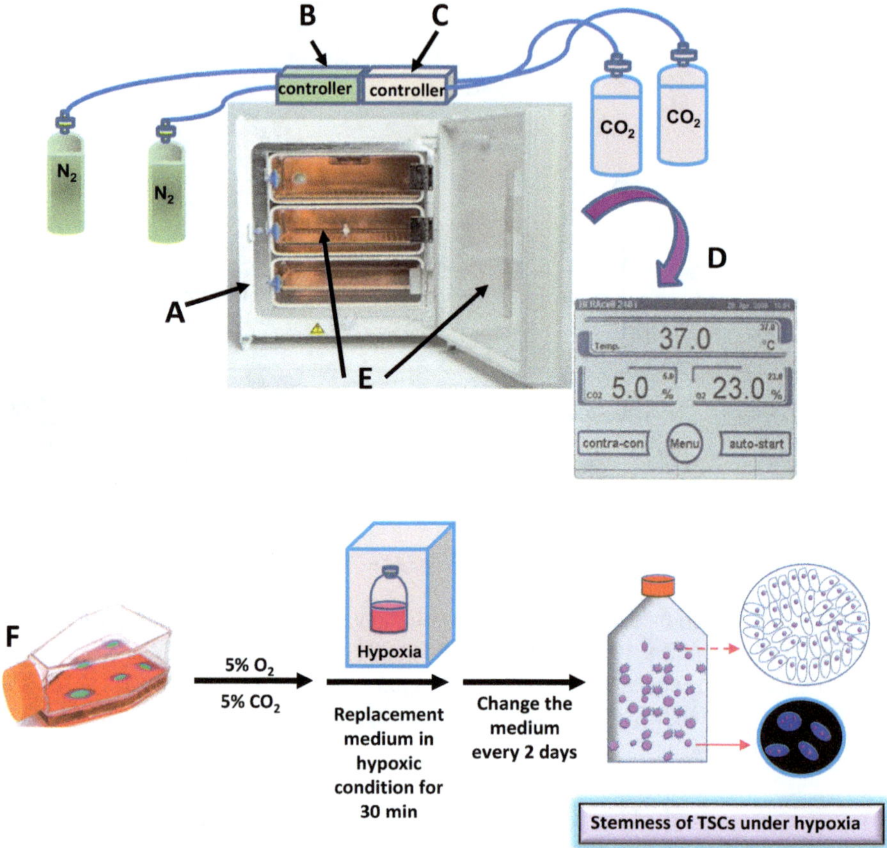

Fig. 5 Maintenance of TSCs in hypoxic culture conditions. (**a**) Trigas hypoxic incubator. (**b**) Nitrogen (N_2) gas controller connects to two nitrogen tanks and switches the supply of gas automatically from the first tank to the second tank when the first tank is empty. (**c**) Carbon dioxide (CO_2) gas controller connects to two carbon dioxide tanks and switches the supply of gas automatically from the first tank to the second tank when the first tank is empty. (**d**) Oxygen concentration controller is located in the door. (**e**) Three isolated chambers separated by three doors maintain hypoxic condition. (**f**) A brief protocol for TSCs cultured under hypoxic condition

7. Change the medium every two days under both hypoxic and normoxic culture conditions (Fig. 5f) (*see* **Note 9**).

8. Determine the hypoxia effect on the stemness of TSCs by colony formation, stem cell marker expression, and multidifferentiation potential (Fig. 5f).

9. To determine the multidifferentiation potentials, the TSCs at passage 1 are seeded in 6-well plates with the density of 2.4×10^5/well and cultured with basic medium (DMEM with 10% FBS and 1% P/S) either in 5% O_2 or 20% O_2 at 37 °C overnight. Then change various differentiation media at the next day and culture the TSCs in various differentiation media for 21 days with seven changes.

10. To assess adipogenic potential, TSCs are cultured in adipogenic induction medium that consists of basic growth medium supplemented with dexamethasone (1 μM), insulin (10 μg/mL), indomethacin (100 μM), and isobutylmethylxanthine (IBMX) (0.5 mM). The medium is changed every 3 days. After 21 days of culture, the adipogenesis is assessed by performing Oil Red O assay for staining lipids, immunostaining for detecting PPARγ expression, and qRT-PCR for measuring adipocyte-related gene expression.

11. To determine chondrogenic potential, TSCs are cultured with chondrogenic differentiation medium consisting of basic growth medium supplemented with proline (40 μg/mL), dexamethasone (39 ng/mL), TGF-β (10 ng/mL), ascorbic 2-phosphate (50 μg/mL), sodium pyruvate (100 μg/mL), and insulin–transferrin–selenous acid mix (50 mg/mL). The medium is changed every 3 days. After 21 days of culture, the chondrogenesis is evaluated by performing Safranin O assay for staining proteoglycans, immunostaining for detecting collagen type II, and qRT-PCR for measuring chondrocyte-related gene expression.

12. To evaluate osteogenic potential, TSCs are cultured with osteogenic induction medium that consists of basic growth medium supplemented with dexamethasone (0.1 μM), ascorbic 2-phosphate (0.2 mM), and glycerol 2-phosphate (10 mM). The medium is changed every 3 days. After 21 days culture, the osteogenesis is determined by performing Alizarin Red S assay for staining calcium deposits, immunostaining for detecting osteocalcin, and qRT-PCR for measuring osteocyte-related gene expression.

4 Notes

1. The activity of enzymes (collagenase and dispase) differs among the lots. The digestion solution contains at least 600 units/mL of collagenase (200 units/mg) and 7.2 units/mL of dispase (1.8 units/mg). The enzymes dissolved in PBS are filtered through 0.22 μm syringe filter before use.

2. Before dissecting the tendons, clean up the tendon areas by shaving the hairs and spraying the skin of the tendon areas with 70% ethanol.

3. The cells isolated from paratenons and core tendon tissues are different. To isolate TSCs, the blood vessels, membranes (paratenons), and fat tissues should be removed as much as possible. The core portions of the tendons are cut into small pieces (<1 mm × 1 mm × 1 mm) and used for TSC culture.

4. Seed 1 μL of the single-cell suspension (1 cell/μL) into a well of 96-well plate and culture the cells until the confluence. In some wells, more than one colony may occur because more than one cell is seeded. In order to purify the stem cells, the individual cell colonies are collected by locally applying trypsin to each colony under a microscope.

5. The antibody concentration and reaction time depend on the activity of the antibody. The listed reaction time and dilution are based on the fresh antibodies. The storage temperature and conditions will affect the activity of the antibody.

6. The antibody is added to the cell solution and mixed well by pipette. The reaction time depends on the concentration and activity of the antibody.

7. The serum used for washing buffer depends on the second antibody. However, FBS can be used for all secondary antibody reaction.

8. The replacement medium should be kept at hypoxic condition for at least 30 min, but not longer than 60 min.

9. The cells grow faster in hypoxic condition than in normoxic condition, and the medium also changes pH in hypoxic condition faster than in normoxic condition. In order to culture the healthy cells the medium should be changed every 2 days.

References

1. Bi Y, Ehirchiou D, Kilts TM, Inkson CA, Embree MC, Sonoyama W, Li L, Leet AI, Seo BM, Zhang L, Shi S, Young MF (2007) Identification of tendon stem/progenitor cells and the role of the extracellular matrix in their niche. Nat Med 13(10):1219–1227. https://doi.org/10.1038/nm1630

2. Zhang J, Wang JH (2010) Characterization of differential properties of rabbit tendon stem cells and tenocytes. BMC Musculoskelet Disord 11:10. https://doi.org/10.1186/1471-2474-11-10

3. Zhou Z, Akinbiyi T, Xu L, Ramcharan M, Leong DJ, Ros SJ, Colvin AC, Schaffler MB, Majeska RJ, Flatow EL, Sun HB (2010) Tendon-derived stem/progenitor cell aging: defective self-renewal and altered fate. Aging Cell 9(5):911–915. https://doi.org/10.1111/j.1474-9726.2010.00598.x

4. Rui YF, Lui PP, Li G, Fu SC, Lee YW, Chan KM (2010) Isolation and characterization of multipotent rat tendon-derived stem cells. Tissue Eng Part A 16(5):1549–1558. https://doi.org/10.1089/ten.TEA.2009.0529

5. Zhang J, Wang JH (2014) Prostaglandin E2 (PGE2) exerts biphasic effects on human tendon stem cells. PLoS One 9(2):e87706

6. Zhang J, Wang JH (2013) Human tendon stem cells better maintain their stemness in hypoxic culture conditions. PLoS One 8(4):e61424

Intravital Imaging to Understand Spatiotemporal Regulation of Osteogenesis and Angiogenesis in Cranial Defect Repair and Regeneration

Xinping Zhang

Abstract

Angiogenesis plays a critical role in skeletal repair and regeneration. Our understanding of the intricate relationship between osteogenesis and angiogenesis at a repair site has been hindered by the lack of an effective approach that allows tracking of bone healing and neovascularization simultaneously at a high spatiotemporal resolution in living animals. To overcome this barrier, we have recently established a cranial bone defect window chamber model in mice that enables high-resolution, four-dimensional imaging analyses of bone defect healing using multiphoton laser scanning microscopy (MPLSM). The windowed defect model confers imaging of the defect through both micro computed tomography (microCT) and MPLSM in vivo, facilitating lineage tracing and longitudinal analyses of osteogenesis and angiogenesis in repair. The windowed chamber model further permits insertion of cellular implants or bone graft materials, aiding in spatiotemporal analyses of the interactions between biomaterials and vascular microenvironment in living animals. In this chapter, we describe the improved technique for establishing the chronic cranial defect window chamber model for long term imaging as well as the imaging analysis protocols for quantitative analyses of osteogenesis and angiogenesis at the site of bone defect repair.

Key words Intravital imaging, Osteogenesis, Angiogenesis, Bone defect repair, Lineage tracing, Spatiotemporal analyses

---

1 Introduction

Cranial bone defect healing is driven by coordinated osteogenesis and angiogenesis at the site of repair. Current analyses of bone defect healing heavily rely on histology, which is destructive and challenging to provide a three-dimensional perspective of the defect healing. A microCT-based approach has been used for quantitative and qualitative analyses of osteogenesis and angiogenesis during bone repair and regeneration [1–3]. Although successful, microCT-based analyses do not have the spatial and temporal resolution to correlate neovascularization with cellular differentiation during bone defect repair and reconstruction. Furthermore, microCT-

Shree Ram Singh and Pranela Rameshwar (eds.), *Somatic Stem Cells: Methods and Protocols*, Methods in Molecular Biology, vol. 1842, https://doi.org/10.1007/978-1-4939-8697-2_17, © Springer Science+Business Media, LLC, part of Springer Nature 2018

based method lacks the resolution to visualize microvessels less than 20 μm in diameter. Since bone defect repair is a dynamic process controlled at multiple spatial and temporal scales, establishing the capability to analyze progenitor cell dynamics and their interaction with neovasculature is important for advancing our knowledge of bone defect repair and for further optimizing progenitor cell and material-based approaches for enhanced repair and reconstruction.

Multiphoton laser scanning microscopy (MPLSM) has emerged as a superior in vivo imaging modality for analyses of thick tissue in living animals. MPLSM offers confocal-like imaging quality with enhanced imaging depths and reduced photo damage. Multiphoton microscopy further permits morphological and functional analyses of neovasculature with unprecedented advantages of high spatiotemporal resolution, minimal invasiveness and 3D capability [4, 5]. Compared with other existing modalities (e.g., electrodes, MRI, Doppler), multiphoton has the resolution to image structure and function of a single vessel, enabling the evaluation of the impact of local vascular microenvironment on selective donor stem/progenitor cells. In addition to imaging nonlinear fluorescence excitation, multiphoton microscopy can be used for imaging collagen matrix through second harmonic generation (SHG) [6, 7]. The unique capability of this technology to allow simultaneously visualizing donor cells, ECM as well as the surrounding vascular network has made it extremely useful in bone tissue repair and engineering applications. Combined with an adequate in vivo bone-healing model, multiphoton microscopy could be especially powerful for in vivo imaging of complex cell–cell and cell–matrix interactions, with added advantages for simultaneously performing functional analyses for cellular motility and vascular ingrowth in a nondestructive, noninvasive, and dynamic fashion [8].

In this chapter, we describe the techniques for the establishment of a chronic window chamber model for visualization and longitudinal analyses of osteogenesis and angiogenesis in bone defect healing. The cranial window chamber model in mice has been previously reported for analyses of brain cell function and tumor-associated neovascularization [9, 10]. The model was modified to meet the need for long-term tracking of defect healing via intravital imaging.

2 Materials

2.1 Creation of a Cranial Defect Window Chamber Model in Mice

1. Col2.3GFP transgenic mice (Jackson Laboratory, Cat No. 013134).

2. Ketamine and xylazine.

3. A stereotaxic frame.

4. Mouse heating pad.

5. Alpha-MEM medium containing 1% penicillin and streptomycin, 1% glutamine.

6. Aryl-ether-ether-ketone (PEEK).

7. Cyanoacrylate glue.

8. Tungsten vanadium inverted cone burs, various sizes.

9. High-speed microdrill.

10. Eight mm round glass coverslip in #1 thickness.

11. Bone cement.

2.2 Imaging and Image Analyses

1. Olympus FV1000-AOM multiphoton imaging system (Olympus), equipped with a Titanium:Sapphire laser (Spectra-Physics), a C-Apochromat ×10/0.45 (Zeiss), and a ×25/1.05 (Olympus) water immersion objective.

2. Q-tracker 655 nanocrystals (ThermoFisher Scientific).

3. Amira software developed by Visualization Sciences Group (VSG, Burlington MA).

4. Image J with plugins for simple image processing.

3 Methods

3.1 Surgical Procedures for Establishing a Chronic Window Chamber Model for Cranial Defect Repair

1. Anesthetize Col2.3GFP mice via intraperitoneal injection of 0.2 mL/20 g body weight of a mix of 20 mg/mL ketamine and 3 mg/mL xylazine in 0.9% NaCl (*see* **Note 1**). The Col2.3GFP mice express Enhanced Green Fluorescent Protein (GFP) gene under the control of the 2.3 kb rat procollagen, type 1, alpha 1(Col1a1) promoter, which allows for visualization and tracking of the proliferation, differentiation, and migration of osteoblasts in living animals.

2. To control any acute pain induced by the surgery, preemptive analgesia, buprenorphine (0.5 mg/kg SQ) should be given prior to surgery and after surgery every 12 h for 3 days.

3. The hair should be thoroughly shaved over most of the scalp using a conventional electric razor.

4. Set a heating pad on low (~35 °C) and place it under the mouse on the stereotaxic frame to maintain body temperature.

5. To stabilize the mouse head for surgery, the animal should be raised to the level of the ear bars mounted on the platform of stereotaxic frame. First lock the nose clamp and incision bar to mount the head on the stereotaxic device and then lock the two ear bars into the external auditory meatus. Tighten both

sides of the bars to make sure the mouse head is not movable and stably fixed onto the stereotaxic platform. Optionally apply ophthalmic ointment or mineral oil to the eyes to protect them from accidental spills.

6. Following stabilization of the mice with iodine surgical scrub, make a midline scalp incision using microsurgical tools. Clear the soft tissue and excessive fluid on the skull with Q-tips.

7. A customized Polyetheretherketone (PEEK) spacer with a circular opening of 6 mm (diameter) in its center is glued on the skull with a thin layer of cyanoacrylate glue. Press it for a few minutes to ensure the plate is tightly glued to the skull and secured it with a small amount of glue around the edges.

8. Seal the custom-made plate on the skull with a mix of bone cement and glue at about 1:1 ratio.

9. Wait 10–15 min until the metal plate is firmly attached to the skull.

10. Wash the open chamber with extensive sterile PBS with antibiotics (*see* **Note 2**).

11. Use a tungsten vanadium inverted cone bur attached to a high-speed microdrill to create a full thickness defect in the parietal bone of mouse calvarium. The size of the defect can be controlled using a 0.9 mm or 1.8 mm size Busch inverted cone bur which generates a 1 mm or 2 mm full thickness defect, respectively, in the parietal bone. Avoid suture and regions with visible vasculatures (*see* **Note 3**).

12. Carefully remove the debris of bone and clear the area with an air blower, avoiding damaging the dura and the vessels on the dura.

13. Thoroughly rinse the opening with room temperature alpha-MEM medium containing 1% penicillin and streptomycin, 1% glutamine.

14. Cover the clear dura with a thin layer of a-MEM medium containing 1% penicillin and streptomycin, 1% glutamine and keep the wound wet.

15. Carefully place a custom-made circular glass coverslip on top of the custom-made plate to cover the open-skull region.

16. Seal the edge of the circular coverslip onto the custom-made plate with a thin layer of glue.

17. Allow the mouse to wake up and recover on a warm pad.

18. Return the mouse to vivarium and keep each mouse separately in a cage to avoid fight during the first few weeks after surgery.

19. Once established, the mouse with a window chamber can survive over a 6-month period with no visible signs of distress (Fig. 1) (*see* **Notes 4** and **5**).

3.2 Imaging of Osteogenesis and Angiogenesis

1. MPLSM. An Olympus FV1000-AOM multiphoton imaging system, equipped with a Titanium:Sapphire laser, a C-Apochromat ×10/0.45, and a ×25/1.05 (Olympus) water immersion objective, is used for live imaging of the cranial defect healing. Images are acquired at 512×512 pixels, 0.2 ms pixel dwell with the laser tuned to 780 nm. The fluorescence of GFP and second harmonic signals (SHG) is collected with a 517/23-nm and a 390/20-nm bandpass filters (Semrock),

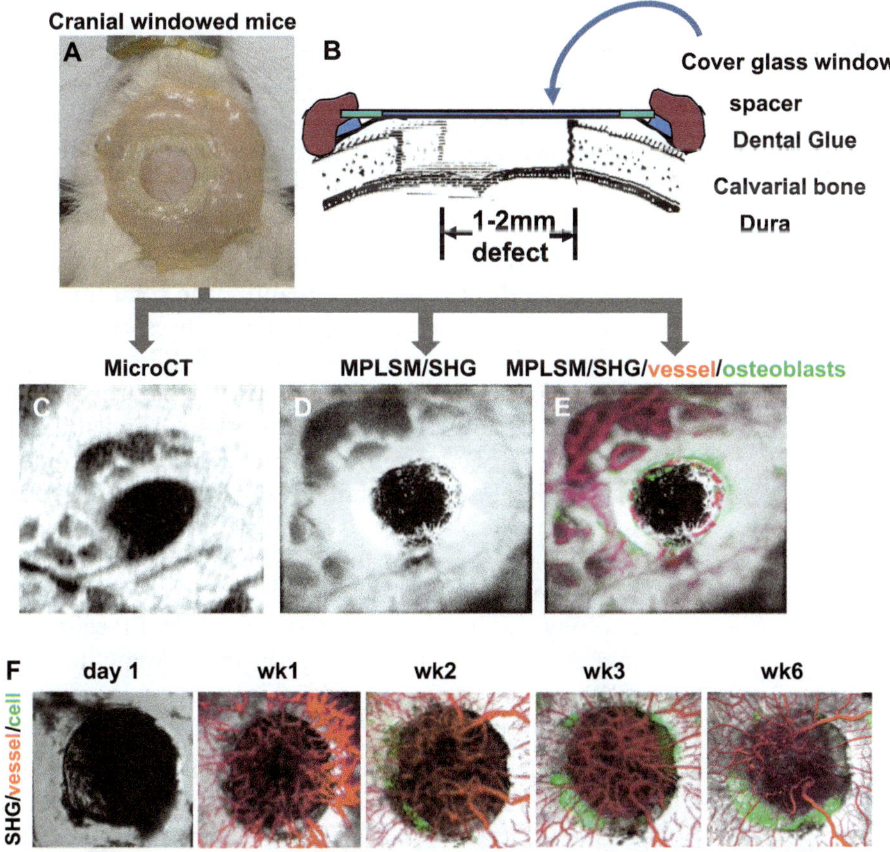

Fig. 1 Superior view photograph of a Col 2.3 GFP mouse with the instrumented head piece exposing a cranial defect is shown (**a**). Schematic illustration of the model (**b**). Windowed mouse is subjected to both micro-CT scanning (**c**) and MPLSM (**d, e**). Three-dimensional reconstructed images of the healing defect derived from microCT scanning (**c**) and SHG microscopy (**d**) are presented to illustrate the remarkable agreement between the current standard and our new MPLSM approach. Longitudinal tracking of bone defect healing in Col2.3GFP transgenic mice over a 6-week period to illustrate osteogenesis and angiogenesis (**f**). SHG, white; vessel, red; osteoblasts, green

respectively. To visualize the blood vessel network, Q-tracker 655 nanocrystals are dissolved in sterilized water according to the instruction from the manufacturer and injected intravenously via retro-orbital venous sinus into mouse circulation. Far red fluorescence from nanocrystals is detected using a far red bandpass emission filter (Semrock, 655/40 nm). Using the ×10 water immersion objective lens (Zeiss), a 1.3×1.3 mm multichannel z-series stack with 5 mm steps is obtained. The z-series stack allowed 3D reconstruction of the defect up to a depth of 300 mm (*see* **Note 6**). To characterize collagen matrix propagation and Col2.3GFP (+) cell dynamics, the defect can be repeatedly imaged over a 3-month period by MPLSM. To ensure accuracy of the analyses, all imaging parameters, including laser power, PMT voltages and compensation should be maintained constant throughout the entire experiment.

3.3 MPLSM Image Analyses

3.3.1 Quantification of SHG Propagation and Col2.3GFP Cell Dynamics in Cranial Defect Repair

1. The multichannel 2D slice viewing and 3D reconstruction of the defect are conducted using Amira software. Based on the SHG microscopy, which provided the contour of the defect region, a region of interest (ROI) can be established in a time-series, referencing back to the circular defect region at the day 1 following surgery.

2. To quantify SHG propagation within the ROI, a global thresholding method is used to quantify the total number of SHG (+) voxels above an appropriate threshold in time-series images. The percentage volume of SHG occupying the total volume of the defect can be used to depict the propagation of collagen bone matrix within the defect. Similarly, the volume of Col2.3GFP (+) cells occupying the defect regions within defined ROI can be obtained by quantifying the total number of green fluorescent voxels above an appropriate threshold across the time-series images. SHG signal from bone matrix can be easily distinguished from soft tissue by its distinct morphology and its association with Col2.3GFP cells during healing. A schematic illustrating the measurements of SHG and GFP (+) cells can be found in Fig. 2. All measurements can be conducted using a combination of Image J and Amira.

3. To examine the extent of bone defect healing, the area of the defect is measured in ImageJ by tracing the circular defect region in the 2D images obtained from MPLSM. Since we are using a Busch inverted cone bur to create the defect, the shape of the defect is circular or nearly circular. Based on mathematical calculations, the radius of a circle can be approximated using the area of the circular region, i.e., $r = \sqrt{\text{Area}/\pi}$). The mean of the cell advancing distance from the edges of the bone is determined by calculating the difference of the outer

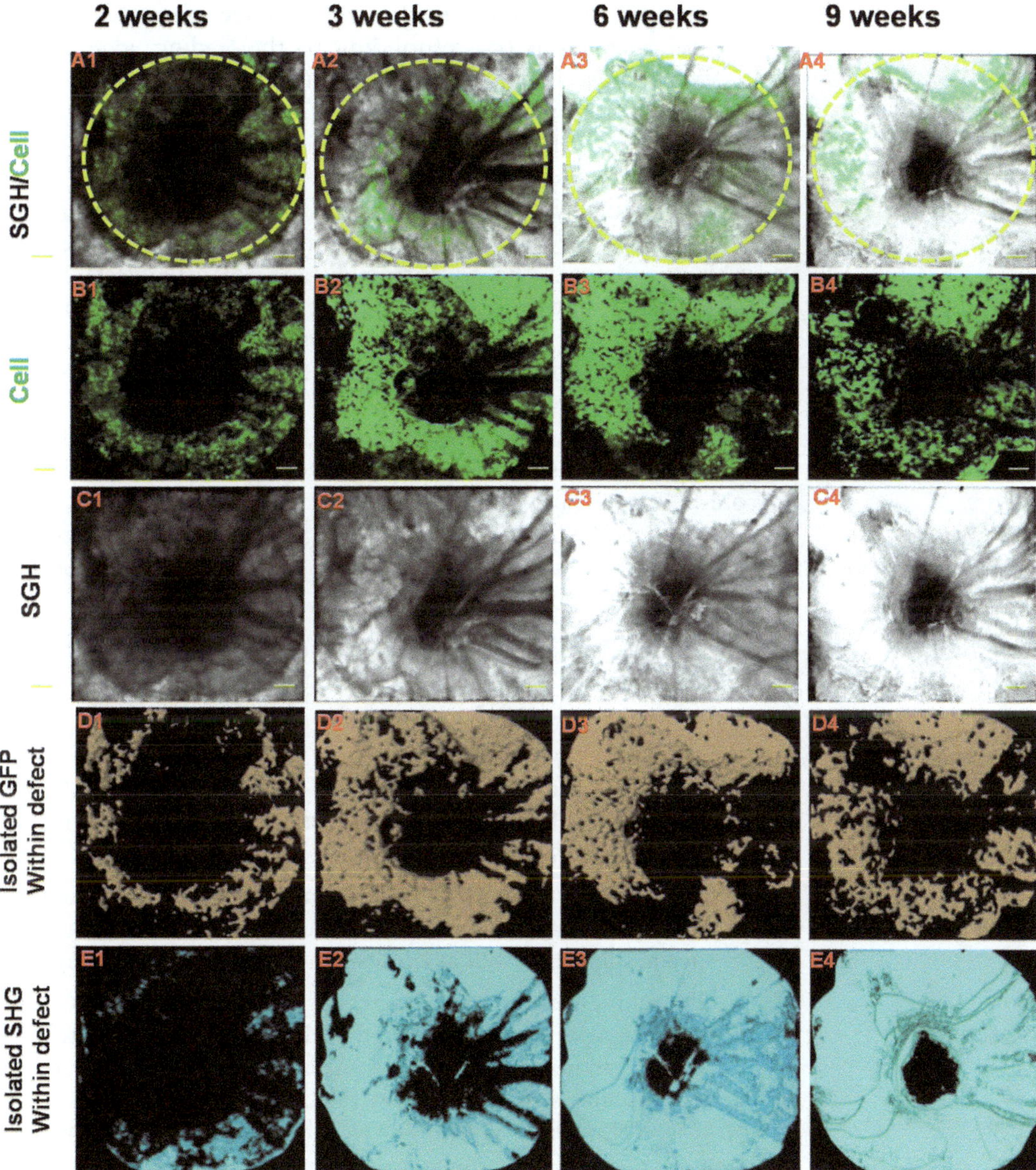

Fig. 2 Quantitative and volumetric analyses of GFP and SHG signals in cranial defect healing. A bony defect was created in Col2.3GFP transgenic mouse as previously described. The defect healing was imaged over a period of 9 weeks using MPLSM. Propagation of SHG and differentiation of Col2.3GFP (+) osteoblasts were illustrated in panel (**a**) (SHG and GFP + cells), (**b**) (GFP + cells) and (**c**) (SHG, white). Signals of GFP and SHG were isolated within the defect region by proper thresholding across all samples. Volume of GFP (+) cells (**d**, orange) and SHG (**e**, cyan) were further calculated based on isolated signals

radius (corresponding to the original edge of the defect) and the inner radius (corresponding to the leading edge of the expanding cells). The original radius of the defect at day 1 is calculated as $r\_0 = \sqrt{A\_0/\pi}$, where A_0 is the area of the defect at day 1. The area of the defect region that is not occupied by the cells at each indicated time point can be used to approximately calculate the radius of the inner circle as: $r\_inner = \sqrt{(A\_inner/\pi)}$, where A_inner is the area of the void in the defect. The advancing distance of osteoblasts at the indicated time points is then given by $r\_0 - r\_inner$. The difference is further normalized by the radius of the original defect and plotted as a function of time.

3.3.2 Quantitative and Histomorphometric Analyses of Neovascularization at the Site of Cranial Bone Defect Repair

Quantitative and histomorphometric analyses of neovasculature can be performed simultaneously with volumetric quantification of Col2.3GFP cells and SHG. A schematic for quantitative analyses of blood vessels is illustrated in Fig. 3. All analyses including segmentation, skeletonization, and filament editing are performed using available modules in Amira.

1. A multichannel imaging stack that includes the entire defect as illustrated by SHG and the surrounding vascular network is reconstructed in 3D format using the Votex Function available in Amira.

2. To analyze the vascular network, a region of interest (ROI) is defined to include all the vessels associated with bone defect as determined by SHG microscopy.

3. The vascular network is isolated using Amira Segmentation Editor to obtain a segmented vascular image, excluding non-specific signals as well as noises generated during imaging.

4. The segmented vascular image stack is used to perform volumetric analyses available from the MaterialStatistics function in Amira: vascular volume (Vasc. Vol.), total volume (T. Vol.), and vascular volume fraction (Vol. Fract.) (i.e., the ratio of Vasc. Vol. over T. Vol.).

5. Using Amira's AutoSkeleton module, which implements a distance-ordered homotopic thinning operation, the segmented 3D vascular network is further skeletonized to generate a line-based network that is topologically equivalent to the original network [11, 12]. The skeleton can be superimposed on the original image to assess the relative accuracy of this method.

6. The final skeletonized vessel network can be obtained by manually retracing of the skeletons using Amira's Filamental Editor to remove false or redundant segments.

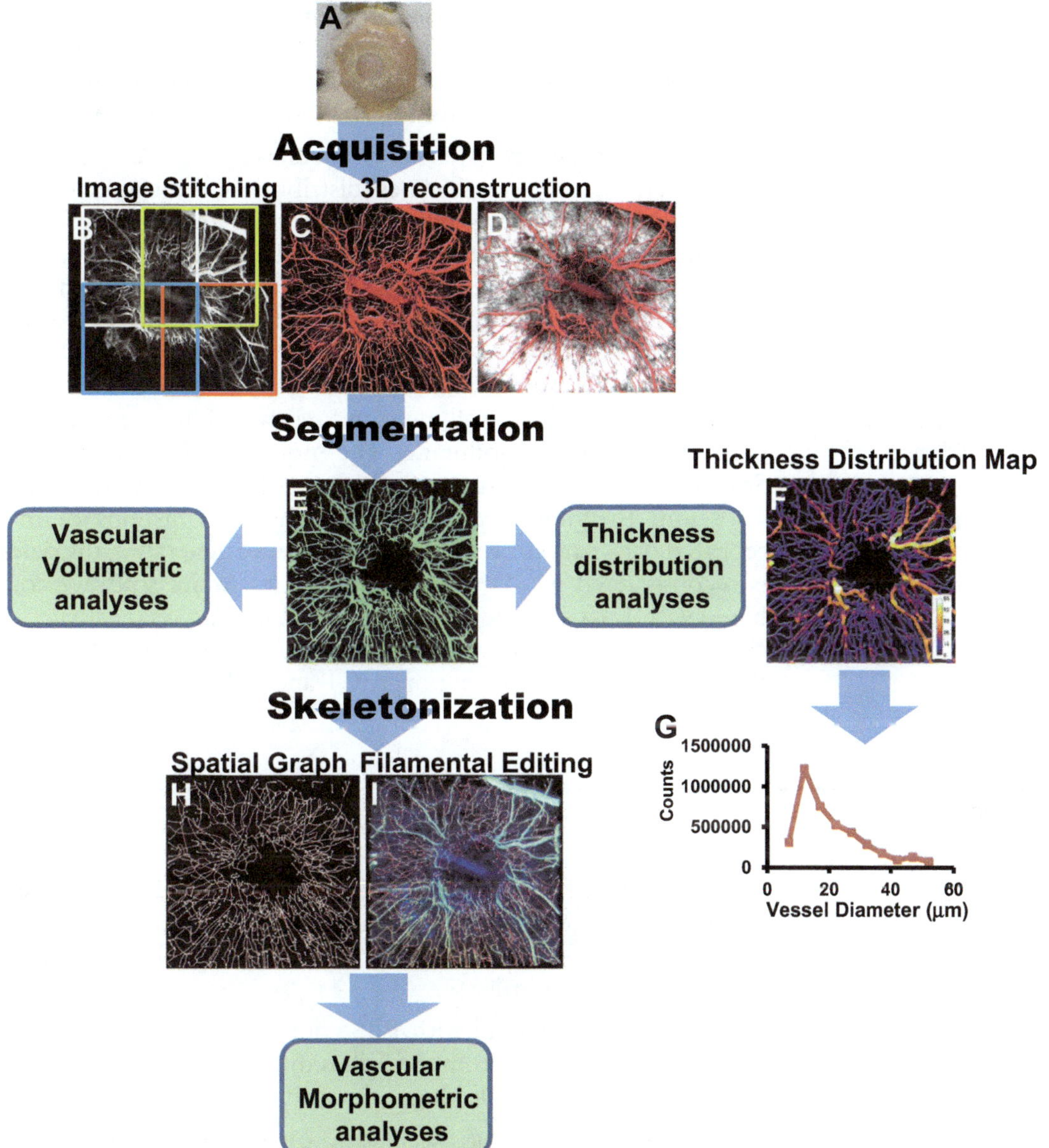

Fig. 3 Simultaneous imaging of SHG (white) and vessels (red) allows 3D reconstruction of the healing defect and vasculature at 8 weeks postsurgery (**a**). The total vasculature within the window (**b**) was segmented using the SHG volume as region of interest (ROI) to isolate the vessels in the new bone (**c**). These vessels were skeletonized (**d**), and superimposed on the original image of the vessels (**e**). From binarized 3D images, a vessel thickness distribution map was obtained of the ROI (**f**), and used for morphometric analyses (**g**). The table lists quantification data from four quadrants of the defect

7. Based on the skeletonized network, the number of vascular segments (NV), total vessel length (T. Length), and vessel length fraction (L. Fract.) (i.e., ratio of vessel length to total volume) can be read from SpatialGraphStatistics in Amira.

8. To analyze the average vessel thickness (Avg. Vess. Th.) and the associated vessel thickness distribution, the segmented vascular image stack is imported into the Image J. Using an ImageJ plugin developed by Robert Dougherty [13], vascular thickness and thickness distribution can be output in excel spread sheet. The complete process for quantification of vascularity is illustrated in Fig. 3.

3.3.3 Longtudinal Live microCT Scanning

Mice are sedated with isoflurane and restrained in a custom-made chamber containing isoflurane. The mouse skull is scanned by a Scanco Viva CT40 system (Scanco Medical) at a voxel size of 17.5 μm. Imaging is performed on the same groups of mice repeated over 3 months. From the 2D images generated, the defect is reconstructed in 3D and analyzed using Amira software combined with Image J. The rate of the defect healing was evaluated by calculating the area of the defect closure over a period of 3 months. We recommend performing microCT and MPLSM on alternating days/weeks.

4 Notes

1. The mix of ketamine and xylazine will ensure deep anesthesia for about 1 h. We suggest weighing animals before surgery in order to provide an adequate amount of anesthetic agents to the surgical animals.

2. Washing the wound with PBS containing antibiotics such as penicillin and streptomycin will reduce infection and improve the success of establishing a clear window for imaging.

3. Damaging suture or vasculature on the dura could lead to excessive bleeding, which could result in the failure of constructing a clear window for imaging. If bleeding occurs, press the area with sterilized Q-tip for several minutes until bleeding stops.

4. During the first week, some window chambers could become a bit cloudy due to the inflammatory infiltration from the wound. Most of the cloudy windows will become clear in week 2 or 3.

5. The improved window system provides a stable and clear window for long term imaging. Compared with published cranial window models, the current model allows for repeated imaging over months with about 80% successful rate.

6. The maximum penetration of laser scanning in bone tissue is ~250 μm, which is close to the full thickness of calvaria bone. By adjusting the laser power, the imaging resolution and depth can be further improved.

Acknowledgments

This study is supported by grants from the Musculoskeletal Transplant Foundation, NYSTEM N08G-495 and N09G346, and the National Institutes of Health (R21 DE021513, RC1 AR058435, AR067859, R01DE019902, R21DE026256, P50 AR054041, and P30AR061307).

References

1. Duvall CL, Robert Taylor W, Weiss D et al (2004) Quantitative microcomputed tomography analysis of collateral vessel development after ischemic injury. Am J Physiol Heart Circ Physiol 287:H302–H310

2. Duvall CL, Taylor WR, Weiss D et al (2007) Impaired angiogenesis, early callus formation, and late stage remodeling in fracture healing of osteopontin-deficient mice. J Bone Miner Res 22:286–297

3. Zhang X, Xie C, Lin AS et al (2005) Periosteal progenitor cell fate in segmental cortical bone graft transplantations: implications for functional tissue engineering. J Bone Miner Res 20:2124–2137

4. Lecoq J, Parpaleix A, Roussakis E et al (2011) Simultaneous two-photon imaging of oxygen and blood flow in deep cerebral vessels. Nat Med 17:893–898

5. Hoover EE, Squier JA (2013) Advances in multiphoton microscopy technology. Nat Photonics 7:93–101

6. Brown E, McKee T, diTomaso E et al (2003) Dynamic imaging of collagen and its modulation in tumors in vivo using second-harmonic generation. Nat Med 9:796–800

7. Osswald M, Winkler F (2013) Insights into cell-to-cell and cell-to-blood-vessel communi-cations in the brain: in vivo multiphoton microscopy. Cell Tissue Res 352:149–159

8. Sidani M, Wyckoff J, Xue C et al (2006) Probing the microenvironment of mammary tumors using multiphoton microscopy. J Mammary Gland Biol Neoplasia 11:151–163

9. Holtmaat A, Bonhoeffer T, Chow DK et al (2009) Long-term, high-resolution imaging in the mouse neocortex through a chronic cranial window. Nat Protoc 4:1128–1144

10. Madden KS, Zettel ML, Majewska AK et al (2013) Brain tumor imaging: live imaging of glioma by two-photon microscopy. Cold Spring Harb Protoc 2013:237–240

11. Cassot F, Lauwers F, Fouard C et al (2006) A novel three-dimensional computer-assisted method for a quantitative study of microvascular networks of the human cerebral cortex. Microcirculation 13:1–18

12. Pudney C (1998) Distance-ordered homotopic thinning: a skeletonization algorithm for 3D digital images. Comput Vis Image Underst 72:404–413

13. Dougherty R, Kunzelmann KH (2007) Computing local thickness of 3D structures with ImageJ. In: Microscopy and microanalysis meeting, Ft. Lauderdale, FL

Chapter 18

Beating Heart Cells from Hair-Follicle-Associated Pluripotent (HAP) Stem Cells

Robert M. Hoffman and Yasuyuki Amoh

Abstract

The neural stem-cell marker nestin is expressed in hair follicle stem cells located in the bulge area which are termed hair-follicle-associated pluripotent (HAP) stem cells. HAP stem cells can differentiate into neurons, glia, keratinocytes, smooth muscle cells, and melanocytes in vitro. Subsequently, we demonstrated that HAP stem cells could affect nerve and spinal cord regeneration in mouse models. We subsequently demonstrated that HAP stem cells differentiated into beating cardiac muscle cells. The differentiation potential to cardiac muscle is greatest in the upper part of the mouse whisker follicle. The beat rate of the cardiac muscle cells differentiated from HAP stem cells was stimulated by isoproterenol and inhibited by propanolol. The addition of activin A, bone morphogenetic protein 4, and basic fibroblast growth factor, along with isoproternal, induced the cardiac muscle cells to form tissue sheets of beating heart muscle cells. Under hypoxic conditions, HAP stem cells differentiated into troponin-positive cardiac-muscle cells at a higher rate that under normoxic conditions. Hypoxia did not influence the differentiation to other cell types. This method is appropriate for future use with human hair follicles to produce hHAP stem cells in sufficient quantities for future heart, nerve, and spinal cord regeneration in the clinic.

Key words Hair follicle-associated pluripotent (HAP) stem cells, Nestin, Cardiac muscle cell, Beating, Neuron, Nerve regeneration, Pluripotent, Hypoxia

1 Introduction

The neural stem-cell marker nestin is expressed in hair follicle stem cells located in the bulge area which are termed hair-follicle-associated pluripotent (HAP) stem cells [1]. HAP stem cells can differentiate into neurons, glia, keratinocytes, smooth muscle cells, and melanocytes in vitro [2, 3]. Subsequently, we demonstrated that nestin-expressing stem cells could affect nerve and spinal cord regeneration in mouse models [4].

HAP stem cells originate in the bulge area (BA) and migrate to the dermal papillae (DP) [5, 6]. HAP stem cells from the DP and BA cells differentiated into neuronal and glial cells after transplantation to the injured sciatic nerve and spinal cord and enhanced injury repair and locomotor recovery within 4 weeks [5–7].

Shree Ram Singh and Pranela Rameshwar (eds.), *Somatic Stem Cells: Methods and Protocols*, Methods in Molecular Biology, vol. 1842, https://doi.org/10.1007/978-1-4939-8697-2_18, © Springer Science+Business Media, LLC, part of Springer Nature 2018

In Gelfoam® histoculture, HAP stem cells from mouse whisker follicles formed nerve-like structures which contained β-III tubulin-positive fibers [8]. The growing fibers had growth cones on their tips expressing F-actin, indicating they were growing axons. These results suggest a major function of HAP stem cells is for growth of the follicle sensory nerve [8]. HAP stem cells can be cryopreserved with full function [9].

Yu et al. [10] isolated HAP stem cells from the bulge area of human hair follicles. The human HAP (hHAP) stem cells gave rise to myogenic, melanocytic, and neuronal cell lineages after in vitro clonal single cell culture. In addition, the hHAP stem cells differentiated into adipocyte, chondrocyte, and osteocyte lineages. Neuronal differentiation of hHAP stem cells induces increased expression of neuron-associated genes. Neuronal cells differentiated from HAP stem cells persisted in mouse brain and retained neuronal differentiation markers after transplantation [11].

We subsequently demonstrated that HAP stem cells differentiated into beating cardiac muscle cells. The differentiation potential to cardiac muscle is greatest in the upper part of the mouse whisker follicle. The beat rate of the cardiac muscle cells was stimulated by isoproterenol and inhibited by propanolol [4].

2 Materials

2.1 HAP Stem Cell Isolation, Culture, and Spinal Surgery Components

1. Animals: Nestin-driven GFP (ND-GFP) transgenic mice for BA culture, RFP transgenic mice for DP culture, nontransgenic nude mice for spinal surgery.

2. HAP stem cell sources: BA and DP of whisker follicles.

3. Dissection or surgery equipment: stereomicroscope, surgical knife, fine forceps, fine scissors, retractor for mice, syringe needles (10 mL, 5 mL, 1 mL syringes), 31G insulin syringe.

4. Tissue and cell culture plates or flasks: tissue culture dishes (60 mm), collagen I-coated 12-well and 48-well plates, collagen I-coated 25 cm² flasks.

5. Culture medium: Dulbecco's modified Eagle's medium/ F12; 10% fetal bovine serum (FBS); 1% N2; 2% B27; 200 mM L-glutamine; 0.025% ITS +3; 20 ng/mL epithelial growth factor, 20 ng/mL basic fibroblast growth factor. Subculture medium is similar to the culture medium but free of serum.

2.2 Isolation and Dissection of Vibrissa Hair Follicles

1. DMEM/F12.

2. Gentamicin.

3. FBS.

2.3 Immunofluorescence Staining [4]

2.3.1   Primary Antibodies

1. Anti-βIII-tubulin monoclonal (1:500, Tuj1 clone; Covance, San Diego, CA).

2. Anti-glial fibrillary acidic protein (GFAP) chicken polyclonal (1:300; Abcam, Kinxton, Cambridge, UK).

3. Anti-keratin 15 (K15) monoclonal (1:100; Lab Vision, Fremont, CA).

4. Anti-smooth muscle actin (SMA) monoclonal (1:200; Lab Vision).

5. Anti-cardiac toroponin T monoclonal (1:500; GeneTex, Hsinchu City, Taiwan).

6. Anti-desmin rabbit polyclonal (1:25; Cell Signaling, Tokyo, Japan).

7. Anti-nestin monoclonal (1:200, rat401 clone; Millipore, Tokyo, Japan).

2.3.2   Secondary Antibodies for Immunofluorescence

1. Alexa Fluor® 568-conjugated goat anti-mouse (1:400; Molecular Probes, Eugene, Oregon).

2. Alexa Fluor® 488-conjugated goat anti rabbit (1:400; Molecular Probes).

3. Alexa Fluor® 568-conjugated goat anti-chicken (1:1000; Molecular Probes).

2.3.3   Secondary Antibodies for Fluorescence-Activated Cell Sorting (FACS)

1. Goat anti-mouse IgG H&L phycoerythrin (1:500; Abcam).

2. Goat anti-chicken IgY biotinylated (1:1000; R&D Systems, Minncapolis, MN).

3. Brilliant Violet 421™ streptavidin (1:1000; BioLegend, San Diego, CA).

2.4 Testing the Effect of Isoproterenol and Proterenol on Cardiac-Muscle Cells Differentiated from HAP Stem Cells

1. Isoproterenol (Sigma-Aldrich) (200 nm).

2. Propranolol (Sigma-Aldrich) (200 nM).

3. Video microscopy (ScopPad-500, GelleX, Tokyo, Japan).

2.5 Immunofluorescence Staining

1. Anti-cardiac toroponin T (cTnT) mouse monoclonal antibody (1:250, Gene Tex, Hsinchu City, Taiwan).

2. Blocking buffer (4% BSA, 0.5% Triton X-100, 0.04% NaN_3 in PBS).

3. Goat anti-mouse IgG conjugated with Alexa Fluor 568® (1:400, Molecular Probes, Eugene, Oregon).

4. 4′,6-diamino-2-phenylindole, dihydrochloride (DAPI) (Molecular Probes).

5. LSM 710 microscope (Carl Zeiss, Germany).

2.6 Flow Cytometry [12]

1. Anti-cTnT mouse monoclonal antibody (1:200).

2. Anti-smooth muscle actin (SMA) mouse monoclonal antibody (1:200; Lab Vision, Fremont, CA).

3. Phycoerythrin (PE)-anti-mouse CD31 rat monoclonal antibody (1:200, BD science).

4. Goat anti-mouse IgG H&L phycoerythrin (1:500; Abcam, Cambridge, UK).

5. FACS Verse (BD Bioscience).

6. FACS suite™ software (BD Bioscience).

2.7 Immunofluorescence Staining and FACS of hHAP Stem Cell Colonies and Cells Differentiated from hHAP Stem Cells [14]

1. Anti-SSEA1 mouse monoclonal IgM (1:100, BioVision, Milpitas, CA).

2. Anti-SSEA3 rat monoclonal IgM (1:100, Millipore).

3. Anti-SSEA4 mouse monoclonal IgG (1:100, BioLegend).

4. Anti-Nanog goat polyclonal (1:100, R&D).

5. Anti-Oct3/4 goat polyclonal (1:100, R&D).

2.7.1 Stem Cell Markers

2.7.2 Anti-nestin Antibodies [14]

1. Alexa Fluor® 594-conjugated goat anti-mouse IgM (1:400, Molecular Probes).

2. Alexa Fluor® 594-conjugated goat anti-rat IgM (1:400, Molecular Probes).

3. Alexa Fluor® 568-conjugated goat anti-mouse IgG.

4. Alexa Fluor® 568-conjugated donkey anti-goat IgG (1:400).

5. Alexa Fluor® 568-conjugated goat anti-rabbit IgG along with DAPI (Molecular Probes).

2.7.3 Primary Antibodies for Differentiating Cells

1. Anti-nestin rabbit polyclonal (1:50, IBL, Gunma, Japan).

2. Anti-βIII-tubulin monoclonal (1:500, TUJ1 clone; Covance, San Leandro, CA).

3. Anti-glial fibrillary acidic protein (GFAP) monoclonal (1:200, GA-5, Lab Vision, UK).

4. Anti-GFAP chicken polyclonal (1:300, Abcam, UK).

5. Anti-keratin 15 (K15) monoclonal (1:200, Lab Vision).

6. Anti-smooth muscle actin (SMA) monoclonal (1:400, Lab Vision).

7. Anti-cardiac troponin T (cTnT) monoclonal (1:500, GeneTex, Taiwan).

2.7.4 Secondary Antibodies for Immunofluorescence	1. Alexa Fluor® 568-conjugated goat anti-rabbit IgG (1:400, Molecular Probes, Eugene, OR).

2.7.4 Secondary Antibodies for Immunofluorescence

1. Alexa Fluor® 568-conjugated goat anti-rabbit IgG (1:400, Molecular Probes, Eugene, OR).

2. Alexa Fluor® 568-conjugated goat anti-mouse IgG (1:400, Molecular Probes).

3. 4′,6-diamino-2-phenylindole, dihydrochloride (DAPI) (Molecular Probes).

2.7.5 Secondary Antibodies for FACS

1. Goat anti-mouse IgG H&L phycoerythrin (1:500, Abcam, Cambridge, UK).

2. Goat anti-chicken IgY biotinylated (1:500, R&D, Minneapolis, MN).

3. Brilliant violet 421™ streptavidin (1:500, BioLegend, San Diego, CA).

3 Methods

3.1 C57BL/6-mice

1. Use C57BL/6 mice (CLEA Japan, Tokyo Japan) to isolate the vibrissa hair follicles.

2. All animal experiments are conducted according to guidelines [4].

3.2 Isolation and Dissection of Vibrissa Hair Follicles

1. To isolate vibrissa hair follicles from C57BL/6 mice, expose the upper lip containing the vibrissa pad and cut the inner surface under anesthesia.

2. Dissect whole vibrissae hair follicles under a binocular microscope and pluck them from the pad by pulling gently by their neck with fine forceps.

3. Wash the isolated vibrissae in DMEM/F12 with 50 µg/mL gentamicin.

4. Do all surgical procedures under a sterile environment.

5. Surgically separate the vibrissa hair follicles into three parts, upper, middle, and lower under a binocular microscope [4].

6. Resuspend the divided the vibrissa hair follicle parts into fresh DMEM containing 10% fetal bovine serum (FBS), 50 µg/mL gentamicine, 2 mM L-glutamine (Gibco), 10 mM HEPES [4].

3.3 Differentiation of HAP Stem Cells from Different Parts of the Whisker Hair Follicle

1. After 4 weeks culture, transfer the separated parts of the whisker in DMEM with FBS to DMEM/F12 without FBS. After 1 week culture in DMEM/F12 without FBS, the cells will grow out from the follicle to form hair spheres [14] containing HAP stem cells.

2. Transfer to DMEM with 10% FBS after 2 days, and the hair spheres will begin to differentiate and 1 week later, the HAP stem cells will differentiate into multiple types of cells [4].

3.4 Testing the Effect of Isoproterenol and Proterenol on Cardiac Muscle Cells Differentiated from HAP Stem Cells [4]

1. Treat cardiac muscle cells differentiated from HAP stem cells with isoproterenol (Sigma-Aldrich) (200 nm) and propranolol (200 nM).
2. Record the beat rate per minute of the cardiac muscle cells by video microscopy (ScopPad-500, GelleX, Tokyo, Japan).

3.5 Immunofluorescence Staining [12]

1. Incubate the cells with anti-cardiac toroponin T (cTnT) mouse monoclonal antibody (1:250, Gene Tex, Hsinchu City, Taiwan) in blocking buffer (4% BSA, 0.5% Triton X-100, 0.04% NaN_3 in PBS) at room temperature for 3 h.
2. Then incubate the cells with goat anti-mouse IgG conjugated with Alexa Fluor 568® (1:400, Molecular Probes, Eugene, Oregon) and 4′,6-diamino-2-phenylindole, dihydrochloride (DAPI) (Molecular Probes) in blocking buffer for 2 h at room temperature.
3. Visualize the images using a LSM 710 microscope (Carl Zeiss, Germany) or equivalent.

3.6 Flow Cytometry [12]

1. Incubate the cells with anti-cTnT mouse monoclonal antibody (1:200) and anti-smooth muscle actin (SMA) mouse monoclonal antibody (1:200; Lab Vision, Fremont, CA) as the primary antibody and phycoerythrin (PE)-anti-mouse CD31 rat monoclonal antibody (1:200, BD science).
2. Use secondary antibodies goat anti-mouse IgG H&L phycoerythrin (1:500; Abcam, Cambridge, UK).
3. Analyze the cells by FACS Verse, then analyze by FACS suite™ software (BD Bioscience).
4. Repeat flow cytometry analyses in triplicate.

3.7 Intracellular Calcium Imaging [12]

1. Wash the cells with HEPES Ringer solution at 37 °C, and load with Fluo-3AM (Molecular Probes) for 20 min at 37 °C.
2. Measure the beating cardiac cells or sheets with the Fluo-3-AM fluorescence (excitation at 508 nm and emission at 527 nm) every 150 ms with the LSM 710 microscope.

3.8 Isolation and Culture of hHAP Stem Cells [14]

1. Obtain surgical specimens of normal human scalp skins from patient for isolation of hHAP stem cells.

2. Take the surgical specimens of normal human scalp skin from patients without systemic disease, who have given informed consent.

3. Perform all the experiments according to the Declaration of Helsinki guidelines, in compliance with national regulations for the experimental use of human material.

4. Cut scalp skin containing the hair follicle pad with its inner surface exposed to isolate whole hair follicles.

5. Dissect the scalp hair follicles under a binocular microscope and split off from the pad using a surgical knife.

6. The specimen size from scalp skin should be approximately $0.5 \times 0.5 \times 0.5$ cm.

7. Isolate 80 ± 26 whole hair follicles per patient.

8. Perform all procedures under sterile conditions.

3.9 Efficient Generation of hHAP Stem Cells from the Upper Part of Human Hair Follicle [14]

1. Suspend the upper part of hair follicles in fresh DMEM containing 10% fetal bovine serum (FBS), 50 µg/mL gentamycin, 2 mM L-glutamine, and 10 mM HEPES (MP Biomedicals, Santa Ana, CA) in 6-well flat-bottom cell culture plate in order to induce differentiation.

2. Four weeks after culture of the upper parts of human hair follicles in DMEM with FBS, treat the growing cells enzymatically with Accumax (Innovative Cell Technologies, Inc., San Diego, CA) to detach.

3. Transfer the detached cells to nonadhesive culture dishes with DMEM/F12 containing 2% B-27, 5 ng/mL basic fibroblast growth factor (bFGF) without FBS.

4. One week after culture in DMEM/F12 medium without FBS, growing cells form many hHAP stem cell colonies.

5. Two days after transfer to DMEM with FBS, hHAP stem cell colonies start to differentiate.

6. Two weeks after switching to DMEM containing FBS, hHAP stem cell colonies differentiate into multiple types of cells.

7. Use differentiated cells from hHAP stem cell colonies for immunofluorescence staining.

3.10 Immunofluorescence Staining and FACS of hHAP Stem Cell Colonies and Cells Differentiated from hHAP Stem Cells

1. Immunostain the hHAP stem cell colonies for stem cell markers as described above.

2. Test stem cell markers (*see* Subheading 2).

3.11 Statistical Analysis

1. Express the experimental data as the mean ± SD.

2. Perform statistical analyses with the unpaired Student's t-test [13].

3.12 Results

3.12.1 The Differentiation Potential to Cardiac Muscle Cells Is Greatest in the Upper Part of the Hair Follicle

1. The mouse vibrissa hair follicle was divided into three parts (upper, middle, and lower part), and suspended each part separately in DMEM containing 10% FBS. Two weeks after culture all three parts of the hair follicle cells differentiated into troponin- and desmin-positive cardiac muscle cells. The number of cardiac muscle cells differentiating from the upper part of the hair follicle was significantly higher compared to the other parts (Fig. 1). All three parts of hair follicle could produce troponin-positive cardiac muscle cells as well as βIII-tubulin-positive neurons, K15-positive keratinocytes, smooth muscle actin-positive cells, and GFAP-positive glial cells. The differentiation potential to form all these cell types was greatest in the upper part of the hair follicle [4].

3.12.2 Isoproterenol Increases the Spontaneous Beating Rate in Cardiac Muscle Cells Differentiated from the Hair Follicle

1. The beating rate of cardiac muscle cells differentiated from the HAP stem cells ranged from 51.3 to 66.6 (n = 10; average 51.3 ± 14) beats/minute and increased significantly by 130.3% with isoproterenol treatment. However, propranolol reduced the isoproterenol-induced increase in the beating rate by 99.2% [4].

3.12.3 Cardiac Muscle Cells Differentiated from Hair-Spheres

1. Hair spheres which grew out of the upper follicle differentiated into troponin- and desmin-positive cardiac muscle cells as well as nestin- and βIII-tubulin-positive neurons, GFAP-positive glial cells, K15-positive keratinocytes and actin-positive smooth muscle cells [4].

3.12.4 Isoproterenol Directs the Differentiation of HAP Cells into Cardiac Muscle Cells

1. Hair follicles cultured in inducing conditions supplemented with isoproterenol alone, or the combination of isoproterenol, activin A, BMP4, and bFGF, had increased differentiation to cardiac muscle cells compared with nonsupplemented conditions [12].

2. The corresponding number of differentiated cTnT-positive cardiac-muscle-cells increased in the hair follicle cultures supplemented with isoproterenol alone or isoproterenol, activin A, BMP4, and bFGF compared with nonsupplemented medium [12].

3.12.5 The Combination of Isoproterenol, Activin A, BMP4, and bFGF Induced HAP Stem Cells to Form Beating Cardiac-muscle Tissue Sheets

1. Upper parts of the mouse whisker follicles in Matrigel culture were supplemented with isoproterenol, activin A, BMP4 and bFGF once every 4 days. We observed beating sheets layered over the hair follicles by day-10 (Fig. 2). Immunostaining of the beating cardiac-muscle tissue sheets showed that the cardiac-muscle cells were distributed within the sheets [12].

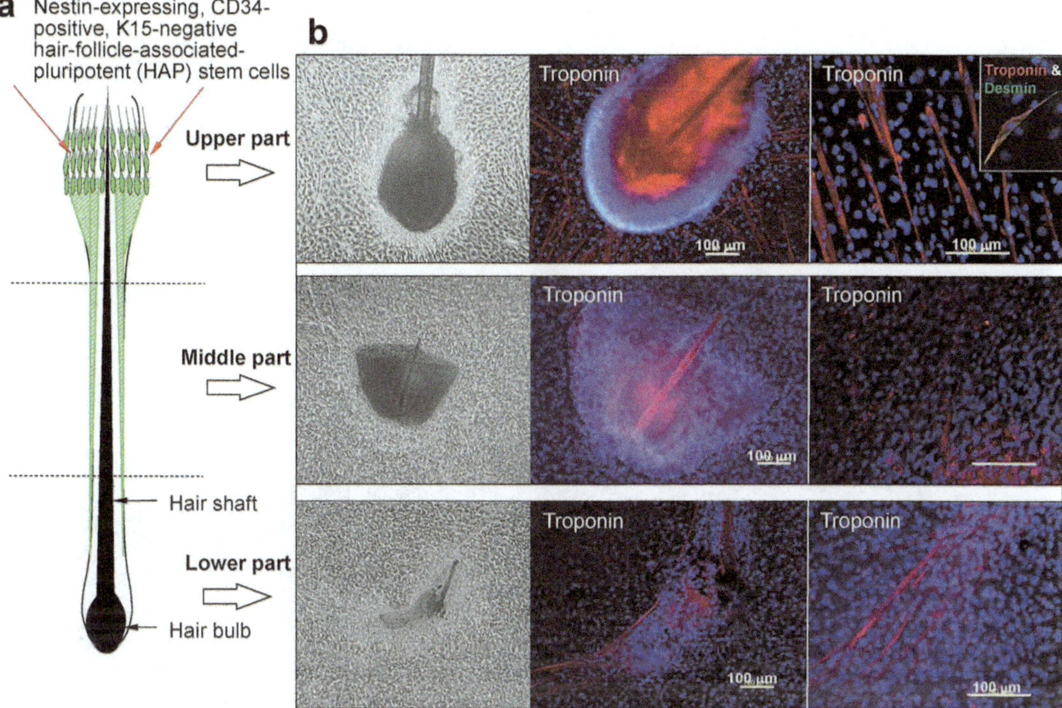

Fig. 1 (**a**) Schematic of the separated mouse vibrissa hair follicle. The upper part of the vibrissa hair follicle contains nestin-expressing, CD34-positive, K15-negative hair-follicle-associated-pluripotent (HAP) stem cells. (**b**) The mouse vibrissa hair follicle was separated into three parts (upper, middle, and lower part). Each hair follicle part was separately suspended in DMEM containing 10% FBS. (**c**) Two weeks after culture, cells from all three parts (upper, middle, and lower) of the hair follicle, incubated in DMEM with 10% FBS, differentiated into troponin- and desmin-positive cardiac muscle cells. The number of cardiac muscle cells was significantly higher in the upper part compared to the middle, and lower part of the hair follicle. Bars 100 mm [4]

2. Cardiac-muscle tissue sheets contained 39% cTnT-positive cardiac-muscle-cells; 22% CD31-positive vascular endothelial cells; and 30% smooth-muscle-actin-positive cells. Intercellular Ca^{2+} was observed in the beating cardiac-muscle tissue sheets using Fluo-3 [12].

3. Cultures of 5 upper parts of hair follicles placed on the bottom of glass culture dishes in medium were supplemented with either the combination isoproterenol, activin A, BMP4, and bFGF, formed beating cardiac-muscle tissue sheets at 65%. In contrast, cardiac-muscle tissue sheets were not formed in the nonsupplemented cultures or in cultures supplemented with isoproterenol alone [12].

3.12.6 Hypoxia
Enhanced
the Differentiation of Hap
Stem Cell
to Cardiac-Muscle-Cells

1. Under hypoxia conditions, cardiac-muscle-cells were significantly increased compared to normoxia ($P = 0.027$). Hypoxia did not increase differentiation to other cell types [13].

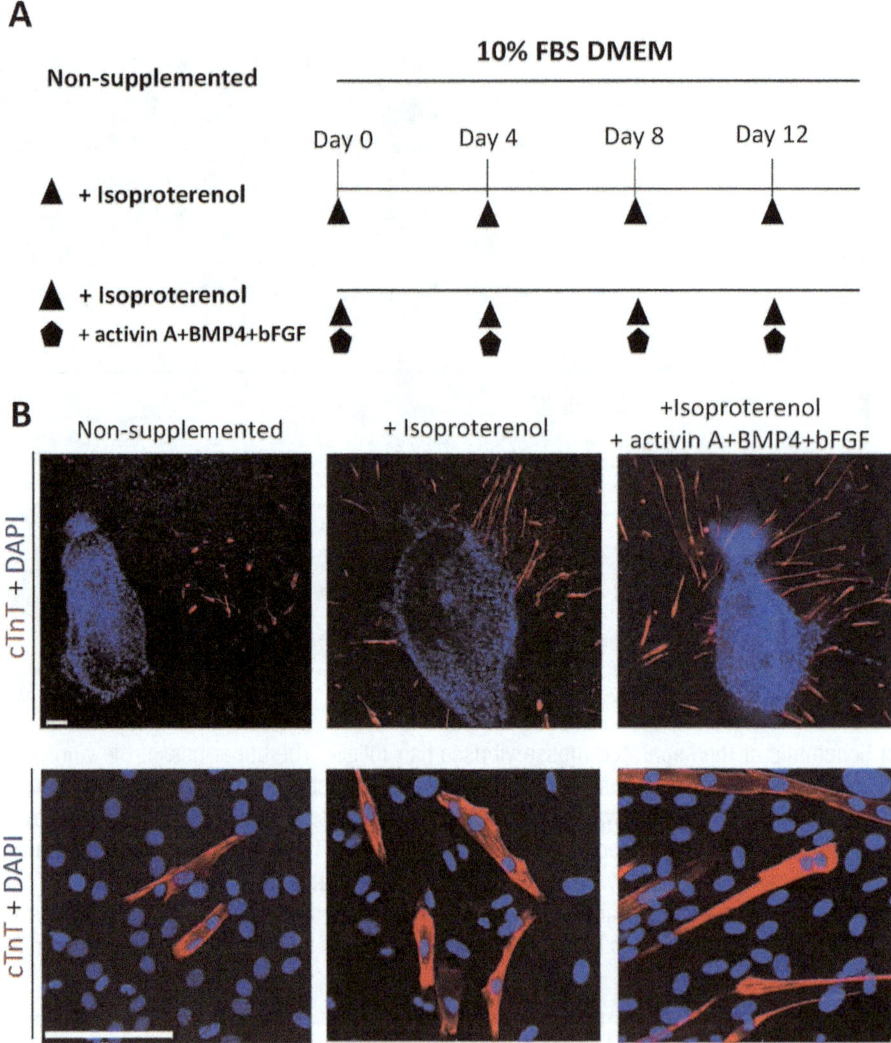

Fig. 2 Isoproterenol directs the differentiation of HAP cells into cardiac-muscle-cells. (**a**) Schematic diagram of inducing conditions for cardiac-muscle-cell differentiation from the upper part of vibrissa hair follicles. Hair follicles were cultured for 14 days in 3 conditions in DMEM medium with 10% fetal bovine serum (FBS): unsupplemented; supplemented with isoproterenol; or supplemented with the combination of isoproterenol+activin A + BMP4 + bFGF; once every 4 days. (**b**) Immunostaining of cardiac muscle cells differentiated from the upper part of vibrissa hair follicles in culture medium that was nonsupplemented (left panel); isoproterenol-supplemented (middle panel); and supplemented with the combination of isoproterenol+activin A + BMP4 + bFGF (right panel). Cardiac-muscle-cells were observed after 14 days of culture. Red = cTnT, Blue = DAPI. Bars = 100 μm [12]

*3.12.7   Human HAP (hHAP) Stem Cells from the Upper Part of Hair Follicle Can Differentiate into Cardiac-Muscle and Multiple Types of Cells (see **Note 1**)*

1. The upper parts of human hair follicles cultured in DMEM containing 10% FBS (Fig. 3) differentiated into troponin-positive cardiac-muscle cells, nestin- and βIII-tubulin-positive neurons, GFAP-positive glial cells, K15-positive keratinocytes, and smooth-muscle actin-positive smooth-muscle cells in culture (*see* **Note 2**) [14].

3.12.8   Upper Parts of Human Hair Follicles Form hHAP Stem Cell Colonies

1. Upper parts of human hair follicles were cultured in DMEM containing 10% FBS. Four weeks after culture, the growing cells from the upper parts of the human hair follicles were transferred to DMEM/F12 medium without FBS. One week after culture, the growing cells formed many hHAP stem cell colonies (Fig. 4a). The hHAP stem cell colonies were SSEA1-negative and SSEA3-, SSEA4-, Nanog-, Oct3/4-, and nestin-positive (*see* **Notes 3** and **4**) [14].

3.12.9   hHAP Stem Cell Colonies from the Upper Part of Human Hair Follicles Are Capable of Differentiating into Cardiac-Muscle-Cells and Multiple Types of Cells

1. The hHAP stem cell colonies were switched DMEM containing 10% FBS from DMEM/F12 containing B-27, supplemented with bFGF every 2 days. Two days after switching DMEM containing 10% FBS, differentiating cells migrated away from the nestin-expressing hHAP stem cell colonies. The hHAP stem cell colonies differentiated into troponin-positive cardiac-muscle cells, nestin- and βIII-tubulin-positive neurons, GFAP-positive glial cells, K15-positive keratinocytes, and smooth-muscle actin-positive smooth-muscle cells (Fig. 4B) (*see* **Note 5**) [14].

4   Notes

1. HAP stem cells differentiate into multiple cell types, including beating cardiac-muscle-cells cardiac muscle cells expressing troponin. The differentiation potential to form beating cardiac muscle cells is greatest in the upper part of the hair follicle which is enriched in HAP stem cells above the bulge. Hair spheres, consisting of nestin-expressing HAP stem cells formed from the upper part of hair follicle also differentiated into cardiac-muscle-cells as well as neurons, glial cells, keratinocytes, and smooth muscle cells [4].

2. HAP stem cells are autologous, easily accessible and can be cryopreserved for banking, making them highly attractive for regenerative medicine for heart disease as well as nerve and spinal cord repair [4].

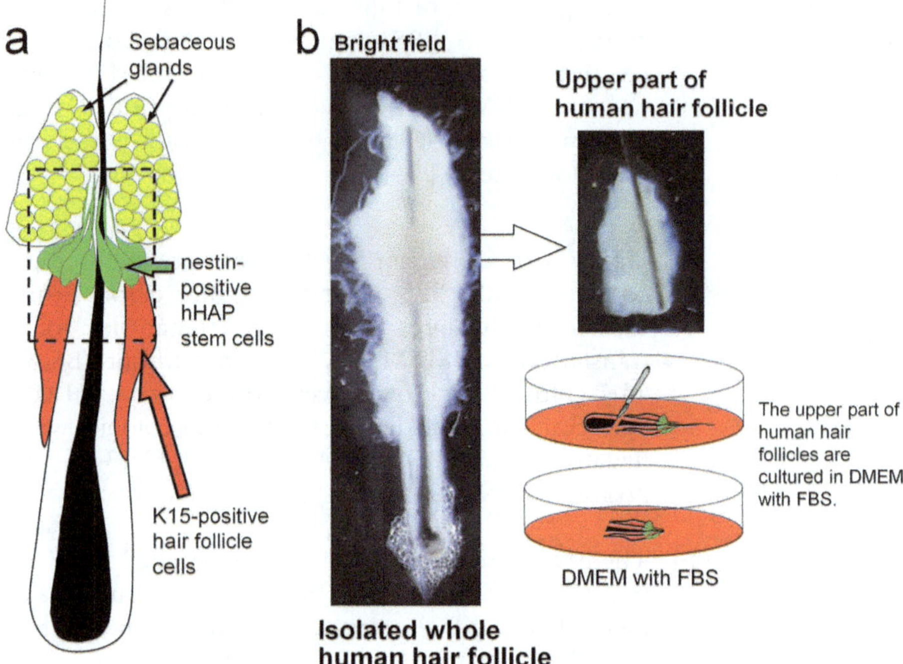

Fig. 3 Isolated human hair follicle and culture of the upper follicle. (**a**) Schema of a human scalp hair follicle shows the location of nestin-positive hHAP stem cells. (**b**) The upper parts of human scalp hair follicles were isolated and cultured in DMEM containing 10% FBS [14]

3. HAP stem cell differentiation to beating cardiac muscle cells was directed by isoproterenol and further additions of activin A, BMP4, and bFGF resulted in the formation of beating cardiac-muscle tissue sheets. The method described here is appropriate for future use of human hair follicles to grow nestin-expressing HAP stem cells in sufficient quantities to differentiate into heart muscle sheets to be used for heart muscle regeneration in the clinic [12].

4. Human hair follicles can produce hHAP stem cells in sufficient quantities for regenerative medicine for heart, nerve, and spinal cord disease and injury [14].

5. HAP stem cells [1–14] are superior over ES and iPS cells for regenerative medicine due to HAP stem cells facile accessibility from any patient, their lack of tumorigenicity, lack of the need to insert foreign genes and lack of ethical problems. It is expected in the near future that hHAP stem cells will be the predominant stem cells used for regenerative medicine [14].

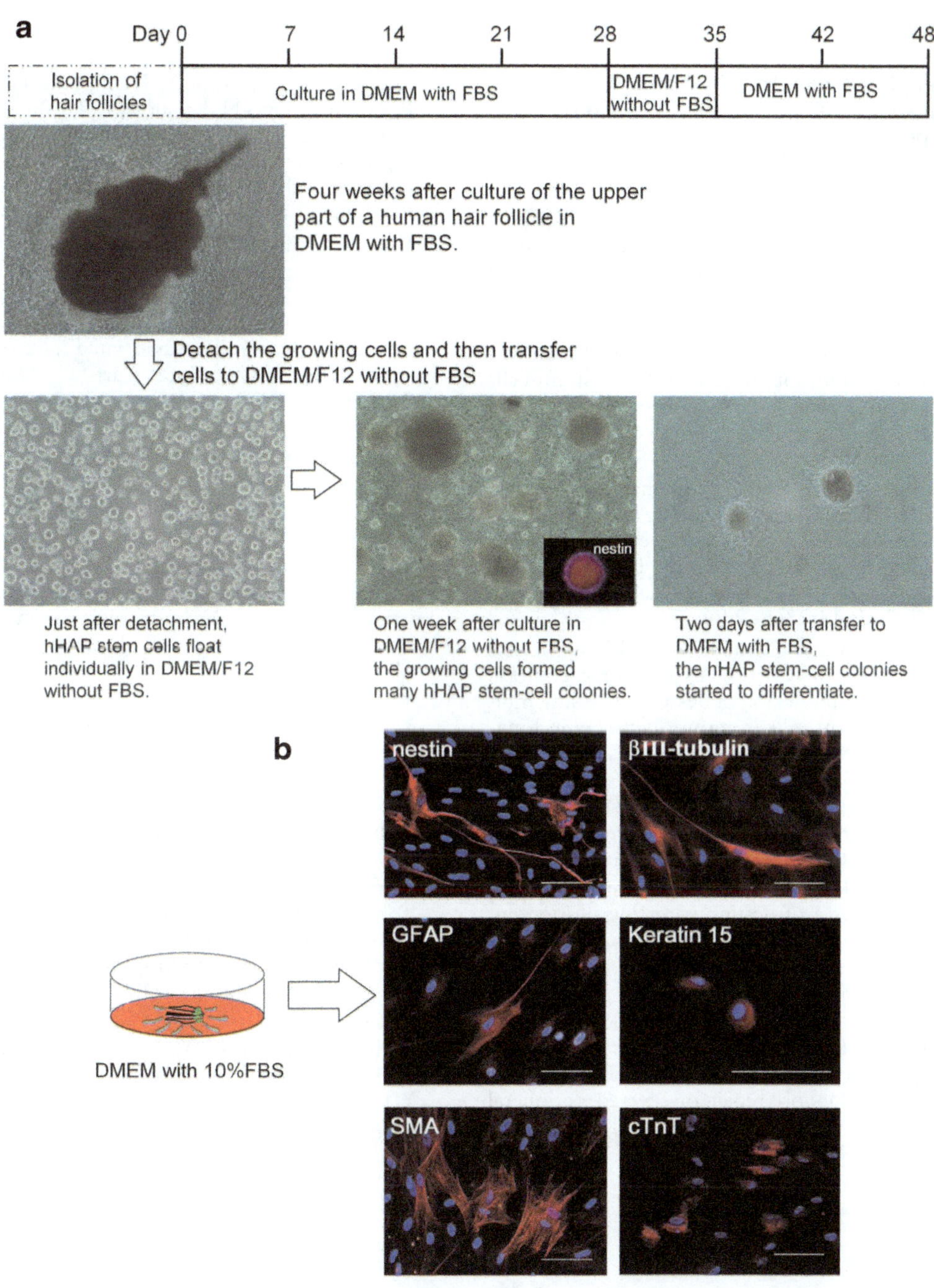

Fig. 4 (**a**) Production of hHAP stem-cell colonies. Human hair follicle culture protocol: isolated upper parts of human hair follicles were cultured in DMEM containing 10% FBS. Four weeks after culture growing cells from the upper parts of human hair follicle were transferred to DMEM/F12 without FBS. One week after culture, the growing cells formed many hHAP stem-cell colonies. Two days after transfer to DMEM containing 10% FBS, hHAP stem-cell colonies started to differentiate. (**b**) Differentiation of hHAP stem cells. Four weeks after culture in DMEM containing 10% FBS, the upper part of hair follicles differentiated into troponin (cTnT)-positive cardiac-muscle-cells, nestin- and βIII-tubulin-positive neurons, GFAP-positive glial cells, K15-positive keratinocytes, and smooth-muscle actin (SMA)-positive smooth muscle cells. Scale bar = 100 μm [14]

References

1. Li L, Mignone J, Yang M, Matic M, Penman S, Enikolopov G, Hoffman RM (2003) Nestin expression in hair follicle sheath progenitor cells. Proc Natl Acad Sci U S A 100:9958–9961

2. Amoh Y, Li L, Katsuoka K, Penman S, Hoffman RM (2005) Multipotent nestin-positive, keratin-negative hair-follicle-bulge stem cells can form neurons. Proc Natl Acad Sci U S A 102:5530–5534

3. Amoh Y, Li L, Katsuoka K, Hoffman RM (2008) Multipotent hair follicle stem cells promote repair of spinal cord injury and recovery of walking function. Cell Cycle 7:1865–1869

4. Yashiro M, Mii S, Aki R, Hamada Y, Arakawa N, Kawahara K, Hoffman RM, Amoh Y (2015) From hair to heart: nestin-expressing hair-follicle-associated pluripotent (HAP) stem cells differentiate to beating cardiac muscle cells. Cell Cycle 14:2362–2366

5. Duong J, Mii S, Uchugonova A, Liu F, Moossa AR, Hoffman RM (2012) Real-time confocal imaging of trafficking of nestin-expressing multipotent stem cells in mouse whiskers in long-term 3-D histoculture. In Vitro Cell Dev Biol Anim 48:301–305

6. Liu F, Uchugonova A, Kimura H, Zhang C, Zhao M, Zhang L, Koenig K, Duong J, Aki R, Saito N, Mii S, Amoh Y, Katsuoka K, Hoffman RM (2011) The bulge area is the major hair follicle source of nestin-expressing pluripotent stem cells which can repair the spinal cord compared to the dermal papilla. Cell Cycle 10:830–839

7. Amoh Y, Li L, Campillo R, Kawahara K, Katsuoka K, Penman S, Hoffman RM (2005) Implanted hair follicle stem cells form Schwann cells that support repair of severed peripheral nerves. Proc Natl Acad Sci U S A 102:17734–17738

8. Mii S, Duong J, Tome Y, Uchugonova A, Liu F, Amoh Y, Saito N, Katsuoka K, Hoffman RM (2013) The role of hair follicle nestin-expressing stem cells during whisker sensory-nerve growth in long-term 3D culture. J Cell Biochem 114:1674–1684

9. Kajiura S, Mii S, Aki R, Hamada Y, Arakawa N, Kawahara K, Li L, Katsuoka K, Hoffman RM, Amoh Y (2015) Cryopreservation of the hair follicle maintains pluripotency of nestin-expressing hair follicle-associated pluripotent stem cells. Tissue Eng Part C Methods 21:825–831

10. Yu H, Fang D, Kumar SM, Li L, Nguyen TK, Acs G, Herlyn M, Xu X (2006) Isolation of a novel population of multipotent adult stem cells from human hair follicles. Am J Pathol 168:1879–1888

11. Yu H, Kumar SM, Kossenkov AV, Showe L, Xu X (2010) Stem cells with neural crest characteristics derived from the bulge region of cultured human hair follicles. J Invest Dermatol 130:1227–1236

12. Yamazaki A, Yashiro M, Mii S, Aki R, Hamada Y, Arakawa N, Kawahara K, Hoffman RM, Amoh Y (2016) Isoproterenol directs hair follicle-associated pluripotent (HAP) stem cells to differentiate *in vitro* to cardiac muscle cells which can be induced to form beating heart muscle tissue sheets. Cell Cycle 15:760–765

13. Shirai K, Hamada Y, Arakawa N, Yamazaki A, Tohgi N, Aki R, Mii S, Hoffman RM, Amoh Y (2017) Hypoxia enhances differentiation of hair follicle-associated-pluripotent (HAP) stem cells to cardiac muscle cells. J Cell Biochem 118:554–558

14. Tohgi N, Obara K, Yashiro M, Hamada Y, Arakawa N, Mii S, Aki R, Hoffman RM, Amoh Y (2017) Human hair-follicle associated pluripotent (hHAP) stem cells differentiate to cardiac-muscle cells. Cell Cycle 16:95–99

Generation of FLIP and FLIP-FlpE Targeting Vectors for Biallelic Conditional and Reversible Gene Knockouts in Mouse and Human Cells

Bon-Kyoung Koo

Abstract

Rapid generation of conditional knockout models in a diploid system is challenging. Recently, CRISPR-FLIP strategy has been introduced, which facilitates the generation of biallelic conditional or reversible gene knockouts in various mammalian cell lines including mouse and human pluripotent stem cells by codelivery of the CRISPR/Cas9 system and a universal intronic cassette—FLIP and FLIP-FlpE. Here, I describe the design and cloning method of FLIP and FLIP-FlpE targeting vectors for conditional and reversible gene knockouts. This method is applicable to mouse embryonic stem cells, human induced pluripotent stem cells, and adult stem cell-derived organoids.

Key words CRISPR/Cas9, Gene targeting, Conditional knockouts, Golden Gate assembly

---

1   Introduction

Gain-of-function and loss-of-function studies have been essential in understanding embryonic development, normal physiology and the pathology of diverse diseases. Genetic loss-of-function studies have become widely available with the introduction of CRISPR/Cas9 technology, due to its robustness and simplicity. Although CRISPR/Cas9-mediated conventional knockouts have been widely reported at high efficiency, even in genome-wide screens, a method to generate conditional loss-of-function models in a simple and time-efficient manner has been challenging and is particularly desirable when studying genes essential for viability. CRISPR-FLIP has been introduced [1], which enables the generation of biallelic conditional and reversible gene knockouts in one single step. This protocol describes in detail how to design and generate targeting vectors.

Shree Ram Singh and Pranela Rameshwar (eds.), *Somatic Stem Cells: Methods and Protocols*, Methods in Molecular Biology, vol. 1842, https://doi.org/10.1007/978-1-4939-8697-2_19, © Springer Science+Business Media, LLC, part of Springer Nature 2018

2   Materials

2.1 Plasmids	1. pUC119_FLIP-puro (Addgene #84538); pUC118_FLIP-FlpE_puro (Addgene #84539).

2.1   Plasmids

1. pUC119_FLIP-puro (Addgene #84538); pUC118_FLIP-FlpE_puro (Addgene #84539).
2. DNA templates.
3. Genomic DNA isolated from target cells.
4. Competent bacteria.
5. Stellar competent cells (#636766, Takara).

2.2   Molecular Cloning Reagents

1. Phusion High-Fidelity DNA Polymerase (#M0530S, New England Biolabs).
2. Betaine solution, 5 M (#B0300-1VL, Sigma).
3. PCR Nucleotide Mix (#C1145, Promega).
4. Agarose (#A9539-500G, Sigma).
5. MinElute PCR Purification Kit (#28006, Qiagen).
6. QIAprep Spin Miniprep Kit (#27106, Qiagen).
7. Qiagen Plasmid Midi kit (#12145, Qiagen).
8. T4 DNA Ligase (#M0202M, New England Biolabs).
9. Restriction enzymes: SapI, EcoRI, NotI, PstI, DpnI (New England Biolabs).
10. Adenosine 5′-triphosphate (ATP) (#P0756S, New England Biolabs).
11. AMP+ plates (LB + ampicillin, 100 μg/mL).

2.3   Equipment

1. Thermal cycler (2720 Thermal cycler, Applied Biosystems).
2. Agarose gel electrophoresis system (AgroPower, A-7020, Bioneer).
3. Water bath.
4. Temperature controlled incubator.
5. Bacterial shaker.
6. Centrifuges.

3   Methods

3.1   FLIP/FLIP-FlpE Targeting Vector Design and Construction

Our cloning strategy is based on SapI restriction enzyme-mediated Golden Gate assembly and involves three steps.

1. First, an intron insertion site (MAGR ($^A/_C$ AG/Pu)) as well as an overlapping (or nearby) CRISPR/Cas9 target sequence is identified in order to design PCR primers to amplify a 300–1000 bp genomic region upstream and downstream of the intron insertion site.

2. Second, gene-specific homologous arms are amplified using the previously designed PCR primers, which also contain SapI sites.

3. Finally, the purified homologous arms are used in Golden Gate assembly to insert them into the plasmid backbone containing the FLIP or FLIP-FlpE intronic cassette to generate the final targeting vector.

3.2 Designing Homologous Arms for a FLIP/FLIP-FlpE Targeting Vector

There are several criteria which must be met in order to identify an appropriate FLIP or FLIP-FlpE cassette insertion site (Fig. 1):

1. To ensure gene inactivation after Cre recombination, the cassette insertion site (intron insertion site (MAGR (^{A}/$_C$ AG/ Pu))) needs ideally to be within the first 50% of the coding region.

2. Insertion of the intronic cassette must leave both parts of the split exon to be over 60 bp long.

3. The intron insertion site must have an overlapping (or at least nearby) CRISPR/Cas9 target sequence, which can be on either of the two DNA strands. Note that target sequences with a PAM sequence (NGG) present in the intron insertion site cannot be used unless silent mutations are generated, as the PAM sequence will not be destroyed after the intron insertion. When the intron insertion site does not overlap with the

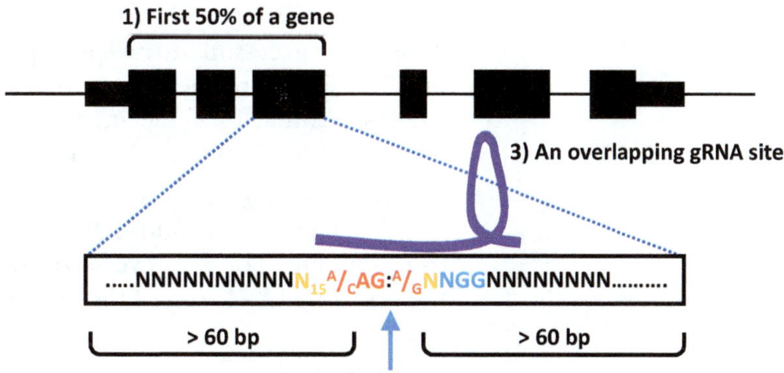

Fig. 1 Identification of an insertion site for the FLIP/FLIP-FlpE cassette. Schematic drawing showing the three main criteria when identifying an insertion site for the FLIP/FLIP-FlpE cassette. First, the intronic cassette is inserted into an exon belonging to the first 50% of the gene, between the G and the ^{A}/$_G$ of the intron insertion site. Second, the intronic cassette is inserted into the middle of the split exon, leaving >60 bp of exon either side. Finally, there needs to be a gRNA target sequence which overlaps with the insertion site. The insertion site is in red, the gRNA target sequence in yellow, the PAM sequence in light blue and gRNA in purple

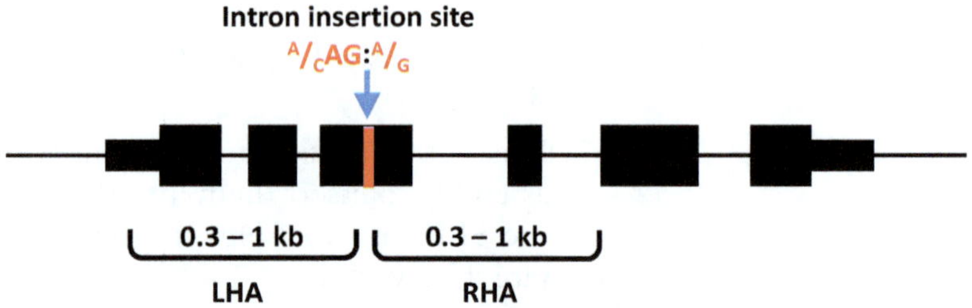

Fig. 2 Schematic drawing of the amplification of the homologous arms. The left (LHA) and right (RHA) homologous arms are amplified from the area around the insertion site and are 300–1000 bp in length

CRISPR/Cas9 target sequence, the introduction of several silent mutations that prevent Cas9 binding needs to be considered.

4. Once both the intron insertion site and the corresponding CRISPR/Cas9 target site have been identified, the genomic sequence upstream and downstream of the intron insertion site can be obtained from the UCSC genome browser (https://genome.ucsc.edu). The upstream region will be used for the left homologous arm (LHA) and the downstream region will be for the right homologous arm (RHA). The size of each homologous arm can vary from 300 to 1000 bp and it is desirable to choose a region devoid of SapI restriction enzyme sites, since this enzyme will be utilized in the Golden Gate assembly (Fig. 2).

5. To ensure successful intron splicing, it is key to insert the FLIP or FLIP-FlpE intronic cassette precisely between the third and the last sequence of the intron insertion site (MAGR ($^A/_C$ AG/Pu)), i.e., between "MAG ($^A/_C$ AG)" and "R (Pu, $^A/_G$)."

6. When designing primers for the LHA, the forward primer (LHA-L) can start 300–1000 bp upstream of the intron insertion site. For the reverse primer of the LHA (LHA-R), the first 3 bp (A/CAG) of the intron insertion site must be the last 3 bp.

7. When designing primers for the RHA, the forward primer (RHA-L) must start from the last sequence (Pu, A/G) of the intron insertion site and the reverse primer (RHA-R) may start anywhere from 300–1000 bp downstream of the intron insertion site.

8. Again, caution should be taken to ensure that the regions used for the homologous arms do not contain any SapI site, as this enzyme will be used during Golden Gate assembly.

9. For optimal primer design, it is recommended to use Primer3Plus (primer3plus.com/cgi-bin/dev/primer3plus. cgi) with the following adjustments: 28 bp optimal primer

Table 1
Primers for PCR amplification of the homologous arms

Primer name	Sequence 5′ to 3′
LHA_L	GTTTAAAC<u>GCTCTTC</u>TGTG(N)$_{24-36}$
LHA_R	GTTTAAAC<u>GCTCTTC</u>TTACCT^{G}/$_{T}$N$_{(21-33)}$
RHA_L	GTTTAAAC<u>GCTCTTC</u>TTAG^{A}/$_{G}$N$_{(23-35)}$
RHA_R	GTTTAAAC<u>GCTCTTC</u>TTTAN$_{(24-36)}$

The final PCR primers are composed of invariable sequences shown below (GTTTAAAC overhang enables efficient SapI digestion, and GCTCTTC is the SapI restriction enzyme site.) linked to a gene-specific sequence of various lengths

length, 65 °C optimal melting temperature (Tm), and a 300–1000 bp product size range.

10. The 28 bp sequences are then linked to invariable overhangs containing a SapI site, resulting in the final primer sequences that will be used for amplification of both homologous arms (Table 1). As a result, the four primers for the amplification of both homologous arms will be:

LHA-L (GTTTAAACGCTCTTCTGTGN(24–36)), LHA-R (GTTTAAACGCTCTTCTTACCT ^{G}/$_{T}$N(21–33)), RHA-L (GTTTAAACGCTCTTCTTAG ^{A}/$_{G}$N(23–35)) and RHA-R (GTTTAAACGCTCTTCTTTA-N(24–36)).

3.3 FLIP/FLIP-FlpE Targeting Vector Assembly

1. PCR amplification of the homologous arms: use a high-fidelity polymerase (e.g., Phusion High-Fidelity DNA Polymerase (#M0530S, New England Biolabs)) and set up two PCR reactions for the LHA and RHA (Table 2) using genomic DNA from mESCs as a template.

2. Run the following program (Table 3) in a thermal cycler.

3. Add DNA loading buffer to 10 µL of the PCR product and visualize on a 1% agarose gel (100 V for 20 min). Check the expected band size.

4. Purify the PCR products with a PCR purification kit and elute the DNA in 15 µL ddH$_2$O.

3.4 Golden Gate Assembly of FLIP/FLIP-FlpE Targeting Vectors

1. Set up the Golden Gate assembly (Fig. 3) using the purified homologous arms and plasmid backbone containing either FLIP (Addgene #84538) or FLIP-FlpE (Addgene #84539) (Table 4).

2. Run the following program (Table 5) in a thermal cycler.

3. Transform 1.5 µL of the reaction product into Stellar competent bacteria.

4. Plate the entire transformation mixture on an AMP+ (100 µg/mL) plate and incubate overnight at 37 °C.

Table 2
Reaction conditions for PCR amplification of the homologous arms

Reagent	Volume (µL)
5× Phusion HF buffer	10
dNTPs (10 mM each)	1
Betaine	10
DMSO	1.5
Genomic DNA template (~100 ng/µL)	1
Primers For + Rev. (10 µM each)	2.5
Phusion HF polymerase	0.5
ddH$_2$O	Up to 50

Table 3
PCR program for amplification of the homologous arms

Temperature	Time
Pre-PCR	
98 °C	30 s
58 °C	30 s
72 °C	1 min 30 s
×2 cycles for next steps	
98 °C	12 s
62 °C	30 s
72 °C	1 min 15 s
×30 cycles for next steps	
98 °C	12 s
68 °C	30 s
72 °C	1 min 15 s
Final steps	
72 °C	3 min
4 °C	∞

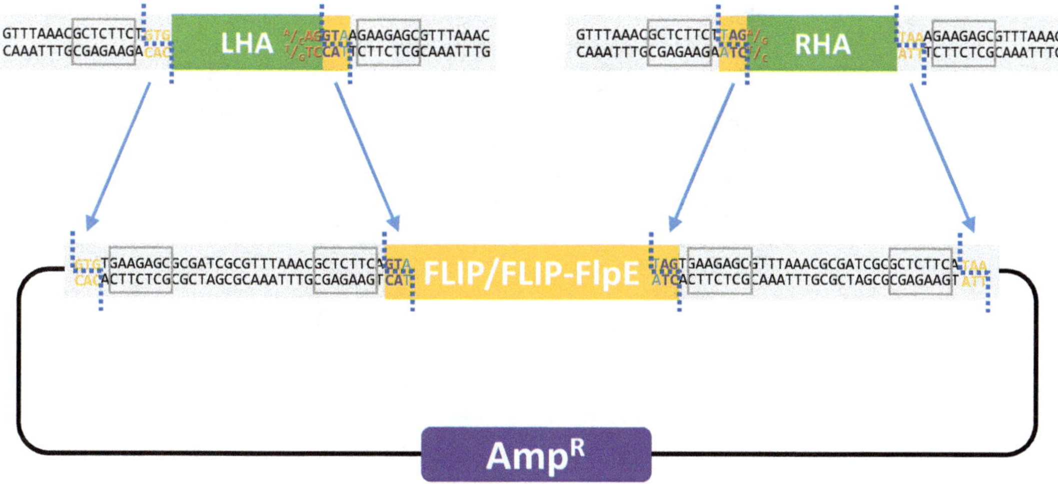

Fig. 3 Schematic drawing of the Golden Gate assembly. The backbone vector containing the FLIP/FLIP-FlpE intronic cassette, the PCR-amplified homologous arms, the SapI restriction enzyme and DNA ligase are mixed in a single reaction. Repeated cycles of cutting (by the SapI enzyme) and pasting (by the DNA ligase) mediates the insertion of the homologous arms into the vector. The recognition site of the SapI enzyme is shown in grey boxes, the cut site is shown in blue dotted lines, the intron insertion site sequence is shown in red, the yellow bases represent the customized SapI overhangs, the purple bases represent the consensus sequence of the splice donor/acceptor regions, and the light blue bases represent less well conserved bases which are still part of the splice donor/acceptor regions. The left homologous arm (LHA) and right homologous arm (RHA) are in green

Table 4
Golden Gate assembly

Reagent	Volume (μL)
10× T4 DNA Ligase Buffer	1.5
10 mM ATP	1.5
SapI	1
T4 DNA ligase	1
Vector backbone (100 ng/μL)	1
LHA (~50 ng/μL)	1
RHA (~50 ng/μL)	1
ddH$_2$O	up to 15

5. Inoculate 5–10 colonies from the bacterial plate and culture overnight in 2 mL of LB media supplemented with ampicillin (100 μg/mL).

6. The following day, purify plasmid DNA using the QIAprep Spin Miniprep Kit.

Table 5
Golden Gate assembly reaction cycle

Temperature	Time
×25 cycles for next steps	
37 °C	2 min
16 °C	5 min
Final steps	
37 °C	15 min
80 °C	10 min
4 °C	∞

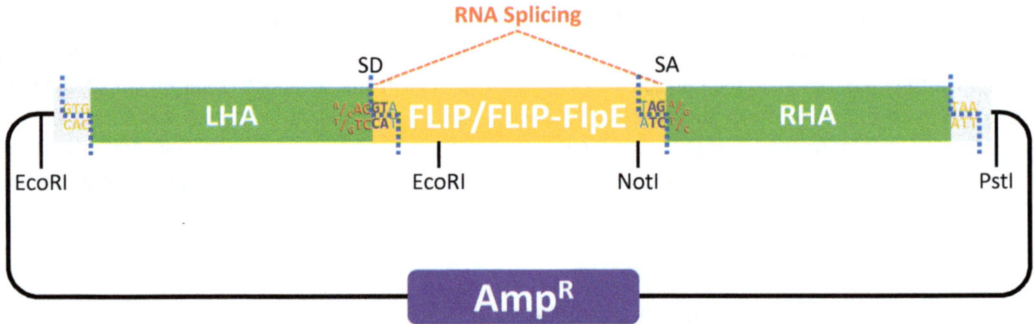

Fig. 4 Schematic drawing illustrating the RNA splicing and restriction sites. The integration of the left homologous arm (LHA) can be confirmed by restriction digest using EcoRI and the integration of the right homologous arm (RHA) can be confirmed by restriction digest using NotI and PstI. RNA splicing will occur in the initial conformation of the cassette from the splice donor (SD) to the splice acceptor (SA), and therefore the cassette will not disrupt gene expression upon insertion

3.5 Confirming Successful Cloning of the Targeting Vector

In the final targeting vector the left homologous arm (LHA) is flanked by two EcoRI sites and the right homologous arm (RHA) is flanked by NotI and PstI restriction sites. Hence, digestion reactions using these enzymes will identify the clone(s) containing the correct configuration of homologous arms (Fig. 4).

1. Perform two individual restriction digest reactions: EcoRI single digest for LHA integration and NotI and PstI double digest for RHA integration. Incubate for 2 h at 37 °C.

2. Add DNA loading dye to the digested samples and visualize on a 1% agarose gel (100 V, ~20 min) to check the expected pattern and confirm the correct insertion of the gene-specific homologous arms (Fig. 5). Note that there could be additional EcoRI, NotI, and/or PstI enzyme site(s) in the homologous

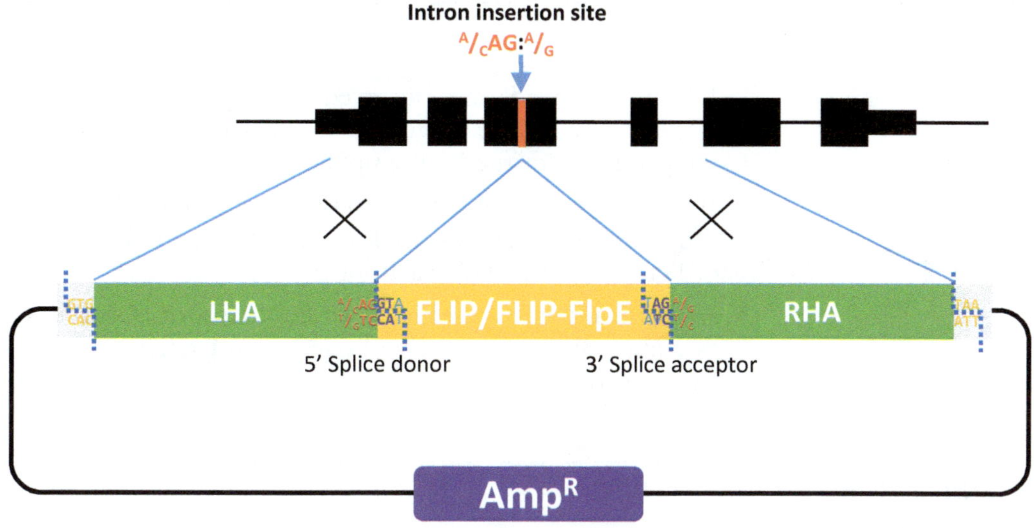

Fig. 5 Insertion of the FLIP/FLIP-FlpE intronic cassette. Schematic drawing showing the insertion of the FLIP/FLIP-FlpE intronic cassette into an exon within the gene of interest. The FLIP/FLIP-FlpE cassette is integrated into the insertion site (in red, $^{A}/_{C}$ AG$^{A}/_{G}$). The yellow bases represent the customized SapI overhangs, the purple bases represent the consensus sequence of the splice donor/acceptor regions, and the light blue bases represent less well conserved bases which are still part of the splice donor/acceptor regions. The left homologous arm (LHA) and right homologous arm (RHA) are in green. The SapI cut site is represented by blue dotted lines

Table 6
Sequencing primers for the FLIP/FLIP-FlpE targeting vector

Primer name	Sequence
5′ of LHA	ATGCTTCCGGCTCGTATGTT
3′ of LHA	TGGTTTGTCCAAACTCATCAA
5′ of RHA	GGCCGCGGTTACAAGACA
3′ of RHA	CTACAGGGCGCGTACTATGG

arms, which may result in additional bands of different sizes. Please consider this when interpreting the results of the restriction enzyme digest.

3. Sequence the clone(s) showing the expected pattern after digestion, using the primers below (Table 6).

4. Once the targeting vector is confirmed, amplify the plasmid by inoculating a large-scale bacterial culture and purify the DNA using a Qiagen Plasmid Midi kit.

4   Notes

1. If multiple bands are observed from genomic DNA PCR amplification of homologous arms, increasing the Tm (meting temperature) for the PCR amplification may help.

2. If no band can be observed from genomic DNA PCR amplification of homologous arms, lower the Tm or add DMSO and/or betaine to the PCR reaction.

3. If there are no colonies observed following Golden Gate assembly, check the quality of competent bacteria with a positive control.

Acknowledgments

This protocol is adapted from Protocol Exchange (2017) doi:https://doi.org/10.1038/protex.2017.026. B.-K.K. is supported by a Sir Henry Dale Fellowship from the Wellcome Trust and the Royal Society (101241/Z/13/Z) and received a core support grant from the Wellcome Trust and MRC to the WT–MRC Cambridge Stem Cell Institute.

Reference

1. Andersson-Rolf A, Mustata RC, Merenda A et al (2017) One-step generation of conditional and reversible gene knockouts. Nat Methods 14(3):287–289

Chapter 20

Analytical Platforms and Techniques to Study Stem Cell Metabolism

Christine Tang, Kevin Chen, Aleksandar Bajic, William T. Choi, Dodge L. Baluya, and Mirjana Maletic-Savatic

Abstract

Over the past decade, advances in systems biology or 'omics techniques have enabled unprecedented insights into the biological processes that occur in cells, tissues, and on the organism level. One of these technologies is the metabolomics, which examines the whole content of the metabolites in a given sample. In a biological system, a stem cell for instance, there are thousands of single components, such as genes, RNA, proteins, and metabolites. These multiple molecular species interact with each other and these interactions may change over the life-time of a cell or in response to specific stimuli, adding to the complexity of the system. Using metabolomics, we can obtain an instantaneous snapshot of the biological status of a cell, tissue, or organism and gain insights on the pattern(s) of numerous analytes, both known and unknown, that result because of a given biological condition. Here, we outline the main methods to study the metabolism of stem cells, including a relatively recent technology of mass spectrometry imaging. Given the abundant and increasing interest in stem cell metabolism in both physiological and pathological conditions, we hope that this chapter will provide incentives for more research in these areas to ultimately reach wide network of applications in biomedical, pharmaceutical, and nutritional research and clinical medicine.

Key words Metabolomics, Mass spectrometry, Nuclear magnetic resonance, Mass spectrometry imaging, Stem cells, Neural stem cells

1 Introduction

Stem cells, the undifferentiated cells capable of self-renewal and differentiation into lineage-specific progeny, hold tremendous promise to help us understand and treat a variety of diseases [1–3]. However, to fulfill this promise and begin to utilize stem cells for organism repair and regeneration, we need to understand the mechanisms that govern their fate and function. Recently, metabolism has emerged as an active regulator of stem cell behavior: stem cells metabolically substantially differ from their progeny,

Christine Tang and Kevin Chen contributed equally to this work

Shree Ram Singh and Pranela Rameshwar (eds.), *Somatic Stem Cells: Methods and Protocols*, Methods in Molecular Biology, vol. 1842, https://doi.org/10.1007/978-1-4939-8697-2_20, © Springer Science+Business Media, LLC, part of Springer Nature 2018

pointing to the critical role of metabolic processes as regulators of stem cell fate and function [4, 5]. To accelerate research in this field and expand our understanding of stem cell metabolism, we can use a variety of analytical platforms and techniques, summarized herein (Fig. 1).

1.1 Stem Cell and Metabolism: A Shift in Balance

Stem cells display metabolic flexibility in response to the demands of their environment and their energy needs to transition from one cell state to another [6]. For example, embryonic stem cells (ESCs), derived from the inner cell mass of the primordial embryo, depend mainly on anabolic metabolic pathways to maintain their energy needs [7–9]. Their main energy source comes from partial breakdown of glucose via glycolysis. In addition, glycolytic intermediates are also shunted to the pentose phosphate pathway to regenerate NADPH cofactors needed to maintain metabolic function. Similar to an early embryo, ESCs have a higher rate of cellular division to generate all definitive cell types of the embryo proper, thus requiring a more immediate need for energy, i.e., ATP. Although catabolic metabolism involving the mitochondria is more efficient to generate ATP, the rapid cellular division restricts the number of mature mitochondrial infrastructure to meet the energy demand. Therefore, anabolic metabolism becomes a better compromise for the energy needs of ESCs. When compared to more lineage-restricted stem cells, such as hematopoietic stem cells (HSCs), neural stem cells (NSCs), and mesenchymal stem cells

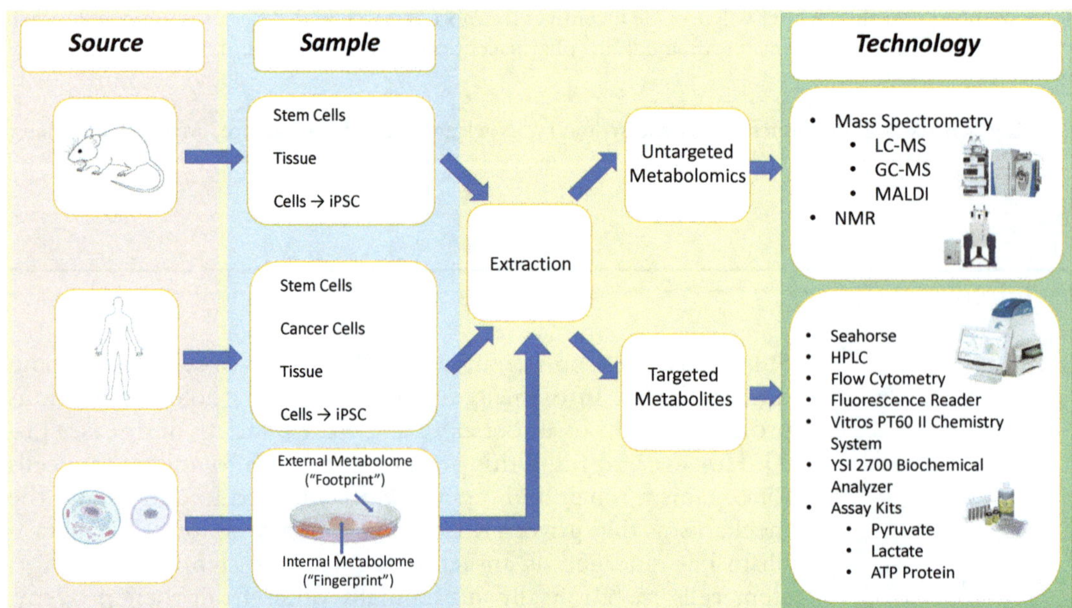

Fig. 1 Metabolism of stem cells can be studied using a variety of analytical platforms. Different types of stem cells can be obtained from a variety of sources. Following extraction, both untargeted and targeted analysis can be done using different technologies

(MSCs) [4, 10–12], the ESC metabolic energy source becomes dominantly catabolic. This is correlated to the number of mature mitochondria, which allocate the energy needs to oxidative metabolism of carbohydrate, lipid, and amino acids. In addition, the regulation of reactive oxygen species via the mitochondria is also implicated for stem cell homeostasis and lineage specification. The reciprocal relationship of anabolic and catabolic means of energy support thus represents the metabolic phenotype of stem cells [4, 10–12].

When stem cells divide to proliferate, the mitochondria are asymmetrically distributed to daughter cells [13]. The one that receives older mitochondria maintains stem cell characteristics, and the other, with younger mitochondria, becomes lineage-specific [14]. After specified lineage differentiation, mitochondrial DNA and emerging energy requirements are gradually increased in support of mitochondrial biogenesis [15, 16]: the spherical and cristae-poor mitochondria of undifferentiated stem cells are transformed into tubular and cristae-rich structures to guarantee sufficient ATP for energy metabolism [16]. Concomitantly, the production of mitochondrial-related key enzymes and mitochondrial reactive oxygen species are increased, and the expression levels of glycolytic genes and the production of antioxidant defenses are suppressed [16–20].

Adult stem cells in the hematopoietic system, bone marrow, hair follicle, nervous system, and skeletal muscles are mostly quiescent [21]. One of the shared features of various quiescent stem cells is the slow rate of cell cycling [4, 21]. One hypothesis is that the slow rate of metabolism, mainly due to low levels of oxidative phosphorylation, prevents reactive oxygen species from causing cellular damage [22]. Characterization of the metabolic state during the quiescent period is hindered by the lack of specific genetic markers for this particular cell state [23]. Due to the slow rate of metabolism, metabolic markers may be the key to distinguish and/ or isolate quiescent stem cells from their metabolically active progeny. Most recently, such strides have been made, identifying lipid metabolism and amino-acid and nucleotide synthesis as well as downregulation of oxidative phosphorylation and urea cycle as key features of the transition from quiescence to proliferation [24, 25].

The metabolic dichotomy of various stem cell fates can also be extended to the metabolism of cancer, in which abnormal metabolic phenotypes also reflect the energy demands of the cell [26, 27]. The most notable metabolic phenomenon is the Warburg effect when cancer cells generate ATP mainly through the glycolytic pathway even in the presence of oxygen [28, 29]. The high rate of glycolysis and lactate production has unveiled many aberrant signaling pathways involving genes that regulate metabolism in cancer. For example, the increased production of 2-hydroxyglutarate, an intermediate of the tricarboxylic acid cycle

(TCA) or the Krebs cycle, has unveiled gene mutations in isocitrate dehydrogenase 1 and 2. This leads to alterations in the α-ketoglutarate-dependent dioxygenase enzyme activity and affects the completion of aerobic metabolic activities seen in malignant tumors such as myeloid leukemias, gliomas, thyroid carcinomas, cartilaginous tumors, and intrahepatic tumors [30].

While the studies of the metabolic changes from one cell type to another focus on the biochemistry of particular metabolic pathways, metabolomics and bioinformatics have been the recent drivers in the discoveries of stem cell metabolic plasticity. To achieve tangible translational results, empirical studies of stem cell metabolism (and metabolic profiling in general) require reproducible and quantitative analytical methodologies; standardized normative data to allow reliable discovery of molecules important for a given experimental condition; and computerized, automated mathematical algorithms for objective analysis of complex datasets [31]. Here, we review the main platforms used to study metabolism and outline methodologies for analysis of some metabolites implicated in stem cell biology.

1.2 Platforms for Metabolic Studies

Today, a wide range of instrumentation is available to analyze, detect, and quantify metabolites, as no single analytical platform is capable of detecting the whole set of metabolites in a biological sample (for review, *see* ref. [31]). Two main platforms are ^{1}H NMR, capable of measuring intact samples but limited to analytes with medium to high concentrations, and mass spectrometry (MS), highly sensitive but less reproducible compared to the NMR [32]. NMR is rapid, nondestructive, and noninvasive; requires minimal sample preparation; and most importantly, it is quantitative (NMR signal intensity is proportional to sample concentration) at a dynamic range of 2×10^5 and very reproducible (coefficient of variation, CV 1–2%) making it ideal for on-site diagnostics and biomarker discovery.

MS is more sensitive than NMR; detecting more metabolites means MS can survey a larger fraction of the metabolome depending on the standard libraries for identification of detected species, but it is intrinsically qualitative [33]. Two main types of MS are gas chromatography (GC) and liquid chromatography (LC). In GC, the two phases are a solid (or immobilized liquid) stationary phase and a gaseous mobile phase, while in LC, both phases are liquid mobile phase and stationary phase of a solid or a liquid covalently coated onto a solid surface. In principle, the stationary phase is contained in a chromatography column through which the mobile phase passes. As the sample in the mobile phase passes through the column, analytes continuously transfer between mobile and stationary phases. Different analytes have different elution and retention times in each phase and this partition is optimized to provide appropriate chromatographic separations.

Internal standards allow relative quantification of the given analyte. Ionization sources create a charged species for each analyte, whether to remove it (electron impact ionization) or to add it (positive ion electrospray and MALDI).

As most of the metabolites are polar in nature (e.g., amino acids, nucleotides, carboxylic acids), their chromatographic separation and analysis can be challenging [34]. Recently, hydrophilic interaction chromatography (HILIC) has become a useful tool for analysis of such polar molecules. HILIC can efficiently separate many classes of highly polar metabolites, including amino acids, nucleotides, nucleosides, nucleotide phosphates, carboxylic acids, and vitamins. The sensitivity and selectivity of the HILIC separation can be greatly enhanced by employing the multiple reaction monitoring (MRM) function of tandem quadrupole mass spectrometers [35, 36].

Finally, mass spectrometry imaging (MSI) is a relatively new technology that provides spatial information on small molecules, enabling an unprecedented capacity to localize lipids, peptides, proteins, and drugs in a tissue of interest [37]. Multiple ionization sources offer a broad range of spatial resolutions that can be used to map the distribution of metabolites in the tissue. A variety of MSI technologies have been developed for analysis of different molecules. Matrix-assisted laser desorption/ionization (MALDI)-MSI can be used to analyze plant and animal tissues, cancer cells, and microbial communities, especially macrobiomolecules such as peptides and proteins. Laser desorption/ionization (LDI)-MSI is more suited for the profiling and imaging of small metabolites.

Here, we present basic methods for metabolite and metabolomics studies using mass spectrometry and NMR, from sample preparation to analysis. While MSI has not yet been used for studies of stem cell metabolism in situ, we outline here the methodology envisioning a variety of applications that could move the field forward.

---

2   Materials

2.1 Sample Preparation and Storage

Stem cells are grown in culture or directly isolated from an organism according to specific protocols; here, we outline the preparation for neural stem cells, which we utilize in our laboratory regularly [1, 38, 39]. In general, for MS analysis you will need about 3–5 million cells (five replicates) and for the NMR analysis, about one million cells (five replicates). Given that NMR analysis does not require any particular preparation of the sample, one may first perform NMR for untargeted analysis and then use the same sample for MS targeted analysis—in such a case, the number of cells needed is determined by the requirement for MS.

2.2 Reagents

1. Phosphate buffered saline (PBS).
2. Neurobasal media.
3. DMEM/F-12 media.
4. B-27 Supplement, vitamin A-free.
5. N-2 Supplement.
6. Poly-L-ornithine.
7. Laminin.
8. bFGF.
9. EGF.
10. Y-27632.
11. SB-431542.
12. Cyclopamine.
13. Dorsomorphin.
14. L-ascorbic acid-2-phosphate.
15. Accutase.
16. Rosette Selection Reagent.
17. Penicillin–streptomycin.
18. Trypan blue, for counting the cells.
19. Acetonitrile.
20. TDFHA.
21. Internal standards for MS (Sigma, St. Louis, MO, or Aldrich, Steinheim, Germany).
22. Internal standards for NMR (TMS, DDS).

2.3 Equipment

1. Centrifuge.
2. Microcentrifuge.
3. Speedvac concentrator.
4. 0.22 μm filters.
5. 2 mL GC-HPLC glass vials with 200 μL conical narrow opening inserts.
6. Ultrasonic bath.
7. For MS experiments: Gas chromatography instrument with electronic flow control to provide constant flow separation. MS ion trap or other mass spectrometers, such as quadrupole, time-of-flight, or triple quadrupole. National Institute of Standards and Technology (NIST) mass spectra library.
8. For NMR experiments: NMR spectrometer (500–900 MHz, Bruker). NMR fingerprinting can be done at any operating frequencies, but the higher, the better. Depending on cryoprobes, one can do proton, carbon, nitrogen and/or phospho-

rus studies. For NMR data preprocessing, use Topspin 2.1, Bruker.

9. Software for analysis: Analyst 3.1.1. or Agilent software for Quantitative and Qualitative analysis (MS data); TopSpin, MetNova, Chenomx, Inc. (NMR data); Simca P, Amix (for multivariate analysis of metabolomics data).

2.4 Mass Spectrometry Imaging

1. Tissue-Tek Mega-Cassette (Sakura; 40 × 25 × 10 mm).
2. 9-Aminoacridine 9AA MALDI Matrix.
3. Red phosphorus (for mass calibration).
4. Methanol.
5. Water.
6. Analysis: HDI 1.4 software.

3 Methods

3.1 Preparation of Human Neural Stem and Progenitor Cells

1. For culturing of rodent neural stem/progenitor cells, please see [39].

2. For culturing human neural stem/progenitor cells, WA09 (H9) human embryonic stem cells (hESCs) are maintained on Matrigel-coated plates in Essential 8 Medium media (E8) [40]. Prepare E8: DMEM/F12 with glutamine and HEPES, 20 μg/mL insulin, 64 μg/mL L-ascorbic acid-2-phosphate, 14 ng/mL sodium selenite, 10.7 μg/mL transferrin, 100 ng/mL human–FGF2, 2 ng/mL human-TGF-β1, and 1% Pen/Strep.

3. To differentiate hESCs into neural stem/progenitor cells (hNPCs), use a variation of the dual SMAD inhibition protocol [41–43].

4. hESCs are first dissociated with Accutase and 2 million cells dispensed per well of Aggrewell plate in neural induction medium (NIM) to form aggregates. NIM is prepared from equal volumes of DMEM/F-12 and Neurobasal medium with 2% B27-supplement, 1% N2-supplement, 2 mM GlutaMAX, and 1% Pen/Strep.

5. During the initial 24 hours of culturing in Aggrewells, use 10 μM Y-27632 to promote cell survival.

6. On the following day, three-quarters of the media should be changed and SMAD inhibition initiated by adding 10 μM SB-431542.

7. 4 μM Dorsomorphin is then added to the culture from day 3.

8. On day 5, aggregates are gently collected, sieved through a reversible strainer and transferred into Matrigel-coated plates

with neural proliferation medium (NPM). NPM is prepared from equal volumes of DMEM/F-12 and Neurobasal media with 1% B27-supplement, 0.5% N2-supplement, 20 ng bFGF per mL, and 20 ng EGF per mL 2 mM GlutaMAX, and 1% Pen/Strep.

9. From day 6, add 10 μM cyclopamine to promote dorsalization. Both SMAD inhibitors and CCP are present in the media until day 9. Media is changed daily until rosette-shaped clusters of HNPCs are harvested with Rosette Selection Reagent on days 12–14.

10. After dislodging, rosettes are incubated in wells coated with 0.2% porcine gelatin to allow the nonneural cells to differentially attach [44]. Floating fraction is collected after 1 hour, transferred into noncoated cell-culture flasks and incubated in suspension overnight.

11. The following day, rounded and floating spheres of hNPCs are plated into Matrigel-coated 6-well plates and propagated until cells form a confluent layer.

12. All cultures are maintained in the presence of 1% penicillin–streptomycin.

3.2 Preparation of Internal Standard Solution for MS

For MS analysis, standard compounds (Sigma, St. Louis, MO, or Aldrich, Steinheim, Germany) should be prepared as stock solutions at 25 mM either in Milli-Q water or in methanol. Standard solutions should be kept in small aliquots at −80 °C. Before analysis, the solutions should be thawed and diluted (*see* **Note 1**).

1. Prepare 1 mL of 10 mM solution for each internal standard, up to 10.

2. Combine them all, to get final 10 mL internal standard stock solution (each compound will be 1 mM final concentration).

3. Take two 10 μL aliquots from the above stock solution into two vials and dry them completely.

4. Reconstitute each vial with 200 μL of methanol–water (1:1), sonicate and combine into a third vial.

5. Rinse again the first two vials with 100 μL of methanol–water (1:1), sonicate and add to the third vial.

6. The final internal standard solution will have 600 μL volume; keep it on ice.

3.3 Preparation of the Sample for MS (see Note 1)

1. Make 26 mL of methanol–water (4:1) solution.

2. Add 600 μL of the internal standard solution to 26 mL of methanol–water (4:1) sample solution to make 26.6 mL final master solution spiked with internal standard.

3. Vortex to mix it thoroughly and keep it on ice.

4. Freeze and thaw cell pellets in liquid nitrogen and water for three cycles, 30s each time.

5. Add 750 µL of ice cold internal standard spiked master solution to the sample vial containing freeze-thawed cell pellets.

6. Sonicate all the solutions (settings: Amp 30%, 20–30 s, three times)

7. Wash rotor with 10 mL de-ionized water and methanol in-between each sample.

3.4 Extraction of Polar/Nonpolar Metabolites for MS (see Notes 1 and 2)

1. Add 450 µL of ice cold $CHCl_3$ (keep it at −20 °C for 10 min) to each sample and vortex it for 10 min at high speed.

2. Add 150 µL of ice cold water and vortex for 2 min.

3. Keep the solutions at −20 °C for 20–30 min.

4. Centrifuge all samples at 4000 rpm for 10 min at 4 °C.

5. Carefully pipette out separate layers (organic and aqueous phase). The top layer (aqueous) is used for LC and bottom (organic) layer is used for GC. If only LC-MS is needed, combine the two layers and used the entire supernatant.

6. Dry the supernatant at 37 °C, 30–45 min.

7. Resuspend each sample in 500 µL of methanol–water (1:1) (vortex—5 min, sonicate—5 min, spin at 5000 rpm—5 min) and filter to separate proteins.

8. Protein separation: precondition the Amicon protein filters by prewashing and centrifuging them with 500 µL methanol–water (1:1) until the entire solution is filtered out.

9. Transfer the resuspended samples into the prewashed filters and centrifuge to collect the filtered samples. Add additional 100 µL methanol–water (1:1) to the filter and centrifuge to collect the remaining metabolites if any.

10. Collect the filtered solution and dry it at 37 °C, 30–45 min. Discard the filters.

11. Resuspend the dried samples in 0.1% formic acid with methanol–water (1:1) (this will be different depending on the mobile phase).

12. Sonicate (5 min), vortex (5 min) and centrifuge (5 min) the resuspended mixture. Split the samples into 2–3 aliquots and store at −80 °C to rerun if needed.

3.5 Analysis of Glycolysis Metabolites

Glycolysis is a main metabolic route for carbon metabolism that leads to production of energy essential for proper functioning of a cell. Glycolysis is one of the metabolic pathways that distinguishes the undifferentiated state from the differentiated state of stem cells, with a dynamic change in mitochondrial morphology and a shift from glycolysis to mitochondrial oxidative phosphorylation as stem cells differentiate into lineage-specific progeny [45–49].

Highly proliferative stem cells convert pyruvate to lactate at high rates to meet energy requirements [50, 51]. All the glycolytic intermediates are phosphorylated compounds except pyruvate. Due to the ionic nature of these metabolites, ion chromatography (IC) is the platform of choice for the analysis of glycolytic intermediates. Another possibility is the use of GC-MS, for which the intermediates have to be derivatized to make them volatile. IC-MS/MS (multiple reaction monitoring, MRM) and GC-MS can be used as complementary techniques [52]. Most of the glycolytic intermediates can be analyzed with the IC method, with the exception of glyceraldehyde-3-phosphate and dihydroxy-acetone-phosphate, which decompose in the column due to high sodium hydroxide concentrations used in the gradient. These metabolites can be analyzed with the GC-MS method.

To measure the glycolytic intermediates, labeled isotopomers are prepared on uniformly ^{13}C-labeled glucose. Cells are harvested and the metabolites extracted: the extract contains ^{13}C-labeled labeled whole metabolome of the cell [53]. The unlabeled intermediate and its isotopolog behave similarly during chromatography separation and coelute. The same amount of the labeled internal standard extract is added to both the calibration standards and the samples. After the elution of the intermediate and its isotopolog, the two compounds are detected and measured by MS. The peak area ratios are computed by dividing the peak area of a given intermediate by the peak area of its isotopolog. The peak area ratios are used for the calibration curves and for quantification.

3.6 Analysis of Amino Acids

Amino acids play an essential role in the overall metabolism of a cell and have been implicated specifically in stem cells as part of the regulatory mechanisms that govern stem cell maintenance [54–58]. Stem cells have elevated levels of the amino acids glycine, alanine, proline, and others [59, 60], and mouse ESCs appear to be highly dependent on threonine metabolism [30, 61]. The stem cell-abundant amino acids may have epigenetic influence on chromatin [62], leading to aberrant stem cell biology [63, 64].

To identify and quantify amino acids in physiological samples, they usually first need to be acidified with sulfosalicyclic acid to remove any intact proteins prior to analysis, but sample type determines the optimal preparation protocol. It is important to note that cysteine (and cystine), methionine, and tryptophan are destroyed during hydrolysis with 6 N HCl. Cysteine and methionine can be determined by oxidation with performic acid, yielding the acid stable forms cysteic acid and methionine sulfone, prior to the standard acid hydrolysis.

Some common contaminants (urea, detergents, high concentrations of buffer/salts, etc.) are incompatible with ion-exchange chromatography and must be eliminated. Also, acrylamide contaminants must be removed from samples purified on PAGE by electroblotting or electroelution. Most biological buffers and detergents, even at low concentrations, will interfere with amino acid analysis and special care needs to be given to sample preparation to ensure quality data.

Underivatized amino acids are prepared by methanol precipitation. Then, ion-pairing reversed-phase liquid chromatography is used for separation of amino acids, and the use of volatile mobile phases allows MS/MS detection. Each amino acid is detected in MS/MS-positive mode by its specific transition.

3.7 NMR Analysis	1. If using neurosphere cultures, dissociate the neurospheres to single cell level. If using monolayer cultures, spin them to collect cells. The cell quantity required to obtain a reliable and reproducible metabolomic profile using the NMR depends on the magnet strength. For our studies, we used 1–5 million cells utilizing 800 MHz NMR [65–67]. Furthermore, all experiments are done in triplicate within the same culture (i.e., on the same day) to ensure reproducibility, with five biological replicates to ensure specificity. Internal standard, absolutely necessary for chemical shift calibration, is run with each experiment (*see* **Note 1**).

2. Centrifuge the sample at 1000 g for 1–2 min at 4 °C to remove the culture media. Use PBS to resuspend the pellet. Add 0.45 mL PBS and 0.05 mL D_2O with 0.5 mM DSS or TMS as concentration standard and chemical shift reference (calibrated to 0.00 ppm). It is very important that cell suspension is homogeneous and sometimes sonication is necessary to achieve this. The sample may be recovered after the spectra collection and stored at −80 °C for several weeks to allow further measurements by NMR or other analytical techniques.

3. Set acquisition temperature for NMR between 26 and 36 °C depending on the type of analysis. Load sample into probe and leave sufficient time to equilibrate. Pay special attention to use powder-free gloves when handling NMR tubes to avoid fatty and other deposits from bare hands onto the glass.

4. Set the RF carrier frequency offset value to the H_2O resonance and determine the water saturation power. Select 1D NOESY-preset pulse sequence (noesypr1d). Record an NMR spectrum. The time required to obtain the consistent spectra depends on the magnet strength and can vary from 5 to 30 min. We recommend that all control solutions should be analyzed in parallel. These include media, conditioned media (also useful for foot printing of excreted metabolites), and any compounds or kits used to dissociate and wash the cells.

3.8 Mass Spectrometry Imaging

1. Dissect the animal to isolate the tissue of choice.

2. Transfer it into the Tissue-Tek Mega-cassette so it sits in one corner of it.

3. Hold the cassette with forceps near the corner where the tissue is placed and keep it in proximity of the liquid nitrogen surface for 30 s (The cassette's corner opposite to the one with the tissue can be used to touch the surface of the liquid nitrogen and control the proximity.).

4. Dip the cassette slowly into the nitrogen and keep it in for 5 s.

5. Retract from the nitrogen for 2 s.

6. Repeat the procedure twice or more (depending on the firmness of the tissue surface).

7. Ensure that there are no cracks on the surface—if there are, the tissue might disintegrate upon processing for MSI. If there are no cracks, loosely wrap the tissue in aluminum foil and then Parafilm and store it at −80 °C or cryo-section immediately.

8. Cryo-section the tissue into 10 μm sections. Do not use OCT or PBS, but mount the tissue using water. Placed the sections onto glass microscope slides. These are kept frozen until ready for matrix application.

9. When ready for the MSI, thaw the slide under vacuum to room temperature. Ensure that there are no water droplets on the surface.

10. An automated sublimation matrix applicator is used to apply the MALDI matrix on the tissue sections. Weigh out 600 mg of 9AA and place it into the heating boat.

11. Set the matrix applicator parameter set point for 9AA and coating duration of 20 min.

12. After the coating process, the resulting coated slides is subjected to a rehydration step in a heated humidifying chamber for 2 min using a wetted filter paper with 1 mL of 10% methanol solution.

13. Prior to loading the slide into the mass spectrometer, the slide is scanned using an EPSON scanner for mapping the interested areas into the HDI 1.4 software. The laser power was set to 250 (arb units) with 300 laser shots per pixel data. The laser raster step was set to 60 μm to match the laser spot size of 60 μm.

14. Load the sample into the mass spectrometer and set the ionization source to negative mode ionization mode.

15. Prior to acquiring the data, the instrument is calibrated for mass accuracy using red phosphorus, calibrating the mass range from m/z 50–2000. After the MS parameters are set,

start the acquisition. For a tissue section of 10 mm × 10 mm, run times is about 130 min.

16. After acquisition, import the raw file into the HDI 1.4 software, where the MS data is processed into a collection of images. After a table of mass lists is generated from the data, a particular mass identifying a metabolite of interest is selected and a heatmap distribution of the signal is displayed. The user draws regions of interest over the generated images to determine signal intensities of particular areas.

4   Notes

1. No good analytical method can compensate for poor sampling or sample handling. As biological samples can easily degrade, storage conditions are very important. Freezing and thawing of samples several times can also degrade them. One of the major problems for the derivatization of analytes is moisture, which can condense on the outside surface of a tube or vial upon taking it out of the fridge or freezer. Thus, wait until the sample is at room temperature before opening the tube. If you need to use only a portion of a sample, thaw it completely, mix it and then divide into several aliquots. Store the stock solution at −80 °C for long-term storage; it will be stable for a 1 year. Always keep samples you are working on ice.

2. Extraction of polar/nonpolar metabolites should be performed in glass tubes instead of plastic vials.

Acknowledgments

We would like to thank the members of Maletic-Savatic lab for helpful discussions on this topic. This work was supported by the NIH R01GM120033-01 to M.M.S.

References

1. Manganas LN, Maletic-Savatic M (2005) Stem cell therapy for central nervous system demyelinating disease. Curr Neurol Neurosci Rep 5(3):225–231

2. Botas A, Campbell HM, Han X, Maletic-Savatic M (2015) Metabolomics of neurodegenerative diseases. Int Rev. Neurobiol 122:53–80. https://doi.org/10.1016/bs. irn.2015.05.006

3. Pati S, Muthuraju S, Hadi RA, Huat TJ, Singh S, Maletic-Savatic M, Abdullah JM, Jaafar H (2016) Neurogenic plasticity of mesenchymal stem cell, an alluring cellular replacement for traumatic brain injury. Curr Stem Cell Res Ther 11(2):149–157

4. Folmes CD, Dzeja PP, Nelson TJ, Terzic A (2012) Metabolic plasticity in stem cell homeostasis and differentiation. Cell Stem Cell 11(5):596–606. https://doi.org/10.1016/j. stem.2012.10.002

5. Ito K, Suda T (2014) Metabolic requirements for the maintenance of self-renewing stem cells. Nat Rev Mol Cell Biol 15(4):243–256. https://doi.org/10.1038/nrm3772

6. Rehman J (2010) Empowering self-renewal and differentiation: the role of mitochondria in stem cells. J Mol Med (Berl) 88(10):981–986. https://doi.org/10.1007/s00109-010-0678-2

7. Zimmerlin L, Park TS, Zambidis ET (2017) Capturing human naive pluripotency in the embryo and in the dish. Stem Cells Dev 26(16):1141–1161. https://doi.org/10.1089/scd.2017.0055

8. Ilic D, Ogilvie C (2017) Concise review: human embryonic stem cells-what have we done? What are we doing? where are we going? Stem Cells 35(1):17–25. https://doi.org/10.1002/stem.2450

9. Desai N, Rambhia P, Gishto A (2015) Human embryonic stem cell cultivation: historical perspective and evolution of xeno-free culture systems. Reprod Biol Endocrinol 13:9. https://doi.org/10.1186/s12958-015-0005-4

10. Gan B, Hu J, Jiang S, Liu Y, Sahin E, Zhuang L, Fletcher-Sananikone E, Colla S, Wang YA, Chin L, Depinho RA (2010) Lkb1 regulates quiescence and metabolic homeostasis of haematopoietic stem cells. Nature 468(7324):701–704. https://doi.org/10.1038/nature09595

11. Gurumurthy S, Xie SZ, Alagesan B, Kim J, Yusuf RZ, Saez B, Tzatsos A, Ozsolak F, Milos P, Ferrari F, Park PJ, Shirihai OS, Scadden DT, Bardeesy N (2010) The Lkb1 metabolic sensor maintains haematopoietic stem cell survival. Nature 468(7324):659–663. https://doi.org/10.1038/nature09572

12. Nakada D, Saunders TL, Morrison SJ (2010) Lkb1 regulates cell cycle and energy metabolism in haematopoietic stem cells. Nature 468(7324):653–658. https://doi.org/10.1038/nature09571

13. Katajisto P, Dohla J, Chaffer CL, Pentinmikko N, Marjanovic N, Iqbal S, Zoncu R, Chen W, Weinberg RA, Sabatini DM (2015) Stem cells. Asymmetric apportioning of aged mitochondria between daughter cells is required for stemness. Science 348(6232):340–343. https://doi.org/10.1126/science.1260384

14. Rossi DJ, Bryder D, Seita J, Nussenzweig A, Hoeijmakers J, Weissman IL (2007) Deficiencies in DNA damage repair limit the function of haematopoietic stem cells with age. Nature 447(7145):725–729. https://doi.org/10.1038/nature05862

15. Cho YM, Kwon S, Pak YK, Seol HW, Choi YM, Park DJ, Park KS, Lee HK (2006) Dynamic changes in mitochondrial biogenesis and antioxidant enzymes during the spontaneous differentiation of human embryonic stem cells. Biochem Biophys Res Commun 348(4):1472–1478. https://doi.org/10.1016/j.bbrc.2006.08.020

16. Chung S, Arrell DK, Faustino RS, Terzic A, Dzeja PP (2010) Glycolytic network restructuring integral to the energetics of embryonic stem cell cardiac differentiation. J Mol Cell Cardiol 48(4):725–734. https://doi.org/10.1016/j.yjmcc.2009.12.014

17. Armstrong L, Tilgner K, Saretzki G, Atkinson SP, Stojkovic M, Moreno R, Przyborski S, Lako M (2010) Human induced pluripotent stem cell lines show stress defense mechanisms and mitochondrial regulation similar to those of human embryonic stem cells. Stem Cells 28(4):661–673. https://doi.org/10.1002/stem.307

18. Zhang Y, Marsboom G, Toth PT, Rehman J (2013) Mitochondrial respiration regulates adipogenic differentiation of human mesenchymal stem cells. PLoS One 8(10):e77077. https://doi.org/10.1371/journal.pone.0077077

19. Urao N, Ushio-Fukai M (2013) Redox regulation of stem/progenitor cells and bone marrow niche. Free Radic Biol Med 54:26–39. https://doi.org/10.1016/j.freeradbiomed.2012.10.532

20. Yanes O, Clark J, Wong DM, Patti GJ, Sanchez-Ruiz A, Benton HP, Trauger SA, Desponts C, Ding S, Siuzdak G (2010) Metabolic oxidation regulates embryonic stem cell differentiation. Nat Chem Biol 6(6):411–417. https://doi.org/10.1038/nchembio.364

21. Jones DL, Rando TA (2011) Emerging models and paradigms for stem cell ageing. Nat Cell Biol 13(5):506–512. https://doi.org/10.1038/ncb0511-506

22. Renault VM, Rafalski VA, Morgan AA, Salih DA, Brett JO, Webb AE, Villeda SA, Thekkat PU, Guillerey C, Denko NC, Palmer TD, Butte AJ, Brunet A (2009) FoxO3 regulates neural stem cell homeostasis. Cell Stem Cell 5(5):527–539. https://doi.org/10.1016/j.stem.2009.09.014

23. Semerci F, Maletic-Savatic M (2016) Transgenic mouse models for studying adult neurogenesis. Front Biol (Beijing) 11(3):151–167. https://doi.org/10.1007/s11515-016-1405-3

24. Lee HJ, Jedrychowski MP, Vinayagam A, Wu N, Shyh-Chang N, Hu Y, Min-Wen C, Moore JK, Asara JM, Lyssiotis CA, Perrimon N, Gygi SP, Cantley LC, Kirschner MW (2017) Proteomic and metabolomic characterization of a mammalian cellular transition from quiescence to proliferation. Cell Rep 20(3):721–736. https://doi.org/10.1016/j.celrep.2017.06.074

25. Knobloch M, Pilz GA, Ghesquiere B, Kovacs WJ, Wegleiter T, Moore DL, Hruzova M, Zamboni N, Carmeliet P, Jessberger S (2017) A fatty acid oxidation-dependent metabolic shift regulates adult neural stem cell activity. Cell Rep 20(9):2144–2155. https://doi.org/10.1016/j.celrep.2017.08.029

26. Krstic J, Trivanovic D, Jaukovic A, Santibanez JF, Bugarski D (2017) Metabolic plasticity of stem cells and macrophages in cancer. Front Immunol 8:939. https://doi.org/10.3389/fimmu.2017.00939

27. Hanahan D, Weinberg Robert A (2011) Hallmarks of cancer: the next generation. Cell 144(5):646–674. https://doi.org/10.1016/j.cell.2011.02.013

28. Beger RD (2013) A review of applications of metabolomics in cancer. Meta 3(3):552–574. https://doi.org/10.3390/metabo3030552

29. Warburg O (1956) On the origin of cancer cells. Science 123(3191):309–314

30. Shyh-Chang N, Daley GQ, Cantley LC (2013) Stem cell metabolism in tissue development and aging. Development 140(12).2535–2547. https://doi.org/10.1242/dev.091777

31. Arnold JM, Choi WT, Sreekumar A, Maletic-Savatic M (2015) Analytical strategies for studying stem cell metabolism. Front Biol (Beijing) 10(2):141–153. https://doi.org/10.1007/s11515-015-1357-z

32. Dunn WB, Bailey NJ, Johnson HE (2005) Measuring the metabolome: current analytical technologies. Analyst 130(5):606–625. https://doi.org/10.1039/b418288j

33. Halket JM, Waterman D, Przyborowska AM, Patel RK, Fraser PD, Bramley PM (2005) Chemical derivatization and mass spectral libraries in metabolic profiling by GC/MS and LC/MS/MS. J Exp Bot 56(410):219–243. https://doi.org/10.1093/jxb/eri069

34. Tolstikov VV, Fiehn O (2002) Analysis of highly polar compounds of plant origin: combination of hydrophilic interaction chromatography and electrospray ion trap mass spectrometry. Anal Biochem 301(2):298–307. https://doi.org/10.1006/abio.2001.5513

35. Bajad SU, Lu W, Kimball EH, Yuan J, Peterson C, Rabinowitz JD (2006) Separation and quantitation of water soluble cellular metabolites by hydrophilic interaction chromatography-tandem mass spectrometry. J Chromatogr A 1125(1):76–88. https://doi.org/10.1016/j.chroma.2006.05.019

36. Bajad S, Shulaev V (2007) Highly-parallel metabolomics approaches using LC-MS for pharmaceutical and environmental analysis. Trends Analyt Chem 26(6):625–636. https://doi.org/10.1016/j.trac.2007.02.009

37. Cornett DS, Reyzer ML, Chaurand P, Caprioli RM (2007) MALDI imaging mass spectrometry: molecular snapshots of biochemical systems. Nat Methods 4(10):828–833. https://doi.org/10.1038/nmeth1094

38. Maletic-Savatic M, Vingara LK, Manganas LN, Li Y, Zhang S, Sierra A, Hazel R, Smith D, Wagshul ME, Henn F, Krupp L, Enikolopov G, Benveniste H, Djuric PM, Pelczer I (2008) Metabolomics of neural progenitor cells: a novel approach to biomarker discovery. Cold Spring Harb Symp Quant Biol 73:389–401. https://doi.org/10.1101/sqb.2008.73.021

39. Ma LH, Li Y, Djuric PM, Maletic-Savatic M (2011) Systems biology approach to imaging of neural stem cells. Methods Mol Biol 711:421–434. https://doi.org/10.1007/978-1-61737-992-5_21

40. Chen G, Gulbranson DR, Hou Z, Bolin JM, Ruotti V, Probasco MD, Smuga-Otto K, Howden SE, Diol NR, Propson NE, Wagner R, Lee GO, Antosiewicz-Bourget J, Teng JMC, Thomson JA (2011) Chemically defined conditions for human iPSC derivation and culture. Nat Methods 8(5):424–429. http://www.nature.com/nmeth/journal/v8/n5/abs/nmeth.1593.html#supplementary-information

41. Li XJ, Zhang SC (2006) In vitro differentiation of neural precursors from human embryonic stem cells. Methods Mol Biol 331:169–177. https://doi.org/10.1385/1-59745-046-4:168

42. Chambers SM, Fasano CA, Papapetrou EP, Tomishima M, Sadelain M, Studer L (2009) Highly efficient neural conversion of human ES and iPS cells by dual inhibition of SMAD signaling. Nat Biotechnol 27(3):275–280. https://doi.org/10.1038/nbt.1529

43. Kirkeby A, Nelander J, Parmar M (2012) Generating regionalized neuronal cells from pluripotency, a step-by-step protocol. Front Cell Neurosci 6:64. https://doi.org/10.3389/fncel.2012.00064

44. Mason RP, Casu M, Butler N, Breda C, Campesan S, Clapp J, Green EW, Dhulkhed D, Kyriacou CP, Giorgini F (2013) Glutathione peroxidase activity is neuroprotective in models of Huntington's disease. Nat Genet 45(10):1249–1254. https://doi.org/10.1038/ng.2732

45. Westermann B (2010) Mitochondrial fusion and fission in cell life and death. Nat Rev. Mol Cell Biol 11(12):872–884. https://doi.org/10.1038/nrm3013

46. Ferree A, Shirihai O (2012) Mitochondrial dynamics: the intersection of form and function. Adv Exp Med Biol 748:13–40. https://doi.org/10.1007/978-1-4614-3573-0_2

47. Simsek T, Kocabas F, Zheng J, Deberardinis RJ, Mahmoud AI, Olson EN, Schneider JW, Zhang CC, Sadek HA (2010) The distinct metabolic profile of hematopoietic stem cells reflects their location in a hypoxic niche. Cell Stem Cell 7(3):380–390. https://doi.org/10.1016/j.stem.2010.07.011

48. Suda T, Takubo K, Semenza GL (2011) Metabolic regulation of hematopoietic stem cells in the hypoxic niche. Cell Stem Cell 9(4):298–310. https://doi.org/10.1016/j.stem.2011.09.010

49. Takubo K, Nagamatsu G, Kobayashi CI, Nakamura-Ishizu A, Kobayashi H, Ikeda E, Goda N, Rahimi Y, Johnson RS, Soga T, Hirao A, Suematsu M, Suda T (2013) Regulation of glycolysis by Pdk functions as a metabolic checkpoint for cell cycle quiescence in hematopoietic stem cells. Cell Stem Cell 12(1):49–61. https://doi.org/10.1016/j.stem.2012.10.011

50. Panopoulos AD, Yanes O, Ruiz S, Kida YS, Diep D, Tautenhahn R, Herrerias A, Batchelder EM, Plongthongkum N, Lutz M, Berggren WT, Zhang K, Evans RM, Siuzdak G, Izpisua Belmonte JC (2012) The metabolome of induced pluripotent stem cells reveals metabolic changes occurring in somatic cell reprogramming. Cell Res 22(1):168–177. https://doi.org/10.1038/cr.2011.177

51. Lunt SY, Vander Heiden MG (2011) Aerobic glycolysis: meeting the metabolic requirements of cell proliferation. Annu Rev. Cell Dev Biol 27:441–464. https://doi.org/10.1146/annurev-cellbio-092910-154237

52. van Dam JC, Ras C, ten Pierick A (2011) Analysis of glycolytic intermediates with ion chromatography- and gas chromatography-mass spectrometry. Methods Mol Biol 708:131–146. https://doi.org/10.1007/978-1-61737-985-7_7

53. Mashego MR, Wu L, Van Dam JC, Ras C, Vinke JL, Van Winden WA, Van Gulik WM, Heijnen JJ (2004) MIRACLE: mass isotopomer ratio analysis of U-13C-labeled extracts. A new method for accurate quantification of changes in concentrations of intracellular metabolites. Biotechnol Bioeng 85(6):620–628. https://doi.org/10.1002/bit.10907

54. Chen G, Wang J (2014) Threonine metabolism and embryonic stem cell self-renewal. Curr Opin Clin Nutr Metab Care 17(1):80–85. https://doi.org/10.1097/MCO.0000000000000007

55. Kilberg MS, Terada N, Shan J (2016) Influence of Amino Acid Metabolism on Embryonic Stem Cell Function and Differentiation. Adv Nutr 7(4):780S–789S. https://doi.org/10.3945/an.115.011031

56. Taya Y, Ota Y, Wilkinson AC, Kanazawa A, Watarai H, Kasai M, Nakauchi H, Yamazaki S (2016) Depleting dietary valine permits non-myeloablative mouse hematopoietic stem cell transplantation. Science 354(6316):1152–1155. https://doi.org/10.1126/science.aag3145

57. Saito Y, Iwatsuki K, Hanyu H, Maruyama N, Aihara E, Tadaishi M, Shimizu M, Kobayashi-Hattori K (2017) Effect of essential amino acids on enteroids: methionine deprivation suppresses proliferation and affects differentiation in enteroid stem cells. Biochem Biophys Res Commun 488(1):171–176. https://doi.org/10.1016/j.bbrc.2017.05.029

58. Antanaviciute I, Mikalayeva V, Cesleviciene I, Milasiute G, Skeberdis VA, Bordel S (2017) Transcriptional hallmarks of cancer cell lines reveal an emerging role of branched chain amino acid catabolism. Sci Rep 7(1):7820. https://doi.org/10.1038/s41598-017-08329-8

59. Casalino L, Comes S, Lambazzi G, De Stefano B, Filosa S, De Falco S, De Cesare D, Minchiotti G, Patriarca EJ (2011) Control of embryonic stem cell metastability by L-proline catabolism. J Mol Cell Biol 3(2):108–122. https://doi.org/10.1093/jmcb/mjr001

60. Washington JM, Rathjen J, Felquer F, Lonic A, Bettess MD, Hamra N, Semendric L, Tan BS, Lake JA, Keough RA, Morris MB, Rathjen PD (2010) L-Proline induces differentiation of ES cells: a novel role for an amino acid in the regulation of pluripotent cells in culture. Am J Physiol Cell Physiol 298(5):C982–C992. https://doi.org/10.1152/ajpcell.00498.2009

61. Wang J, Alexander P, Wu L, Hammer R, Cleaver O, McKnight SL (2009) Dependence of mouse embryonic stem cells on threonine catabolism. Science 325(5939):435–439. https://doi.org/10.1126/science.1173288

62. Burgess RJ, Agathocleous M, Morrison SJ (2014) Metabolic regulation of stem cell function. J Intern Med 276(1):12–24. https://doi.org/10.1111/joim.12247

63. Abraham AB, Bronstein R, Chen EI, Koller A, Ronfani L, Maletic-Savatic M, Tsirka SE (2013) Members of the high mobility group B protein family are dynamically expressed in embryonic neural stem cells. Proteome Sci

11(1):18. https://doi.org/10.1186/1477-5956-11-18

64. Abraham AB, Bronstein R, Reddy AS, Maletic-Savatic M, Aguirre A, Tsirka SE (2013) Aberrant neural stem cell proliferation and increased adult neurogenesis in mice lacking chromatin protein HMGB2. PLoS One 8(12):e84838. https://doi.org/10.1371/journal.pone.0084838

65. Allen GI, Maletic-Savatic M (2011) Sparse non-negative generalized PCA with applications to metabolomics. Bioinformatics 27(21):3029–3035. https://doi.org/10.1093/bioinformatics/btr522

66. Allen GI, Peterson C, Vannucci M, Maletic-Savatic M (2013) Regularized partial least squares with an application to NMR spectroscopy. Stat Anal Data Min 6(4):302–314. https://doi.org/10.1002/sam.11169

67. Peterson C, Vannucci M, Karakas C, Choi W, Ma L, Maletic-Savatic M (2013) Inferring metabolic networks using the Bayesian adaptive graphical lasso with informative priors. Stat Interface 6(4):547–558. https://doi.org/10.4310/SII.2013.v6.n4.a12

INDEX

Shree Ram Singh and Pranela Rameshwar (eds.), *Somatic Stem Cells: Methods and Protocols*, Methods in Molecular Biology, vol. 1842, https://doi.org/10.1007/978-1-4939-8697-2, © Springer Science+Business Media, LLC, part of Springer Nature 2018